the Nature of

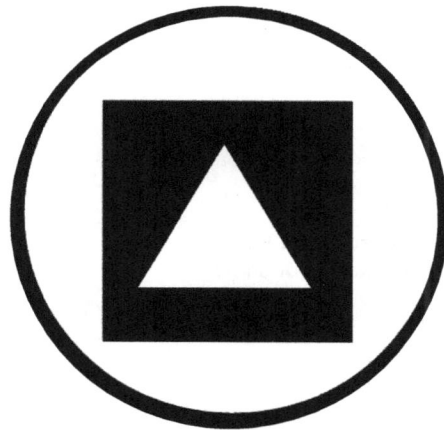

classical chinese medicine
the foundational context to re-unite myriad styles

①

david nassim

The Nature of Classical Chinese Medicine
Published by:
HI Publishing
Stone, Bucks
UK

www.healthinstinct.org

Copyright © 2010 by David Nassim
First published in 2012
Edition 1

ISBN :978-0-9566873-3-3

Editing by Elizabeth Day
Book Design, Illustration by Celine Hogan

*A bird does not sing because it
has an answer;
a bird sings because it has a
song.*

For those investigating the original unity of Classical Chinese medicine

With deepest gratitude to:

Amma Frimpong-Ansah
Douglas Harding
Noguchi, Haruchika 野口晴哉
Masakazu, Ikeda 池田政一
Fukuoka, Maki 福岡真紀
Joan, Victor, Jonathan Nassim
Edward Obaidey
Tony Parsons
Yoshikawa, Akiko 吉川明子

Table of Contents

Section A: Foundation and Constitution

Part 1: Foundational Ideas

Part 2: The Constitution

Part 3: The Male Body-Spirit Combinations

Part 4: The Female Body-Spirit Combinations

Part 5: Relating Cycles of Twelve and Ten to the Constitution

Part 5: Section A — Endnote

Section B: Energetic Anatomy and Physiology

Part 1: Understanding Energetics

Part 2: The Wood Phase

Part 3: The Fire Phase

Part 4: The Earth Phase

Part 5: The Metal Phase

Part 6: The Water Phase

Section C (Book 2): Classical Energy Medicine

Introduction: Principles of Classical Chinese Medicine and Differentiating Appropriate Therapeutic Treatment Approaches

Part 1: Etiology

Part 2: Energetic Pathogenesis

Part 3: Diagnostics

Part 4: Treatment Methodology

Acknowledgments

I would like to thank the following people. Their input has had a tremendous effect, allowing me to see further and ever broadening my approach. Thank you for what you are.

Heidi Able	Jan Diepersloot
Amma Frimpong-Ansah	Gretchen DeSoriano
Carol Alberts	Susanna Dowie
Tommy Boyd	Marian Fixler
Rebecca Bend	Fukuoka, Maki
Michelle Byrne	Simon Fielding
Steve Blair	Christine Grabowska
Jo Bristow	Anthea Grundlingh
John and Hiroko Blazevic	Penny Hamilton
Ingrid Broberg	Andy Harrop
Stephen Birch	Zadie Hasan
Michelle Byrne	Paul Haynes
Melanie Chandler	Jamie Hedger
Emma Chalmers	Andrea Hoffmann
Dianna Cheong	Charles Homonnay-Preyer
Sarah and Rob Carroll	Fiona Hurlock
Amanda Cox	Huide Jin
Anna Couser	Paul Johnson
Fiona Cree	Alan Jansson
Michael Parcell-Davis	Davorah Kadish
Julio da Costa	Aneill Kamath
Siegrid Delaney	Catherine Kato

Vanessa Kempner
Kohga, Chiaki
Jackie Kohnstamm
Andrew Kemp
Oran Kivity
David Lamb
Anna-Maria Lavin
Leslie Anne Lewis
Bob Lloyde
Geraldine McMahon
Anand Marshall
Philip Martin
Masakazu, Ikeda
Julie McBride
Andrew McFarlane
Sam McManus
Hinemoa and Cheyne McSweeney
Marinda Meyer
Kiyoko Montgomary
Iian Montgomary
Audrey Morrell
Tim Mulvagh
Gordon Joslin and Joan Murphy
Hla Myat Saw
Tsutomu, Namikawa
Petra Nannes
Noguchi, Haruchika
Edward Obaidey
Andrew Parfitt
Tony Parsons
Bill Petrie
Alan Plenty
Mike Potter
David Purchas

Mark Preston
Malcolm Reeve
Suzuki, Ritsuko
Michel Rose
Nicky Ryde
Bettina Schriewer
Daniel Schwager
Sophia, Jeremy, and Isabella Smith
Shiro and Smatty
Bill Shirasawa
Ajahn Sumedho
Sugasawa, Fumio
Toby Stephens
Sam Tam
Chris Thile
Alan Traharne
David Towmey
Khai and Rex Tyler
Sandy Richmond
Amy Van Nice
Sallyanne Van Emden
Nicolene Visser
Paul Wan
Snow Wang
Lisa Winston
Daphne Watson
Jo and Jon Whitley
Jackie Whitmore
Yamashiro, Junji
Yoshikawa, Akiko

My Patients: from you all I have learned so much.

My colleagues:
The classes of the London College of Traditional
Acupuncture 2003, graduates of weekend

acupuncture, and herbal medicine groups
The class of the European Shiatsu School 2000
The class of the Japanese Acupuncture and
Moxibustion Skills Association 2002
The class of the Toyohari Association Program
2003
The children of the Homei kindergarden of Japan
Women's University, Tokyo 2003-2006
The staff of the Edward Obaidey Acupuncture
Clinic 2003–2006
The students of the 2007-2008 Classical Oriental
Medicine Study Group, London

And my Family:
Betty Freeman and Avraham Freeman
Lilly Nassim and Nassim Nassim
Joan Nassim, Michel Moncheur
Jonathan Nassim
Victor Nassim
Jack, Adam, Jonathan, Martha, Nami, Ariel, and
Raffi LeRoy
Helen and Bob, Emily and Suzy Gordon
Uncle Matt, Auntie Lena, Betty, and Ron.
Uncle Rafael, Aunty Hilary, Candice, Peter and
Russell

Special Thanks:
I would like to thank those who have taught
me through their sense of being and through
the passion of their enquiry. In particular, Ikeda
Masakazu and Edward Obaidey for passing on
the clarity and wisdom of the Ancients for a new
generation to engage with and explore. This book
is a resultant exploration of deeply considered
resonance from the truth through those who I
was able to make contact with. I am not a spokes
person for any of the teachers I had and I do not
suggest this work is aligned with specific views,
but it is an attempt only, to reveal that which
underpins all of them.

Also, throughout the construction of this work,
the friendship, brilliance, and inspiration of Amma
Frimpong-Ansah, has been a constant vital thread
with which all this work is bound. The Nature
of Amma's consistent intuitive sensitivity and
sense towards the truth of Oneness has drawn
me towards Douglas Harding and Tony Parsons,
whose expressions are utterly transformative.

I would also like to thank Maki Fukuoka, Kiyoko
Montgomary, Audrey Morrell and Sophia Smith
whose friendship and support have always been
greatly nourishing and powerful.

Also, to my editor Elizabeth Day, my proofreader
Ella Tamplin-Wison, and book designer Celine
Hogan. Elizabeth's tremendous task of putting this
work together and Celine's creativity are deeply
important and allow communication of this work:
Thank you.

Author's Note

Why write another textbook on Chinese Medicine? What is so different about this one?

I began studying Chinese Medicine in London. I had previously been interested in Taoist philosophy, and this is what drew me to study medicine. What I immediately noticed within the methods of practise of shiatsu, acupuncture, and herbal medicine is that they were piecemeal in content. The teachers who taught very often could not answer students' questions, and the root philosophy of Oneness, expressed in the Tao Te Ching/ DoToku Kyo/道徳経, * seemed almost completely severed from the practice. There was no unity or natural harmony. I studied various theoretical viewpoints: Traditional Chinese Medicine, Masunaga style shiatsu, and Toyohari style acupuncture. I read and interviewed teachers of Chinese Medicine who upheld different stylistic viewpoints: proponents of Yoshio Manaka, Jack Worsley, Kiiko Matsumoto, and many others. I slowly realised, after a number of years, that all of these methodologies were stylistic viewpoints. None of the practices I saw were focused in unity, void of egoic content, and simply expressed. The bare essential principles of Tao Te Ching were somehow missing.

I was led to Tokyo, Japan, on my quest where I met Edward Obaidey, a teacher I learned a great deal from. His teacher, Ikeda Masakazu, taught that the ancient books of medicine were, at root, one and that it was possible for all methods of practice to be tools of this root philosophy. They suggested that all styles could be understood from this root Oneness. I knew this to be true, as one knows the air within a breath. It was natural. At last, there were the beginnings of peace for me.

---

* (Please note: throughout this book the first time I use a new word of a specifically important ancient text, name or concept it will be written firstly in the Chinese Pinyin then secondly on occasion the Japanese Romanji then thirdly on occasion the ancient character, as the previous "Tao Te Ching" is an example. After the initial use the word will be only in the Pinyin. The Pinyin is used as a main reference point in this book.)

I studied as an apprentice to Mr. Obaidey intensively for three years. During this period and after, I was able to piece together the history and possible reasons why Chinese Medicine has turned towards what it is today, a fragmented and broken edition of the original. What I gained from being amongst practitioners who studied the ancient texts and found unity within them, was total freedom of expression, not limited to or over emphasising egoic style but rather allowing each practitioner to become free of this and so allowing medicine to be practised naturally, style being an inevitable expression or a spoke of a wheel around a silent hub.

In the years that followed, I clarified the theoretical map of the classic literature for myself. I began to understand that the whole of Chinese Medicine, if one understood the classics (i.e., those works of the Han Dynasty), was an echo of the Tao Te Ching, around 500 years after its writing. The simple yet profound unity was present like a thread.

This book is the culmination of 10 years of work. It is the closest rendering I have yet seen in literature of the way to understand the theory of practise directly from the classical material. This is to say that this book attempts to uncover and clarify the very source of Chinese Medicine's theoretical basis. It does so however, from my limited perspective. While I have tried to keep my own renderings and interpretations of the Classical material to a minimum, this book is inevitably an interpretation but one which I hope holds to the core of the simplicity of the principles of yinyang. It is very difficult to make claims of "absolute Truth" and I do not do so, but yinyang is an expression which if the view is broad enough incorporates every phenomena. It is for you the reader to feel whether or not this book expresses such, it was certainly the process by which through me, this book came about. If in areas I have lost this broad view, then, I invite the reader to engage with the material and see what I could not and further render clarity, always from a wider breadth-of-view, not a narrowness. This is not the last word on Classical Chinese medicine but I hope it is a place which springs further editing and rendering into a clearer and clear map of reality with fewer and few words needed.

The study of Classical Chinese Medicine and differentiating it from stylistic medicine today was the key point I needed to tackle for myself. The history of classical to modern is really the history of the ever-expanding and ever-fragmenting mind. If you go back far enough, you will find that all peoples were one and had a sense of this. There was total sense of unity because we engaged as part of the animal kingdom (rather than pretending not to be). As time went on, mind (and brain) and the issue of identification with its contents (mind-identity), expanded, much like heat from a fire rising to the top, as the human being stood upright. From that point forward, humans have been plagued by further and further illusions of a separate, or individuated self, time and space—encountering fear, anxiety, anger, and other emotions, that are the basis of the so-called civilisation we live in.

The fact that this is a great illusion was known by the ancient seers (literally see-er, one that Sees), one of which was Lao Tzu/ Roshi/老子, who was able to cross the seeming divide of the illusion of mind-identity and his ancestral-animal heritage. The Tao Te Ching brought this to society. However, this did not change much. The world moved on and kept focusing more and more deeply on the appearance and the image—fragmenting things ever further. Today, for instance, the CERN particle accelerator requires the energy of a small country to power its insatiable appetite to find the truth. This truth and unity, of course, never left, so cannot be found. They are ever-present as Emptiness and Space but are masked by mind-identity.

This is even true of Chinese Medicine. Even this subject, which has such deep roots in the Tao, has now lost unity. Chinese Medicine now relies on rules, not feelings, on teachers, not first-person interaction, on machine, not Nature, on the self, not no-self, on action, not non-action. As a result, we are left with an industrialized, manufactured product, not a natural medicine able to encompass all phenomena and yet ask nothing in return.

The reason for writing this book is therefore simple: it is to re-establish the connection with the ancient threads that we once knew to be all of reality, so allowing this medicine to still be an Eden within a mirage-desert, so that the mirage does not totally obscure Eden.

David Nassim
2 March 2009
Stone, United Kingdom

Preface

There is a major problem facing Chinese Medicine today, that of breaking its connection with Tao, its root philosophy and lifeblood. This book is a manifesto for reclaiming the ancient roots of the classical (Han Dynasty) period, within a theoretical context, so that Chinese Medicine does not drift further into material-rationalism, dogmatic posturing, and egoic stylist expressions. This book acts as a bridging point to all practitioners and styles of Chinese Medicine. It gives a foundational format with which all can agree, if one senses unity, rather than looking for separations and "special" lineages. All that is required from the reader is openness to the possibility of unity: the very Emptiness at the heart of healing.

The Expansive Mind-Identity and the History of Medicine

The root problem of our time and the reason even for the creation of medicine is founded in the ever-expanding human mind-identity. This, then, becomes the focus of understanding the need for a book such as this. This point is best illustrated by a question I was asked about medical strategies:

Is there a problem in giving a patient a herbal pill for his or her condition rather than using acupuncture or a hands-on modality?

The answer is that if the patient is seeking out a herbal "cure", (which is a particular kind of pathological process), in this period of history he/she will consider the pill a way to fix his or her problem. The pill will become a crutch for the patient, if it is effective, and the patient's condition or way of behaving in the world will not change. In fact, the patient may attempt to continue doing the same thing he or she was doing before—the same behaviour that created the condition in the first place.

So how did that behaviour develop in the patient? I look at Chinese medicine (and modern modern medicine) today, and it is commonplace for such practice to occur. Is this wrong?

It is not a question of right and wrong, but of identification with mental-emotional ideas (mind-identity) or seeing these ideas in context. The Nature of Chinese Medicine in classical times was originally about unity; it was about being one with Nature. By Nature, I mean all that is non-conceptual, or non-abstract; in other words, all that is wild and spontaneous. After all, animals and plants live as one, without recourse to abstractions of thought. In Nature, natural cycles determine life processes. The human mind-identity, on the other hand, attempts to take this on and augment it because of what it thinks it wants; it is a dream of separateness from the world. For instance, the first emperor of China (Qin Shih-huang/秦始皇 ; 259 BC–210 BC) wanted to live forever and directed herbalists to find an elixir that would give him eternal life. This delusional, mind-identity-based idea became the focus of Chinese Medicine for the next several hundred years. Acupuncture and medicines that worked with bodily energy began to decline in importance even at this time (Han Dynasty, 206 BC – 220 AD) at their formative clarity, because they simply offered connection with Nature. To live forever would require Nature to be harnessed for a purpose; massive resources and refinement for an individualistic dream. The elixir is a similar notion to intention around the CERN particle accelerator, and nuclear fusion technologies, but within the human body, a power source that would go on forever. Huge resources from the outside would need to be harnessed in order to stabilise life for the emperor (the Egyptian pyramids, and Qin Shih-huang's own "Terracotta Army" are further examples of this), an impossibility that many emperors (or the like) have come to know at the end of their days.

Again, even in these ancient times just a few hundred years after the end of the classical period of medicine, the thread was beginning to become lost and mind-identity was taking over. Quickly, herbal medicine took the lead. Acupuncture was seen as an adjunct to treatment. Herbal formulas became increasingly focused on symptoms rather than towards the root. There are headache pills, pills for throat problems, pills for phlegm, pills for everything. Mind-identity is not just a modern Western invention. Even in China, the seat of clarity, we see corruption of the Tao of medicine and a loss of connection with Nature. Cities build up, populations run riot, all things distorting Nature, yet are One with it. This is true of all human civilisations.

Today, modern Western medicine, driven by a mind-identity now so fragmented that it cannot see the wood for the trees, is trying to find a cure for symptoms in a similar way. Seemingly broken from Nature, the symptom becomes all-important, just like mind-identity/ the seeming individual, is all-important. The head becomes more important; the feet (or body) less so. We lose our connection with the Earth; we destroy it for our immediate mental requirements, to try to end the anxiety about "our" bodies perishing and changing form. Like a madman so concerned with the "precious" goblet he is grasping, he doesn't see the clear water inside it, that will quench his thirst. The medicine is harsh, associated with death not life, very effective for the acute and physical because it is so aggressive and is anti-biotic rather than pro-biotic.

So, as our mind-identities expanded and fragmented the world we see, our focus became closely identified with symptoms, and symptoms were broken off from their roots. The medicine had to adapt to this and so became a quick fix, a way out. But of course, the problems keep coming back!

The Classical Approach and the Use for this Book

The classical approach is one of non-action "applied" to medicine. This means non-action of the abstractions of mind-identity. It refers to the situation of a person who instinctively heals by connecting, or being one-with, another, who instinctively is healed. The relationship is instinctual. It is about total appropriateness of technique and total efficiency, as is Nature. This book does not go into treatment very much other than to explain the variety of tools and what one can do with them. This book is about the theoretical principles of life upon which all techniques are founded. It is a mapping out of the unity the classical material gives us. Whereas Traditional Chinese Medicine (TCM) is useful for herbalists and not acupuncturists, the Toyohari method is effective for yang deficiency and less so for yin deficiency. Manaka's methodology is often effective for branch treatments but not the root treatment. Classical medicine has no person saying, "This is my way, so follow me". I am not saying that in this book. Classical medicine—at its heart—transcends time. It is part of Nature. It is not from the TCM in the 1950s, or Toyohari of the 1940s, or Masunaga style of the 1960s. It originates from feeling Nature around us. (See Appendix 1 for a history of classical medicine).

This book offers a context, a broad perspective, rather than arguments about which school is right or wrong. If we look from the classical context, we reveal objectivity enough to know when a style is appropriate or not and exclude nothing. Context is the key word in this book, and as far as I have been able to see, context is the very pinnacle of Truth, for in its all-inclusiveness, the Tao provides context for all phenomena, which is what this book attempts to express. This book therefore is for all proponents of any style of Chinese Medicine, for all understanding has the same root and unity, if one is broad enough to encompass all, instead of focusing on parts. The Tao allows us this view. There is no greater need than the need for acceptance and unity. This book attempts to fulfil that need by bringing together practitioners of all styles, like spokes on a single wheel. It is the context and the background.

In this book, we are attempting, for the first time, to really be one with the essence that Lao Tzu talks about, to "have the courage to look for yourself" as Douglas Harding explains, and to experiment and investigate life and experience, rather than to take someone else's word for it. This work attempts to recognise the animal-Nature, the Buddha-Nature, the Christ-consciousness, or simply the Void that we are all part of, and assist us in realising it to be ourselves. Only then, when we are at ease, can we pass this message on to others without ideological or specific intention, in treatment or any aspect of life. The book attempts to move us to a realisation that we can do nothing about mind-identity, and it is the very doing nothing (non-action) that contextualizes mind-identity, seeing the seeming individual for what it is, a mirage no more, no less. No amount or style of training will help us find what we are seeking, for it is already with us, like looking for one's glasses when one is looking through them, or trying to find one's own skin.

In reaching into the depths of the Tao Te Ching and the thread of Nature that it offers, we become able to bridge this seeming distance between Nature and civilisation, and so become curative of dis-ease, through no-self rather than through reputation or command or intention.

This text is not an academic book, because academia is created to further mind-identity. This text is written to render mind-identity neutral and to allow medicine to work of its own accord. This book, I hope, will allow for the relaxation necessary for healing to occur naturally rather than through force.

Hence, what makes this text useful and different from others is that it ends the confusions generated by mind-identity and attempts to take us to what we know instinctively. Please, therefore, feel deeply for yourself. I hope this book serves you on your quest.

Chapter 81

Truthful words are rarely embellished;
Embellished words are rarely truthful.
That which is Naturally-virtuous cannot argue
That which argues cannot perceive Innate-perfection.
Wisdom is not found in extensive intellectual learning
The extensive intellectually-learned are not wise
The Natural-human does not hold back.
Expressing outwards in the world there is great fulfilment
Expressing outwards in the world there is great contentment
Naturalness nourishes all and cannot separate
Naturalness of the Natural-human is awesome and cannot contend.

[Note: I will base the key words, names and book titles that are difficult to translate in Pin Yin, but the first time I use the words in the text I will also provide basic pronunciation in Japanese via Romanji (Romanized phonetic script) and also provide the pictogram character itself so you can investigate and cross reference with other materials easily. There may be instances where I break down a pictogram character itself in which case the Chinese and Japanese pronunciation will be given at this time.]

General Introduction

Chinese philosophy is a massive subject, yet being of simple mind and needing to understand things clearly before I can let them go, I have found that Chinese philosophy is also very simple. It has to be. People today make the world complex. Complexity means fragmentation. It is merely the perspective that fragmented parts are separate that is the difficulty—the belief in brokenness.

We cannot get away from the fact that all things are united. We spend so much time, so much of our lives, involved in reconnecting fragmented parts, and so little time realising the unity of all. Chinese philosophy or its themes are explored in all of the ancient philosophies of Pre-Socratic Greece, Kabbalah, Taoism, Zen, Sufism, Hinduism, Buddhism, Gnosticism, Shinto, Jainism, Bon, and so on. All of these philosophies hold that there is constant unity. We must, of course, differentiate these traditions from their religious counterparts. These are not religious doctrines at their core; these are spiritual truths recognised deeply and all expressing the same thing. It is obvious that any of these philosophies is interchangeable, and those who understand this can give up their label of being a member of a spiritual group as easily as changing a coat.

What we must become aware of is that within the ancient traditions, such as those mentioned above, is a central core of truth. However, this core cannot be entered into using mind-identity. This problem has created the biggest difficulty for human beings thus far in history. Thus is formed the illusion that there are those who know and others who do not know: the teacher and the student. This hierarchy, which appears useful, is really very often a guise for power games and for dissonance to reveal itself. The truth is that the core teachings cannot be taught. It is the place where the teacher ends and also the student ends, literally as the realization is of being fingers of the same hand. So when we are discussing Chinese philosophy, we are looking at a description or dialogue between a seeker and a "no-oneness" answering!

As a result, we must first make our language clear. The section that follows starts to look at the words I use in the book interchangeably, to explore ideas and offer, very simply, an explanation of what is being expressed in the classical works of Chinese Medicine at a core level. What I do not offer is a method of becoming deeply natural, which is as futile as attempting to find a needle in a haystack when the haystack is made of needles, or looking desperately for wood in a forest of trees. The best one can do is to stay at the edge of meaning, just before verbal paradox. Here we see things exactly as they are, and from this instinct or one's true-nature, will know the rest of the way.

Let's start with the concept of qi/ki or energy. This book is about acknowledging a unified Oneness of a sea of energy called qi in Chinese and ki in Japanese. This unified field of accumulating and expanding life-force is what is known as energy. Energy in Modern Western terms is often associated with particular kinds of energy, but the Chinese don't separate and don't fragment. Energy is everything that has form or foundation in the universe. This energy can be expressed in its two unified forms: yin and yang, cooling-accumulation and heating-expansion. We will look more into these ideas later, but the key principle to understand is that energy is like a sea; it is a totally unified quality throughout the universe but lies on that which has no form and does not change. In fact, it is born from that which has no substance at all, something that is underpinning everything but yet is no-thing in itself. We can conceptually call this Wu or Void. Again, these concepts will be talked about in more depth later. However, throughout this work you will see this symbol:

(fig.0.1)

It is essential from the outset that we recognise that energy is represented by the symbol of the Taiji/Taikyoku /太極 or the yinyang symbol above, and it is backed by the Wu or Void, the black background. This in itself holds all the principles required for understanding—the sea of qi underpinned by the Essence, or the mother of life. Please let's go forward with this concept in mind, and we will keep going back to more deeply clarify this picture. Energy is everything, all form and all function; it is the explosion of the celebration of all life happening at this moment. All is Oneness. Now for some further explanations …

There are many explanations for mind. I will offer mine in the context of how I use it in this book. By the term mind-identity, I mean the process of the dream-like projections of past events, furring up the pure connection and acceptance of the world as it is. It is not about saying that this is good, bad, or ugly, but mind-identity without context of itself believes itself to be real, which is the key issue. Like a cloud in the sky, mind-identity constantly changes form and shape but is always in the context of the sky (know it or not!). Mind-identity in the context of this book is a pure "I am cloud" belief, rather than seeing the cloud in the context of the sky and then moving to the position of realising that "sky" (emptiness or Source) and "clouds" make up the two aspects of this full reality—and these are one. Once context is arrived at, Nature takes its course and there is no friction. When mind-identity is apparent, force is often used. Dis-ease occupies the situation, and there is great suffering. Mind-identity as a phenomena incorporates both the identification process held in thoughts and in the body, that which is held in the body we call emotion so mind-identity is the origin of what could be called mental-emotional dis-ease.

Whatever we call the inner consciousness or awareness, light or knowing, that allows us to see the mental activity in context—and therefore not own it and call it a name like "David" or "Anne" or "John"—this It-ness in the background is the Oneness that allows us to know all of the outer manifestations of the world to be part of that Oneness. Though there seem to be myriad forms and separation, this is only the appearance. We are never negating this appearance nor saying style or unique expression are not important aspects. However, if we consider the nature of the reality we live in, 99.9% of the time is take up by the appearance and self-image (mind-identity) and so very little is the recognition of Oneness that simply this is unsustainable in every possible way. This is the dis-ease of humanity. When Oneness is at the root, it occurs like the rest of nature, which is a spontaneous unique expression, arriving without thought of it being "unique" or separate, simply it is as it comes, as it is.

Mind-identity is utterly natural, so we can't really say that there is a split between Nature and human mind-identity. What we can say is that there is a pattern that's happening, and this pattern is the dream or illusion that there is a person here writing this book and another person reading it, when, in fact, these are one and the same Source, having a conversation with itself. The illusion is that there is a world where we live separate lives, and this is reinforced by everything that we see around us, built up from eons of belief in separateness. What is really going on is that mind-identity is not ever seen in context; it is never seen for what it is. A friend of mine suggested to me: "A chocolate bar is only a chocolate bar; it can't do anything other than be a chocolate bar". This is exactly how we need to understand mind. As a divine dividing tool, it divides things, attempts to separate off. That's what it does, but behind it is what it's coming from, and this is Oneness.

Mind-identity is essentially a human behaving as if he or she is nothing but a head. This is true of persons in whom mind-identity manifests much of the time. It is as if everything below the head is far away and they are almost floating. Many academics and students suffer from this! We will consider that mind is a fragment of the human being. Hence the term body-mind will not be used because it implies that mind is separate from the body or spirit. Mind is an expression of both body and spirit. Body-spirit has no words to express itself; it is purely instinctual, like animals or small children. When the sickness of mental-emotional identification comes into the picture, by its nature it attempts to take control of both body and spirit and move them towards identification with separation—as "me", as a separate self. Body and spirit can have differences, just like all of Nature has different and expressive energetic newness arriving from One source. Mind (without its identification with itself) is a tool of body-spirit, just a function of their union, which is sometimes more dominant (mind-identity) and sometimes less so— what we could call the "human condition". The situation of being human seems to bring with it separation as evolutionary development, and natural change. This has to do, I feel, with walking upright and having very little contact with the Earth (see appendix and conclusion sections of this work). However, we now have a function of the body-spirit which has the possibility of consuming the human being within its spiralling. For some, this spiralling has begun to cause change. Larger and larger numbers of people are spontaneously moving to contextualisation of mind-identity and a return to sensing Oneness. Others of us are caught in between the dance of pure mind-identity and contextualised mind, and still others of us are caught in pure mind-identity. Nothing is better or worse here, it is all about a ripening process that cannot be met with "will to force change!", but certainly the strong belief that everything is separate from "you" is a "hell" which is the cause of all the suffering in the world today.

The body-Jing I refer to as more condensed energy that involves the constitution and quantity of energy of the body. The body is form or structure, a vessel for the spirit (Shen). This vessel includes the brain, which is not considered to be mind itself, as mind exists in thoughts in the head or in the bodily sense, or in dis-ease/ mind-identity reactive-thinking in the head and emotions in the body. By spirit (Shen), I mean that which permeates the structure of the body and the universe and is the quality of energy of a person. Also, spirit is that activity that is within the Jing-body and so activates it and brings it to aliveness. In reference to all expressions, there is always a notion of a continuum of energetic expression. Hence, what I mean by jingshen, or body-spirit, is really yinyang, or that which is change and has occurred from a canvas of stillness/emptiness, or Wu in Chinese. So Wu-body-spirit means the connection between stillness/nothingness and body-spirit. It is another term for the newborn child or the animal-natural energetic expression.

A sage or seer or Natural Person is usually someone who has reached adulthood. Mind-identity has come into the picture, and there is a realisation that there is an identified self which is not the whole but only part of what is. Thoughts and feelings are placed in context, and this forms the sage or seer, who is simply one who Nature, or Grace, or whatever you want to call it, has affected. But we can't really call this a "person", from their perspective, any further, because the term "I" no longer exists here; it is like a wave recognising it is One with the sea. There is dissolution of the mind-identity-formed self, and as such, the individuality of the person "David" or "Anne" or "John" is simply seen for what it is, a smokescreen enveloped by Emptiness. Once the Void is known to be the Origin of "I", then everything else, all the other manifestations in the world, become less absolute and more like highly expressive forms of life and colour—or as Tony Parsons puts it, "life happening". Mind-identity in this context is a dis-ease of the spirit's interaction with the body. The body is the vessel; the spirit is the activity within the vessel. Hence mind-identity is associated with spirit and yang more than with body and yin.

The apparent "big issue" for us in today's world is an over-association with the content of mind-identity. This has created much suffering. In fact, the Buddhist term suffering means just this, the manifestation of mind-identity. The place of contextualised mind or acknowledgment of inner Empty-Completeness is liberation from this overheated-contracted state and therefore a place where the spirit and body unite in unimpeded flow. One can think of mind-identity as being the barrier or resistor in the circuit, stopping the body and spirit from engaging to their full ability, or at least realising what was always there.

Chinese philosophy (and all of ancient philosophy and indigenous medicines) developed in order to counterbalance the effect of an ever-expanding mind-identity dis-ease. The medicine has developed in order to treat it. If one is aware, it means that one is as an animal: one with exterior, not bound within the inside, yet not attaching to the things of the world. This makes up the two aspects of Truth: the water—the fullness (everythingness), the detached and objective, the unconditional-ness; and the fire—the emptiness (nothingness), unity, intimacy, or love,….. hence together: unconditional love. Mind-identity is the aspect of the system that sees itself as a thing separate from the outside and therefore forms fragmentation in action and in being. This is mind-identity. The spirit is different and works through inspirational change; it is a force of Nature.

Definitions of Terms Used throughout the Text

Considering the previous and further explanations, I offer the following definitions (also see Glossary):

Mind-identity (yang + yang/ warped yang)= ego = fragmented focusing (forcing a narrowing of view) = seeing form/matter as separate units. Belief in broken and whole, rather than Oneness. Mind-identity has an interest in the future (based on a projected past) and the past, which can be called the mind-attached-state. The illusion of separation takes over. Mind-identity causes suffering. It is about duality, right and wrong, and intentionally doing or undoing something. Karma. Ideas of Karmic consequence. Individuated soul, reincarnation, personally owned "past-life", reactionary expression based on past-based fantasy, emotion (or mind-identity expressed in the body) of all types (tends to be found more in women), Attachment/ contraction/ desire/ grasping, mental analysis and constructed dogma and idealisms of all types (tends to be found more in men). A sense of pointlessness and depression or forced and focused intention based on belief. Contraction-tension-heated

Contextualised mind (Wu-yinyang/ health) = connection to the super-consciousness/collective consciousness = unity of body to spirit = function/energy all unified. This equals connection to the original stillness from which the change of the universe is derived (acceptance of broadening of view). Contextualised mind has an interest only in the present moment. This can be called Clear-awareness or Being. Contextualised mind is the end of suffering. There is no-one to suffer. All identity with suffering has been lost. There is no resistance, which is non-duality or Oneness. There are no concepts at all; broken and whole do not exist, only Oneness. It is about awakening from the illusion of separation. It is about non-doing or letting go, the spontaneous movement, the order of the universe, everything and nothing accepted as One. Seeing Karma as an illusion, looking past the mirage of past experience. Seeing all past experience as the past of No-One simply past/ memory, response and sensitivity, life flowing through, feeling without attachment into emotion or mental constructions. Utterly spontaneous and without requirement for point or direction or individuated intention. Relaxed-cooled.

Using terms from modern psychology, an egotistical formation is one that sees parts, whereas that which sees the underlying mind-identified reality is that which is unity. Notice that mind-identity requires force or impetus (yang), whereas the role of contextualised-mind is acceptance (yin) of the way it is, which is unity. Yang doesn't have to move towards fragmentation as we can see in the natural world around us, but the mind-identity phenomena in humans is an aspect which is like a dis-ease or a pathological change towards death for the humans species and so yang within humans has a strong tendency to be associated with mind-identity. Healthy yang is always deeply anchored by yin.

Freedom or liberation from suffering takes place when there is realisation of the fact that there is no-one to be freed and that the mind-identity game of going to extremes (good/bad, right/wrong, true/false) is a drama of the identification process with a "something" that we call "you" or "me". All of these are mental expressions. For every polar action mind-identity attaches to, there will be an equal and opposite reaction, because all energetic expressions create polarity in the change that is occurring in the yinyang of the universe. As long as this is accepted, it works without resistance, but if attachment to an absolute, unchanging state arises, this will cause suffering.

Suffering involves placing energy into an illusion. This is not chosen by anyone. In fact, there is nothing chosen at all, ever; it only seems like it. You may feel that choosing to "become liberated" is the best cause of action, but this in itself is not chosen (who is the "you" that "chooses"?) and doesn't necessarily lead to liberation. Very often, it forms identification with the idea of "you" being the person that is seeking to find what already is. This is the big difficulty in the misunderstandings of most of the messages of the ancient seers. The broader the viewing angle, the closer to stillness one comes, and the realisation occurs that there is no-one. All is unity. We must understand that what is sometimes called "no-mind" is not the end of the thought processes but the end of the identification of a person called "me" who in reality doesn't "exist". All of life, even the illusion of separation, is occurring in the perfection of Oneness whether it knows it or not! With the realisation that there clearly isn't a person there within you, many of the thought processes die down, for there is much less to say. Awakening is not better than non-awakening. It is just the end of suffering. Suffering is not good or bad; it just is. The illusion is not good or bad; it just is … until it isn't!

This book is aligned with the work of Douglas Harding and Tony Parsons, whose work is beyond my inadequate description. However their message is realising that there is no place to get to, life is just here as it is, or simply just seeing what is. When the seeker ends, the awakening occurs. It's as simple as that. The point is to stop looking,…or carry on….but at least to see the world from where you are at this moment. This culminates in the viewing of the "original face" which is the essence of Zen and all the sages, old and new. In this book, I may talk about de-focusing or contextualised-mind, but this is not to say it is "achievable". It is simply an attempt to explain something with the best choice of words. It means freedom from the absoluteness of "me" or "mind-identity" or "self" in the emptiness, viewing the world as it is—or as Douglas Harding puts it, "The Head-less State".

Hindu culture describes ultimate reality in terms of negations: neti neti, literally "not this, not that". Such an expression is a letting go of all mental categorising in order to know reality. Openness to this allows revelation of the true Nature of things without needing a thunder clap or earth-shaking drama, but with the obviousness and bareness of being wherever you are, doing whatever you are doing. There is no practice involved, just a shift in perception which, as with mind-identity's arising and blossoming and eventually dissolution, is a purely natural phenomenon. This place of openness is Buddha's "Middle Way", although there really is no "way" to it, as it is never due to working hard at attempting this "goal" because there is no-one seeking in reality. The key is that mind-identity will constantly arise and attempt to make something into a "way" or a path. This is why all of the religions of the world fall short of the mark, which is in fact right back at the very "Centre". When extremes of change die away, one is left in the middle of all things, as a singularity where everything is happening. Douglas Harding points out that the view from the first-person perception is Singular, meaning it is the One view that we are all seeing through, like different windows in the same house. If one considers a spinning wheel, the still point is always at the origin of movement, the centre or middle. One allows both extremes to exist together, and therefore finds a broader perspective, which incorporates the two poles upon a still framework/background. Action is no longer confined to change based on past mental attachments, on fragmentation, and narrowing and blurring the awareness of now-ness, but is able to feel and act, becoming an expression of Nature and at one with all things. This is called right action, as expressed by J. Krishnamurti, or Christ's "righteousness". The same principle permeates all Chinese culture and forms of art.

Right action, or action from the contextualised mind (Wu wei, unintended action or empty, natural, action) alludes to the central principle of all truly important messages. It is also impossible to "teach" this. In reality, everything around us is still, as time and space cannot exist without the distorted perspective of mind-identity which we see life through. We constantly become attached to particular mental patterns. It is very much like playing a film inside your own head, while simultaneously the senses experiencing reality (i.e. that without the add-on images of the past). This makes for a highly confused picture, because one will be reacting to events now, based on the internal film, as well as the event itself, now. If, in addition, there are several films playing at once, this can be almost impossible to live with, a constant madness of existence. Everything today is deciphered through the cognitive process, very rarely through the feeling of being present, now, and then knowing instinctively what to do. All dis-ease, and even the concept of dis-ease itself, is attributed to mind-identity and its illusions. Beyond illusion is merely the here and now, and within this is the stillness of the truth of being. Here, there is freedom from the illusion of constant movement and the idea of absolute states of life and death. At the core of things, there is no-one to live or die. There allows freedom from the suffering that arises from impermanence and change. Through the acceptance of change, we accept the fact that mind-identity is an illusion. We accept that time does not exist. We therefore accept that the I, as self, cannot exist. Then we accept that the space-less and time-less origin is all that there is and that we are one and the same with this. Change in the universe becomes a detail. As it is placed in context, we are free, for we see the prime illusion that there was never any-one there to be free. The self was an illusion. The belief in "self" as absolute reality, dies.

The so-called moral attitude or created principles of mind-identity follow a pattern in most religious (non-spiritual/non-truth) associated contexts and often the social laws we seem bound to. Good and bad create a dilemma, because who is the objective judge? Ultimately, this draws us to how one interprets religious philosophy. Seen through the eyes of the fragmented mind, a classic work, such as the Sutras or the Bible, seems to show clear lines of good and bad, right and wrong, heaven and hell. However, it is so easy, through these eyes, to fragment the stillness of objectivity to subjective status, and in most situations, this is what occurs. There are few who can see clearly what is appropriate, objectively, because they have gone beyond the belief in the separate state. Chinese philosophy allows us to perceive an objective view through its non-moralist, naturalist, and simple principles. It shows us that in a universe where all eventualities can be seen as relations of yin and yang (everything-ness, view from yang), founded on stillness (nothingness, view from yin), morality is truly a creation of the subjectivity of mind-identity and informs dis-ease.

When we consider the meaning of unity, there are no boundaries to be had. If we call this unconditional love, or if we call it Void or nothingness, or everything-ness, it is all equally meaningless. But the judging mind-identity considers some words better than others, some things better than others. A hierarchical expression is developed from this where some people are "more equal" than others, and this just doesn't concur with Unity. Unity is found in torture and pain and terror and tyranny, just as it is found in flowers and trees and sex and caring and Nature. Nothing is excluded. If we take the mental state of a murderer as an example, he wishes to kill in order to control his fears. If he kills, he may need to kill again and again; he may believe that if he kills all of life, he will kill out his own self and finally be free and at peace. This is a form of love he tries to find—love through killing of all life, killing away his self. The other way is controlling ownership; the big boss gets bigger and bigger, and his company becomes larger and larger until he owns more and more and more, until he is in control of the government, industry, everything. He expands to engulf the world, and then the solar system, and then everything in the universe. If it is all in him, then he is safe; he doesn't have to fear because he has formed Oneness.

Even the so-called "worst" of human behaviour is actually an attempt to form Oneness, even if it doesn't look like it or it seems utterly misconceived. Still, the power at the root of it is the universe; it is never the mind-identity of man, who only believes he is making the choices. We are in a situation of budded awareness, budded consciousness, which is about to bloom through Nature's process, but if the mind-identity attempts to control this state, it just ties it up before blooming, which resists the process. Since this is also the divine way, this is a time that is about resistance. It will break through, however that occurs , which could be massive human "disaster" and death just as it could be entropy towards Oneness with our environment. It will occur just as clouds break when the air currents settle after the resistance of a storm force. All is one, no matter how it may seem to mind-identity to be otherwise. This is the key and only understanding of non-duality, found in Taoist understanding.

Hence, seeing mind-identity in this context of Chinese philosophy, we can see the dis-ease and the cure:

Dis-ease = fragmentation, subjectivity, mind-identity, and absolute life and death as states. Pain with suffering and the belief in separate individuals. Belief that one is broken and needs to be whole, though people often believe they need help or external influence to find this wholeness. Caught in the illusion of separation or being "an individual". Contracted-tension-heated.

Cure = Nature/Truth/unity, objectivity, contextualised mind, and acceptance of the continuum of change on a background of stillness. Pain without suffering—therefore, not pain but being. Death of the self and re-birth of life simultaneously. The realisation that the self is an illusion and what is left is pure life or nothingness. The end of the interplay between broken and whole, as both are seen as an illusion. Realisation/acceptance/dissolving. One was never broken but always was and ever has been within

Oneness. Relaxed-cooled.

Let's be clear about the following four terms, as they will be used throughout the text:

Mind — In the context of this work, mind by itself means mind as part of body-mind or body-spirit. The mind aspect is an action of the spirit through the body. It's the expression of the spirit as spontaneous thought forms unattached and inspirational, including many past images occurring spontaneously as remembrances connected to events occurring now, but never more than images on a screen, without them being a tangible reality in themselves. This is the child-mind and can be also understood as the contextualised-mind (see below). This means the Mind of Wu.

No-Mind — The presence of no thought, and a return to the original emptiness, the un-manifested. It means Wu.

Mind-identity — This is dis-ease of mind; it is the dis-ease of the spirit, that of identifying with the smoke or cloud of thought that manifests within the human expression. This is the game of mind-ego, the illusion of mind being separate. Mind-identity is suffering, which means the idea of being separate, and so pain of whatever kind occurring to a separate individual, which is just not the case. Rather, it is occurring as an expression of Oneness. Everything is Oneness; this is hidden from the mind-identity state, which is a heated-contracted state of energy.

Contextualised mind — Freedom or liberation from the perception that one is mind-identified; enlightenment. This is has the same meaning as "mind" within the text, but it also considers something that happens later in life than in childhood, after a process of being within the mind-identified state and then letting go or dissolving from this. This too means the Mind of Wu. It is also the cooled-expanded/ exploded open energy expression; there is no energy trapped in the body-spirit in a state of heated-contraction associated with idea of "self".

It is as simple as that. No matter what tradition you endeavour to practise, or how complex the meditation or practice or medicine, Truth cannot be taught. Of course, it does not matter either way. Life brought about the mental condition of suffering, and so it will end it. "You" cannot. "You" never did or will have a choice in the matter. Practice in this case doesn't make perfect; it makes for an identification with the idea of perfection and reinforces the process of "self". So then what is there to do? In the allowance of this question to remain unanswered and in the allowance of mind-identity to seek and scrabble to make a foothold where there is none, there is a possibility of something else to be observing it all. From this place, things are done because they are enjoyed and/ or is natural and for no other reason.

    The process of differentiation is vital in Chinese philosophy and is stated in the great Classic, the Tao Te Ching, when the seer within speaks and says: "So the natural-person takes this, and leaves that". It means to see the essence of stillness and unity and to keep one's eye on it. It was not just Lao Tzu who had this clarity. The following is an excerpt From Plato's work, the Timaeus:

"Every diagram and system of number and every combination of harmony, and revolution of the
stars, must be made manifest as the "One Through All" to him who learns in the proper way. And it
will be made manifest, as we say, "a man learns by keeping his gaze on Unity." (quoted in: Harvey C.
& S., 1999)

As students/ Observers of life, differentiation to see through to the simple from the seemingly complex, which is another way of saying, to be authentically our self, gradually occurs. There seems to be little choice in the matter of how one approaches it, other than to see the unity behind the fragments. It is important to differentiate what can be taught and what cannot. A method or skill can be taught, a way of analysis, theory - all of this can be taught, this is all associated with mind-identity and usually formatting a construction to see things through. However, if the skill is nothingness or being-one, pure-observation, not having a self, this is anti-teaching, it is neither method nor observation-explanation. It is non-teaching, and as such, any method will always engage the same attempt by mind-identity to tackle the problem. Therefore, the only true teaching of being-ness or unity can be a non-teaching, an imagined iron wall that mind-identity cannot pass through. This is why we must be very careful when looking at interpreted works of a "master", or spiritual works and take from them only what resonates with Truth in a broader context. A super-structure of mental ideas have often been added to the basic principle in these works, as mind-identity immediately turns one into two; this is often what so-called disciples did of expressions of others who could "see". However, methods that claim to lead to loss of self may actually hinder one from realising non-action, rather than help. The mind-identified way is one of contraction and heat; it pertains most strongly to yang, and is fundamentally complex although it believes itself to be simple. The scientist will "simply" explain a fragment of knowledge or "his" idea of something but fail to realise that his focus and individuation leads to complexity. It may seem like a simple explanation because he has narrowed his field of view so much or, disregarded so much that in his idea there are now only 2 or 3 points of focus not the seeming billion fragments of what presents. Considered a "brilliant teacher", he only teaches that which he knows. The contextualized mind is fundamentally yin and without direct action or teaching; it is "in acceptance of...". It accepts the seeming billion fragments as one and so sees a pattern in them. This is why, for example, quantum theory requires and through its own unfolding, explains a change of mind from Newtonian-materialism (yang/ separatism) towards a "mystic" or in fact yin understanding, if it will ever transpire to be the "theory of everything" that is searched for. Hence yin teaches by example but not by "being an example", just by being what it is, a constant direction to Centre. This seems, to the yang individuated state, to be very complex and also threatens to dissolve his individuality, for it accepts all aspects. However she holds to the universal underbelly and understands fundamental unity and as such is not in contention with the other "fingers of the hand-of-God", so to speak.

So it is with this premise that we start the investigation into Chinese philosophy: there is nothing highly complex that we need to understand with our mind. It is something we investigate with far more than just a mind. As we learn it, it changes us, to become whole and unified within ourselves, and partway through the project, we find ourselves in a situation where we know we are doing what we need to do, like me writing this book, and perhaps you reading these words.

If you have encountered non-you for a moment or two in the stream of mind that makes up the ego's belief in itself, you have seen the reality which underlies the fragmented movement of mind-identity, the moving stream of thought we inhabit. Mind is not the problem, but a symptom of the belief in the content of mind being self or ego (mind-identity). The Truth in the Stillness of the background seems a long way from mental activity and yet has been with us all, from the beginning. It is here that we can diagnose a patient, paint a flower, think, create a song, create food, build a house, walk, breathe, or do whatever, with freedom of belief in mind-identity, living "in the stillness of seeming change". It is important to realise the fundamental difference between what is out there in the world—literally the view out from where you are sitting now, viewing the words on the page in front of you and perhaps a desk, a chair, arms and legs—and the "view" or sensing of the inward direction. The inward direction is like a child. It isn't three-dimensional like the outer world, but two-dimensional because there is a headless-ness; there is no head in the way when we observe, hence you are looking out of emptiness into the world. Wittgenstein wrote that "the subject is not in the world"; this subject is the true I or I am. It is the paradox of being and not-being, nothing and everything, both happening at the same time, and nothingness enveloping all that is in movement. This level of obviousness is the root of the whole of Chinese medical philosophy; it is the foundation from which everything else is looked at. When we discuss anything other than the core emptiness (that which is a capacity for all things), then we are discussing the world. Most of this book is a discussion of two-ness, yinyang, but we are constantly in touch with the fact that the two derives from Emptiness or Wu at the root. Mind-identity yields, when life is ready for it to yield, just as a bud flowers when it is right to flower, or a fruit is picked when it is ripe. Until then, one can do no more to accelerate the process than simply not seek the answer.

On each page of the classics of Chinese Medicine in the classical Taoist expression is a perspective, a view of the Tao. The Way isn't a way; it is the naturally unravelling expression of Stillness (Wu) that is sometimes near, sometimes far, sometimes broad, sometimes more focused, but always attempting to show the same thing from a different context, massaging mind-identity to be flexible rather than rigid, full, and hard. For Mind to be soft and unrestricted, is to perceive stillness as everything. The essence one absorbs through reading is not another piece of information, or another technique, or a list of ingredients or points. Rather, one relaxes to absorb the principle innate within the words, and this principle is unity, and unity is close to the Tao. In a sense, you let go more, rather than accumulate, it is the opposite of what is normally done in "learning something", as in fact this is about sensing innate instinctual sense which is already there behind the mists. The words of the classics act as a wall to mind-identity. You cannot penetrate the deep notion that the principle is arrived at through the words.

From this we can understand that a principle is something that underpins, and hence is static. However, if it is a principle of the Tao, it must incorporate change, and thus, when I say principle, I mean the changing truth, the static principle that everything changes. The root is stillness or the void or Wu. This will be explained further, but again keep your eye on the unity of what it means, rather than the specific words involved.

Once mind-identity understands that on its own it can go no further, then please read on, for by the end of this book, mind-identity can be seen as part of the whole, not as the whole of the whole, at least intellectually speaking. Our culture is one of collecting and holding on to sensation (extremes), and this is what makes the process of understanding the words easy, but non-doing very "difficult". However, it is only mind-identity that believes it is difficult, in order to validate itself further. Breathing is easy, or a heart beating, so mind-identity doesn't bother with it …

The main difference between the ancient East and modern West is that the modern Western mind is associated more with fragmentation, with an egocentric standpoint, and with a philosophy and science that are products of this, which reinforces it. This we can describe as being inside looking out. The ancient Eastern point of view adheres more to the outside and inside joining and becoming unified (inner looking). It is about attempts to take God's perspective (both in the macrocosm and microcosm) and see this in objective stillness, exterior openness, rather than by being within and looking out. The narrower of the two is the modern West, and so it creates the narrowness of fragmented science, philosophy, politics, and religious practice, all with this mindset.

The ancient Eastern understanding is broad in its outlook and therefore is about seeing context and perspective, opening (rather than focussing mind-identity), and integration of all aspects of the being. Notice, therefore, why ancient Eastern principle absorbs the modern West, but the modern Western principle can only be said to be a part of the picture. It is for this reason that I have decided to explain and explore, in this text, using entirely ancient Eastern concepts. I have left out all methods of attempting to bridge ancient Eastern and modern Western understanding because this is impossibility (I have looked at points at which modern Western science is moving towards the Classical perspective in Appendix 2- page 859). Many have tried and failed. One cannot look at the whole from a fragmentary point of view. The ancient Truth can only be approached with an opening outwards from the modern contracted-fragmented perspective, it is spontaneously and instinctually accepted. The nature of nature is yin, it cannot be bartered with or found after long seeking, it is dissolved into. Actually to understand in this way is not only cognitive, cognitive is the by-product not the organ of this sense, it is common-sense which is acknowledged, something that is the most difficult thing for the individualistic, modern Westerner, who wants to "have his own opinion", rather than dissolving/ letting-go into a background Oneness. This book will not pamper to these modern terms.

Modern Western methodology breaks things up through the nature of its mental focussing. This relates, in part, to the constitutional differences between Western and Eastern peoples, making it literally harder to see a broader perspective, especially for modern Westerners. Ancient cultures do not use this methodology and view things as a whole, for this anchors and contextualizes the seeming individual. It is therefore imperative that you look at this text in this way. For example, if I say gallbladder in the text, you should assume not only the organ in a physical sense, but also, incorporated into the nature of gallbladder are all the superficial/yang region tendons of the whole body. Similarly, by tendons, I do not mean simply the physical substrate, but the energy of holding and drawing inwards and accumulation of the sour flavour (we will learn to do this). The point is: this is a textbook about ancient philosophy, not modern fragmentary thinking. My suggestion is that one uses differentiation of background from foreground, rather than simply an eclectic approach, which really is about being so overwhelmed by parts that one can't see the whole. In this way, one can identify what causes fragmentation and thereby see it objectively. What causes fragmentation from the whole is a pathology emerging. It is focused on, out of context.

Modern biomedicine is a fragmentary medicine that creates a legacy of fragmented patients. Be aware of the view/ words you use to explain things, which often dictate the way mind-identity in you is working to break things up. The words in this book, I hope, draw things together.

Modern Western physiology is founded entirely on the notion of the brain being the controlling factor. The upper body is not only focused on to "find happiness", but also to understand, somehow, the root of dis-ease and how to change it. This is because Modern Western physiology is based on the assumption that we are our brain-minds, that we are our heads. Quite the opposite is true. In fact, the head is merely the flower of the structure. The roots are in the body and organs and legs of the structure. Hence all we find in the head are things triggering themselves. It is actually the body that feels and can respond directly, without the brain's perpetual controlling involvement. This is the Natural State. The point is to take a broader view. The head is above, and the legs are below, so how can it possibly be that the head is the controlling factor? It is like suggesting we put the last tile on the roof before we work on the foundation. There is always another biological theory that expresses how another part of the body is, in fact, connected to the pituitary or pineal gland. Had they thought that perhaps the whole body is one, with the inside as well as the outside? No, because they are focusing with their minds in a fragmented state, rather than feeling unity within and without.

Chinese philosophy developed differently from, perhaps, any philosophy in the world. From prehistoric times, Chinese people have been in China, where they have been without invasion but have had much turmoil and change. Chinese understanding has developed from the roots of humankind. No other large civilization of cultural heritage, with the exception of the Indian culture (and many smaller groups of indigenous peoples), can say the same, up until the modern day, when the Maoist revolution practically destroyed the cultural heritage of China, in one fell swoop. This was fed by modern Western philosophy! This many-thousand-year-old culture and the nature of the Chinese people, which could understand and build on what came before, has formed a medical culture, that is not so much about a defined term, such as medicine but in fact is the health-instinct. We investigate, through the ancients, a map of intuition and learned cultural experience about what it is to be a human and what it means to be unified and connected to the universe, at one with everything. There is nothing to fix or to do. It is simply the realization to let go of "learning" and to simply be, thereby ending the process of abstraction that gets in the way.

A group in Japan, who studied both Classical Chinese Medicine and Modern Western medicine, decided to make a clear differentiation between the broad methodologies they were being exposed to. They decided to draw all phenomena into two categories: (a) made by Heaven or Nature (although all things, including mind-identity, are natural, and everything formed from it is also natural); and (b) fragmented by the human mind-identity. In this way, they could see what was truly Nature and what was "natural" for mind to split things up, although "natural" does imply "unnatural"—which of course does not exist in reality! This is the process we will be undertaking in this book, in order to get to the roots of what differentiates modern Western mental cognition and the culture of the ancient East, which is often traditionally based in pure observation of Nature. Differentiation itself is a process of mind, yet it is a movement back into the context to see what Is—to understand something deeply and not to confuse it with that which looks similar but isn't, for example the natural bitter taste of chicory versus the super-charged extracted bitter taste of Aspirin. Also, mind as a tool of Oneness is very different from the dis-ease state of mind-identity. Please differentiate the key concepts of this book: mind-identity/dis-ease versus contextualised mind/cure-medicine.

If abstract cognition is the dis-ease, then this Classical expression of medicine or perspective shift in unlearning/ unravelling to be one's self, is the cure. It is one of the greatest of all philosophies. It not only outdates most other cultural philosophies, it also agrees with all of them and adds clarity and unity to them, thereby absorbing them all within it. The Tao is a life manifesto. The classical period of Taoism gave us a glimpse of a cultural root that had no morality or set way of being; it was an anarchy of harmonious being, which essentially is what Nature is. Classical Taoism is simply a profound and clearly rendered voice for all the naturalism in the world, whether seemingly destroyed by religious doctrine, invasion, war, or whatever. Still today, the voice of the Tao can be heard as a call back to Nature.

Another important point is the invention of the joke. Comedy is about the most potent form of medicine. All jokes are about breaking what mind-identity goes into, something that it cannot conceive of. Notice how animals do not laugh. They have no-one to laugh about; they are already free. The deeper that humans can truly laugh at their total separation from unity and how unbelievably trivial it all is, the better.

Acceptance/dissolving/relaxing/ripening/decaying

When we talk about "acceptance of unity" or "Oneness of things", in Chinese this refers to Wu, the background Oneness and Stillness, the Mother that all life is from. The best word to describe the process of acceptance is dissolving. Why? Because dissolving is a totally non-forceful/non-acting yin mode of being. It is a purely natural process that does not impose structure or order, which are forms of mental conformity. Hence, whenever I talk of acceptance throughout this work, the words dissolving or to dissolve can be used in its place, if this is more comfortable to the reader. If one can see acceptance as the process of ice melting, dissolving from the iceberg into the sea around it, or better, lava from a volcano cooling off and dissipating its heat into the sea, this is a beautiful way of looking at the hardness of the egoic mind-identity dissolving back to the still unity beneath; silhouettes are another good image of this. This will be explored time and again, through the text. However, that acceptance-dissolving expression is the best way to observe the end of mind-identity. These same words can be described in other ways: drop away, let go, open from heated-contraction. All of these are the same. They are about losing one's self, the loss of one's identity as the identity of life in a box, separated from everything. When we begin to open to the idea that there is no-one, then we dissolve mind-identity into a joke, and there is freedom/ no-action. From contraction-tension-heating there is a transition to total relaxation-freedom-and cooling or calming.

There is no method to this, simply a question: What do "you" feel? In this question lies all of the answers. The Essence/ Source of life cannot feel and see the world as it actually is and views from the emptiness (headless-ness) that we actually are, and thinks , all at the same time. Hence, to notice the body and the energy within the body, draws us naturally back home to our-non-self! There is no method in it and this is why the Tao has no mental interpretation, does not get involved in words or phrases, and does not torture the body through contortion or focusing mind-identity, in order to get some result. Instead, it attempts to ask the question: what do you feel? Then the spontaneous movement occurs. This is like Taiji Chuan, which is intended natural movement. This also occurs during the stillness of Qi Gong. These forms were derived from natural movement. In the states that we learn them from teachers, they are often contrived and structural. However, what they represent is something that is close to natural movement. And this is what we are doing. However, as soon as something becomes a practice, it loses its potency; it becomes part of the mind-identity structure and is not felt. This is why any structure or method will yield no liberation from suffering. We leave the question open-ended and simple: what do "you" feel right now?

Before you continue with the rest of the book, have a look at the appendix, which will give you a brief history of Chinese philosophy and differentiate the periods when various texts appeared.

Remember, the key reason for this work is to explain a principle to you, in every conceivable way, in the context of Chinese Medicine. There is, in fact, only one root principle, which is yinyang in the context of stillness. I will be constantly reiterating this. Once you have the gist of it, you can apply it to anything and everything yourself, and this work will become totally useless and confining for you. Then please give it to someone else, burn it, or use it as a doorstop or whatever, but please do not attach yourself to this book or the words in it. Please look beyond this and find that which resonates with the essence of Source within you, and so to freedom.

Keep clear the following notions as they have been described in the previous text:
- Wu —Void or Stillness
- Qi/Ki/yinyang, the sea of energy
- Mind-identity/unconsciousness/ego/suffering/someone
- Contextualised mind/consciousness/end of sufferance/no-one
- Mind/spontaneous thought with no attachment
- Body — Jing
- Spirit — Shen
- Body-spirit/Jing-Shen (yinyang)
- Unity/cure
- Fragmentation/dis-ease
- Principle
- Differentiation
- Acceptance/dissolving/relaxing/ripening/ decaying

No-oneness/ non-self is that still timelessness which is before change, and that which change dissolves back into. Change is yinyang. Change is the coating of stillness, which changes with the seasons. Embracing both the worlds of Nothing and Everything, sees the paradox of the Universe to be As It is.

Section A

Foundation and Constitution

"The subject is not in the world".
—Ludwig Wittgenstein

Part 1: Foundational Ideas

1.1: The Study of the Classics, and "You"

i) Before the Word, Was Stillness

In the pages that follow, we will be looking into the Chinese classics as the foothold for our mental understanding of Nature. However, first we should pay homage to what is truly at the root of all such literary expressions—intuitive, instinctual sense. The ancients, whose literary activity reached its efflorescence during the Han Dynasty, were connected to their own ancients, who in turn, through shamanism, had handed down the root principle by which to live. This fundamental sense can be found throughout the classics. The root principle is, of course, the Oneness that unites all. This is the unity of Wu (Void), God, Love, the Un-manifest—or however one wishes to designate our Origin. Thus, this is not only a Chinese understanding, but is also expressed in the philosophy of the West. Parmenides (475 BC), who lived around the same time as Buddha and Lao Tzu, stated that "nothing comes from nothing". However, Heraclites, who also lived in the same era, was drawn into the illusion that change is all that exists, that Wu (Void) is not the foundation. The Parmenidian and Pythagorean tradition, which possibly included Socrates and Plato amongst others (although hidden within their work, because to utter lineage of this kind would be sacrilege, see: Internet Reference 1- bibliography) dominated, although it became diluted with each consecutive generation.

The understanding of the nature of the universe of change (post-heaven; see page 32) forms the main theme in the ancient classics, as well as in the medical classics. Their foundation is that we are all of the same Origin, and that this Origin, in relation to the universe, is still—without the constant change or cycle of life and death. In fact, the change we perceive is only due to the function of adherence/ attachment of the mind to images of the past, so seeing what is present as a change from the past. The Origin is stillness in relation to movement. Hence, our instinctive connection to this allows us to see objectively and to feel at peace, no matter what is occurring subjectively in the world. The form we take, the spirit energy that moves us, as well as the combinations of these, will play out according to natural rules, including the impediment and abstraction of mind-identity and its dissolution.

The background root is called Wu in Chinese. A life in connection with Wu is called Wu wei, which means "without-preconceived or intended (mind-identified) action" or simply "empty-action". The ancients who wrote the Han Dynasty texts using symbols, used as few words as possible. It is therefore on this basis that we must view everything written in the classics during this limited period in history. The books themselves are of no intrinsic value. Their words are only signposts to the truth of total unity and are from the Origin of stillness. As we will discover, the classics of the true Origin are the ones that express the reality of Wu wei.

The words used in such books are always simple. They are ways to acknowledge the unexplainable. Words attempt in various ways to describe experience, which experience itself outstrips. We derive meaning from words due to collective experience of the past. So when we are using words, we are engaging mind-identity and thus engaging with the past. This can draw us into delusion, or alternatively, draw us into the present. The classics (and, in our own age, On having No head by Douglas Harding and As It Is, by Tony Parsons, See annotated bibliography) amongst other works, can help us to see what Is and can at least show us what is inherently true about the predicament of mind-identification which we, as humans, find ourselves within. Books often draw us into imaginary worlds and abstractions of the imagination, expansion. There is nothing wrong with this, but ask yourself: can you put mind-identity down when you put the book down? It often doesn't occur because mind-identity has been lit/ignited/ expanded, rather than questioned/cooled/realised. With the ancient works of Taoism and the other works mentioned in this text, mind-identity is loosened. One loses interest in that as life's focus, for each time you read these books, there is less of "you", rather than more accumulated, which is the opposite of how we learn—an unlearning, in fact.

If we look into the nature of words on a page, such as the one you are now reading, the black represents yin and, therefore, in its nature is shadowy and indistinct. The white of the page highlights the shadows we put in. It is as if the white is secondary and the shadow of the words is primary. The words are uncovered by the white. Hence words are interesting, as they, in themselves, point to something that is behind. The white makes the words stand out, gives them outline or definition from the world of shadow. This is the old understanding of black behind, white foreground. The origin of lettering is actually carving a piece of stone, or forming indentations in a material to form dips, which create shadows. This again is the clear picture of writing. We are actually carving away the whiteness of the world to get to the blackness—the Stillness behind the page. We are attempting to reach Stillness, form and yin, and a relationship to Wu. At the time of the classics, the written word had profound significance, was of "spiritual" significance, as were most written documents of the ancient world. Today, words mean very little, and we forget that they attempt to reveal the underlying or the very origin of things, through absence, rather than addition.

In ancient China, calligraphy, painting, and artwork representing forms, are drawn in a special way. First, accuracy is not the intention, because nothing is absolute. So there is no reason to create specific accuracy in representation. Second, the shadows are drawn on the page. The shadows are the ink or the paint, and the actual outline image is not often drawn in Chinese and Japanese artwork, until it is influenced by the West. The reason for this is a cultural difference. Some ancient Eastern peoples understood that everything is fundamentally based in Stillness and unity. Why, then, draw outlines to differentiate and separate? For the ancients, the way to view was always from Stillness (a looking inwards), looking outwards. For the Modern Western mind, separation is everything. Modern Westerners seek to find wholeness through acquisition, which is their way of trying to end the anxiety of existence. They destroy that which they cannot obtain, or they offer illusions to bow down to. When borders are created on maps and outlines drawn, one knows this is the work of mind-identity, for no such borders exist in reality. Here there is a looking –outwards only.

Further, we can notice the three-dimensional perspective of much of Modern Western art, especially paintings (pioneered by the ancient Greeks but taking hold mainly in the 1400's) where a flat two-dimensional canvas is converted into something which has the supposed perspective of distance within it. Ancient Eastern art, uninfluenced by the modern West, has no such expression; the renderings are utterly two-dimensional, as flat as the canvas. A child's perspective, too, is without depth; it is totally flat to the canvas. This is because for a young child, time and space do not exist. Depth of field and perspective is something thought about, not directly experienced. This too is the ancient Eastern (and universal indigenous cultural) expression, a development from the child's drawing to a more ornate expression, but still in two dimensions. Original seeing, or seeing the world from one's true nature, is very much about the actual, the obvious: people very far away are actually very small, and when they come closer, they have grown in size. This sounds ridiculous, the uttering of a child's view, but if we view things from the perspective of the Stillness within, this is always our actual experience of the world and so is our primary basis to understand the world. The dimensions of time and depth are later additions to our timeless knowing of what actually is, now. Lewis Carroll often played on the difference between the adult (mind-identity looking outward only) and the child (deeper, looking inward, AND looking outward) reality in his books Alice in Wonderland and Through the Looking Glass. These important works show us the perspective we lost as children; they move into a world of time and space, where we are on a journey away from our self, rather than rooted in the central experience of emptiness. This is also expressed in the work of Douglas Harding (e.g. Harding, D. E., 2006. On Having No Head: Seeing One's Original Nature).

The fact that Japanese is one of the very few languages that has its original written form and presumably grammatical notation, dominated by women, gives it a rare, expressive insight. (see: Internet Reference 2- bibliography) Words are used as fill-ins for the silence of being, rather than the other way around, as in the yang-mind-identity-filled modern West. The yin quality of Japanese expressions is difficult for modern people (especially Westerners) to understand because it is not a language that is mind-identity-orientated or mentally expressive. There is very little use of the subjective notion of "I". It is not a subjective, looking-outwards only language; it is a looking-inwards language at its core. It is one that allows for the gaps between words and the intention behind sentences to emerge. This shows us how words can be used from a totally opposite perspective from our own: to hint at meaning—and for silence to invoke understanding, rather than to further obscure meaning, by use of a lot of expressions and thereby a lot of mind-identity. (It is true, though, that Silence itself, when it is about withholding expression, is much "noisier" than natural sound, like the rainforest or cicadas which express sound spontaneously from Oneness. As always it is the quality of an expression that is its clarity; yang-expression can be clear or distorted, a lot or a little.).

It is grammatical juxtaposition that gives words meaning. This seems reasonable because we see in all languages, particularly in Japanese, that communication on this basis is possible. However, if the feeling aspect of communication is heightened by its lack of a focused, narrow description in words, much is left to the intuitive experience of the present situation. This renders words more efficient and makes them valuable rather than overused. This is also why the classics of medicine in China are so revered. They are valued not only for what they say, but for what they imply. One must try to sense the feeling of the author to know and understand her/his work, and this means that one must be as mentally unbound as the author was when writing.

The Japanese and Chinese languages and cultures are utterly different from what they were before modern Westernization destroyed so much of them, driven by its mind-identity of dis-eased masculine energy. However, the more ancient forms of these languages show a feminine approach to ideas. This can give us understanding through triangulation of metaphors, using few words, whereas many words say very little, have superficiality, and often lead to too focused and specific mental constrains to render a generalized or contextual viewpoint. Few words said with grace and quality can be powerful and direct the senses. As one goes further in understanding the picture of the Ancient and the East in relation to the Modern West, we find a key point which is that the Modern expression is really a child (yang) held within the mother, or background (yin) of the world. The roots and foundations of universal sense and clarity is something hidden to modern Western peoples, only because we are too narrow in focus and can't see what is simple all around us; this sense is lost. The Ancient people all over the world, exemplified by the Ancient Chinese, are an expression of the root and origin and represents an acknowledgment of the background, the earth, the mother which has been forgotten in the mists of masculine mental identification. The reason for why mental identification has taken over?....natural growth....., but the result is the "child" now contends to kill its "mother"! An impossibility for the background Oneness or "Mother" is root, the "child" or mind-identity of humans therefore attempts to kill-itself, seeing it and it's Mother as separate from an idea of "self".

If one can possibly listen to words in the way one does to musical notes—hearing the sounds as if they were a bird's song, without understanding the exact meaning, one can get the general feeling, the essence of language. As Wittgenstein says, "Essence is expressed by grammar" (Wittgenstein, L. 1953). If one is to engage with mind-identity, one loosens connection with this song that uses the words—with the broad perspective behind the words. This is how we get lost in mind-identity and in fragmentation rather than hearing the context. The latter allows us to deeply understand, through the experience of hearing the song of someone, to gain insight. Usually the song is far more important than the specifics of the language, as far as true understanding at a deep level is concerned. Words on a page are shadows of form and yin—indistinct. If one looks at words and expression as outlines to the absolute, one is thereby training mind-identity to fragment itself further. This is why Lao Tzu writes poetry. Poetic imagery, with its lack of boundaries and its use of metaphor, adds even further indistinctness to the words on the page. This is the writing of the Tao. The words do not form borders or outlines, but give an impression. To this extent, we can say that all visual expression in art is impressionistic, or that "realism" does not exist.

ii) The Classic Texts: The Classics of the Tao, within the context of this book

Reading the classics is an important part of the training of students of Ancient Eastern philosophy. How we do this depends on the degree to which we wish to delve into the subject. Some scholars will learn the classics in the original languages and study the ancient characters written during the Han Dynasty. In Japan, the Chinese classics have long held an important status (see Appendix 1-page 762 for histories), and so translations into Japanese are among the best in the world.

In the last 100 or so years, translations have been made into English. However, English is quite different from ancient Chinese, which was originally written and understood in pictograms-symbolism (today in China modern characters have replaced or simplified ancient ones in many cases). In ancient Chinese—and in ancient Japanese too, (to the extent that Japanese borrows ancient Chinese pictograms)—deeper understanding of general-feeling or notions, not only cognitive sense, behind words is gleaned from metaphorical images through the juxtaposition of the pictograms. Compare this with English and generally modern language, where specific and cognitive meaning is gleaned. Of course it is often true that "pictures tell a thousand words", but it is also true that a thousand people will see something slightly different in a picture. Words make pictures more concrete/specific, which is often why artists refuse to put titles on their pictures to describe their content. Ancient scripts are non-specific in almost all traditions, as they needed to contextualise mind-identity, rather than make points to narrow it, using few words to reach a broader plane and therefore greater understanding. Such languages may seem basic to us. However, the meaning is often far more profound than if it was linguistically specific. This is just as true all over the ancient world and in all texts—for example, in Egyptian hieroglyphics, the Bible, the Koran, and more recently in the poems of mystics such as Rumi.

When looking into works such as these, we must do so with openness. This is not blind faith or a worshipful mental attitude, but one that avoids arrogance and narrowness: looking for things that you want to find and thereby finding them and losing the broad perspective. If we look towards these works, we must do so in the light in which they were written, which is most probably from people, whose clarity of being was of the utmost; otherwise, what they had to say would have been lost or would not be relevant today. The first chapter of the Su Wen (the first book of The Yellow Emperor's Classic of Internal Medicine: Su Wen or Fundamental Questions), for example, immediately makes its relevance apparent. Instead of using the analytical mind to break something into parts, we have to look at this synthetic work as attempting to bring everything together again, which is why it is held in such reverence—much like an intuitively channelled, often called "religious" work. These works do have a connection to our Centre, so reverence to the degree that this is "curative", is a given. They speak of the roots of the understanding of the shamans and of those connected to the universe in a way we have lost. They are literally the thread of life we have within civilisation, the books that perhaps ends the need for books if they could be understood directly.

Translators have enormous problems putting different cultural meanings into words, especially those from past eras with obscure cultural references. However, the brilliance of most of the important classics is that they are timeless in their quality of understanding—and this can be translated, however roughly. In fact, as time has gone on and the world has opened to cultural differences, translation has become better and clearer. I know this through my study of the practice of Oriental Medicine, in which I have compared translations of the same text. Although I am neither a translator nor a linguist, I have learned much through gaining the thread of metaphorical meaning imbedded in various translations. Being reliant on translators is perhaps not ideal, but as long as one does not rely on a single translation, one can gain a more comprehensive understanding. Often teachers suggest that clarity can only be found through reading the original word. As I have expressed, the ability to "do" something doesn't bring one a step closer to understanding; these two things are very different. I know several linguists that read the Classic texts and practise medicine with "commitment", yet have very little understanding-sense of the Essence, and there are many people who intuitively heal and have no notion of Classic texts. Please bear this in mind before judgement.

The books expounded below are the bases upon which all other works in Oriental medical philosophy are written; they are the keystones for understanding. They are in chronological order and also in order of importance. One needs the foundation of the previous book before one can move on to the next. These books hold the information, but they do so in a way that requires perpetual interest of one's own sensory experience to corroborate and elucidate, or else they are just worthless paperweights. These books tell the story of ourselves, the story of being human, its suffering and the end of suffering.

There is some confusion concerning Taoist literature, which I must point out. Taoism, in its true sense, has no literature attached to it; it is a pure expression of natural being-ness, which is inexpressible in words. In fact, words are the very thing that—once uttered—separate. The Tao is non-dual and as such is the uniform background to everything else. As we will see, the most important works, the Tao Te Ching and the I Ching/ Eki kyo/ 易経 which are the basis of the Taoist philosophy or expression, are really non-works; they are attempting to be the beginning and end of the requirement to write further. Confucianism, on the other hand, is a very different expression. Confucius wrote his works at the same time as Lao Tzu wrote the Tao Te Ching, but they are diametrically different in expression. Whereas Confucius split things up and categorised and wrote numerous works, as well as being involved in political activities, Lao Tzu wrote just the Tao Te Ching, stating that "the Tao that can be named is not the eternal Tao". This clearly shows a non-literature tradition. All of the works after the Tao Te Ching can be considered to be part of the Confucian expression. "Pure" Taoism is rare and not often seen, especially in today's modern world, as is perhaps found only amongst indigenous peoples now. When I refer to Taoism in this text, I am talking about this classical heritage, rather than that augmented by the systemisation ideas of Confucianism (see historical perspectives in Appendix 1). Also, "Taoism" in itself is impossibility. The religious Taoism, that sprung up from the expressed Tao in the Tao Te Ching, did so within a Confucian ideology of holding on to and creating a religious practice out of something which was impossible to practise. This too is really a form of Confucianism, in a way, or at least separatist mental activity (modern materialism is a simple evolutionary step from this ideology). Tao is the root, and so this means the beginning and end. Classical Taoism, therefore, is a very small point in the history of Taoist ideas. Neo-Taoists, of the early part of the first century AD, attempted to bring it back to this, but again created further add-ons to the picture. It is very difficult to just be with what is and leave it at that. This is the nature of the way the human mind expands (see Appendix 1), and perhaps the big joke of this text itself! Anyway....

The Yellow Emperor's Classic of Internal Medicine, which was likely written around 200 BC, although it is often attributed to very ancient times in the Classics, indicating that the origin of the understanding is pre-historic. This could be considered a Taoist work. However, the Yellow Emperor himself is not a "real" character but an archetypical image of the dawn of "civilisation"—cities, town, and culture including medicine. Before this was the heritage of the shamanistic cultural heroes of Fu Xi/ Fukki/ 伏羲 and then Shen Nong/ Shinno/ 神農 (see Appendix 1), who were expressions of the far more Taoist unwritten practices of divination, agriculture, and close connection to the environment. Huang Di/ Kotei/ 黄帝, the Yellow Emperor, was an expression of everything that Taoism was not—a movement towards city, structure, politics, and medicine of all things! Taoism is not medical. Taoism is a pure state of natural being, unadulterated by influence from mind-identification and therefore simply interested in being. No roles can be placed, so a practitioner is therefore very much a Confucian idea. We live in a Confucian world based on a Confucian, individualistic tendency. The idealism of Taoism is far from what we are living; it is also the root of medicine. Numerous people claim to be Taoist, based in Taoism, or have so-called "classical" Taoist connections, due to association with a particular book. This is bound to cause confusion. No book can hold Taoism, and this is what the Tao Te Ching (a book!) is all about expressing: don't try to write it down because it can't be written! No rule or mental concept can hold or behold the essence of what is meant by Tao; this is why it is the most impossible subject to write about!

Taking the above into account, the following will be the way this book looks at the Classical sources and understanding: This book "Classical Oriental Medicine" could very well be called "Classical Taoist Medicine" but due to the fact that this draws attention to the idea that Taoism is "something" and the fact that the whole ideology of Tao would undermine the idea of something called "medicine" as differentiated form any thing else, makes it impossible to give this title to the book. Hence this makes the work Confucian in its ideology of attempting to hold-on to data and to categorize. However this book aims to resonate with the borderline between Taoism and Confucianism, which is a hair's-breadth balancing act, and which are chalk and cheese in their difference. One of them nourishes, the other is for writing on black-boards! What Confucianism. as well as religious Taoism, Buddhism and other religions of this nature , which are all of the same ilk, has achieved is two things: Preservation and distortion. The Confucian type ideology stores the texts about the "master", like the sutra of Buddha held by the monks or the gospels of Christ, held by the Disciples. The point is that this isn't the actual message; it is the preserved teaching. Often it is an interpretation of the teaching. The second aspect is that the words are then distorted to form hierarchical order, and highly yang-masculine dominated control tactics and ideologies. What this book aims to do is to be as close to the clearest expression of the Tao Te Ching and other important works, while noticing the creeping in of the Confucian mind-identified categorization. The constant process is about differentiation of what Confucianism is, and what Tao is not. Throughout, these notions are a hair's breath apart and it is likely that at times, what are Confucian categorizations and ideas will seem to cloud the Tao of clarity. However I will constantly try to connect to this. The book will try to keep to the understanding of the preserved word and the meaning of these words, from the clarity of perspectives today, which gives the ancient expression a new way of being expressed, such as the work of Tony Parsons (see Bibliography). This helps render a pathway through the thicket of Confucianist ideas to see the naturalism, or the foundation, beneath. The annotated bibliography therefore tries to provide links and information which veer away from the Confucian and towards the Classical root. Also the avoidance of Taoism and the various forms and styles of it in a religious expression, which is really another form of Confucianism, as opposed to Natural sense, or Tao, is differentiated constantly. Although it is difficult to walk along this line, the hope is to show that this medicine has no belief system associated and is utterly connected to that which has no name and is part and parcel of natural instinct. As a result this book tries to realise that in the utterance of a single word that moves one from the place of simply being a natural human expression of nature, to being someone who has the "root of knowledge" and holds "the truth", is again the difference between wisdom and knowledge or reality and illusion, for always the notion of separation will be that which creates hierarchical ideology.

I Ching

The I Ching , or Classic of Change is the first classic of Taoist understanding (having understood the above!). Related to the divinatory arts, this work has a lineage directly to the shamanism of the ancients, where the unity of everything was very clear. The study of this unity was achieved using the I Ching as a guide for mind-identity to broaden its perspective, to see beyond the so-called individual—to see oneself or an event in the context of everything-ness.

The origins of the line-images of the I Ching are ancient, relating to the oracle bone readings of shamans from at least 2000–3000 BC. It is difficult to find historical sources; the line images are older than most other aspects of the culture still surviving. King Wen rendered them into 64 I Ching hexagrams around 1000 BC. From this time onwards, this became the fundamental classic and was used by all Taoist scholars as a method of inquiry into the nature of the Tao. The "study" or interest in the I Ching is something all scholars of the classics must take up, but an explanation of it here would fill the entire book. Suffice it to say that this work holds the beginning and end of all other works within it. An understanding of it imparts clarity and the liberation not to need to use it.

The 64 hexagrams of the I Ching, as we will see, have a connection to the pre-Heaven state of the universe, prior to change, expressing the objectivity of the Origin. It is, most accurately, drawn in a square representing yin and Stillness. Because the I Ching is used by/formed by human beings who are part of the post-Heaven (yinyang), or the universe of change, it is used to see the pre-Heaven or the structural context of the world of change. Beneath the post-Heaven universe is the pre-Heaven universe, and beneath this is the enveloping Stillness of the inward-viewed "I am", at the background and centre of all things. This Origin of Stillness is arrived at in the Tao Te Ching.

The difficulty, as with all the root Classics, but in particular with the I Ching, is that the true Taoist- naturalist expression of yinyang, without any add-ons or hijacking from Confucianist ideas of moral justifications, is very hard to find in any commentary in any language. The I Ching is the most deeply yin and also most fragile of the Classics. Originally it was an expression of simply what there is, not what there could be or what there should be, but just what is. Yin was not seen as "inferior" and yang as "superior" but there was an understanding of just how things are, without necessarily much interpretation. It is now almost impossible to find much of this kind of clarity in humans due to the sense of separation and fear of death that go together in our society: the expression of not allowing things to be and making yang better because yang pertains to life and yin to death. Humans pertain to life, but this doesn't mean life is "better". It is therefore important to understand that when looking into the I Ching and Tao Te Ching especially in translation, the essence of the book is sometimes very hidden under the words. In fact the hexagrams themselves and the basic understanding of the trigrams that create the hexagrams can aid us much in attempting to understand the nature of the yinyang expression that has no moral attached. My suggestion is that you steer clear of Confucian commentary (for our purposes of understanding medicine) and the interpretation of this deeply Taoist text.

Tao Te Ching

The Classical Period of Oriental philosophy started around 500 BC, deeply influencing Chinese history. A shift occurred in which spiritualist-based thinking evolved into a more universal picture of energy and its conjoined ebbs and flows in the universe. Instead of numerous gods, spirits, and entities, a move was made to depict reality by unifying these phenomena into a single principle. This move, which had no belief system attached, related all phenomena to their Origin. This we describe as Wu wei—a perpetual connection to the Origin of life. The I Ching philosophy opened the way for this understanding to be heralded by the whole society in the form of Taoism and Confucianism in this next period of China's history. Many described its perfection in wholeness, but none greater than Lao Tzu, a perhaps real, perhaps fictional character, to whom the Tao Te Ching, or often translated as the The Classic of the Way and Virtue is attributed. In this work, from beginning to end, the reader finds repetition of an expression of the unexplainable nature of being, without attachment to past events or to phenomena occurring, but to the unfolding of this authentic being of the self and the unity of all phenomena. It is the second foundational classic of all classical Taoist philosophy and so becomes our basis for revealing the importance of what we study. All scholars of Taiji Chuan, Qi gong (i.e., the internal, physical therapeutic exercise/arts that cultivate energy/health), and traditional arts read it.

This work is not, as most believe it to be, a teaching. Lao Tzu was not attempting to teach Wu wei (non-action), because it is impossible to do so. All that Lao Tzu does is point the direction, by negating that which it is not, but never expressing that which is. This work therefore shows the limitation of the written word to reveal ultimate reality, yet draws us to the brink, where mind-identification is potentially released.

These two works are the basics of Taoism. If we apply the understanding in these two works, we have the possibility of being able to clarify our minds and explore the world with our feelings, backed by or rooted in Oneness. These books are training tools, signposts of observation of the destruction of being and behaving, based on the outer dream of mind, which seems to fragment the world and ourselves. One realises from the context of these works that one is able to subtract rather than add to knowledge, to allow openness, to feel, becoming animal- and intuition-based, rather than reacting to an echo of an event. These works form the core basis of all taught knowledge, as well as of that which cannot be taught.

There are of course other Classics of Taoism, such as the work of Lao Tzu's contemporary, Chuang Tzu. But as with all philosophies, gradually, as one moves from the root, there is a dilution of the essence, and while Chuang Tzu's "Inner Chapters" from his lengthy works (some of which are not attributed to Chuang Tzu, himself) do hold to the main expression of Lao Tzu, still the Tao Te Ching is the base.

(Note: I have quoted the Tao Te Ching many times in this book. The version of this work used is my own and is not a direct translation, but an interpretation of the text that I feel draws to light the Non-dualist understanding of Taoism, beneath its often highly Confucian-dualist interpretation which I feel loses the essence of what is being said. For further interest in this interpretation please see the previously published book :-
Nassim D., 2011, Tao Te Ching by Lao Tzu: The Classic of Naturalness and it's Innate-perfection by "Old Man", Stone: HI publishing.)

iii) The Classic Texts: The Medical Classics
The Classic of Moxibustion (plus other texts of this period)

This work is an aspect of a series of collated writings about medicine, it is the earliest discovered work of the Han Dynasty period of medical classics. The title of the work has been given the archaeological references MSI.A called the "Cauterization Classic of the eleven vessels of the foot and Forearm/ Zubi Shiyi Mai Jiu Jing" and MSI B called the "Cauterization Classic of the eleven yin and yang Vessels/ Yinyang Shiyi Mai Jiu Jing". It was discovered in 1973, in the Mawang Dui tomb, in China. This work comes from the period just before the famous Yellow Emperor's Classic, perhaps around 200 BC. One finds in this work, key descriptions of the meridians of the body, before they were formally arranged into The Yellow Emperor's Classic. Eleven meridians are shown, with their pathways. This is important, as the later Yellow Emperor's Classic describes twelve meridians. The discrepancy is one of the key points in understanding Oriental Medicine and will be discussed in Section B of this book. This point gives the Moxibustion Classic an obscure but important role. It is also filled with shamanic rituals and under-standing, long-since lost, much of which have little relevance other than to the people who performed these treatments at the time. However, it gives us an important glimpse into pre-systematized medicine before The Yellow Emperor's Classic.

The Moxibustion Classic is not the only important work drawn out of the tomb of Mawang Dui; another important work is the Huainanzi/ 淮南子 (or Book of King Huainan, 180-122BC approx.). This work contains numerous accounts of the astronomical, astrological, topographic, and seasonal ex-pressions of Taoist thinking. It also covers other aspects, such as ideas about the origins of life and the human being within the broader picture of the cosmos. It has been drawn into connections with The Yellow Emperor's Classic, as a foundation for acupuncture ideas associated with the Chinese clock cycles (Golding, 2008). It is important to note that this book has also been used to create the ideological foun-dation for other theoretical positions added to The Yellow Emperor's Classic in AD 762 (by Wang Bing/

王 冰). The addition of Wang Bing's material has been considered by most scholars to be his own or his teacher's work and not of the same period as the other aspects of the classic. This is strongly refuted by the Wang Bing schools, which have developed in association with herbal medicine (please see the work of Dr. Arnaud Versluys and Dr. Heiner Frauhauf; also see discussion of the 5 phase and 6 Qi, in Appendix 2). It is important to differentiate Wang Bing's Confucian dialogue from the Taoist lineage, to differentiate its addition much later in history and also its movement towards more externally applied structuralism, rather than intuitive sense, yang not basal yin.

The Huainanzi is a work open to interpretation on numerous levels, as are many of the works of this very early and formative period. The Yellow Emperor's Classic has numerous missing chapters if considered apart from Wang Bing's additional information. This needs to be accepted so that one does not veer away from the central teaching of the Classical Period of Taoism and begin formulating theory that is not rooted in the simplicity of the Han Dynasty material.

The lost scrolls of The Yellow Emperor's Classic must stay lost until truly found, if ever. We need to be clear that in order to view the classics, we have to see them as they are, not trying to add anything to them. If we are to extrapolate, it needs to be without trying to find somewhere in the text to corroborate an idea, but more to see what there is and see the picture developing from this, and then, only then, moving off in a unique expression of this, which is deeply "who you are". Doing this will naturally mean that whatever extrapolations are made, will concur with the root, and be easily seen by others to do so. It will "feel right" to everyone, not just to a few "individuals", because it would not be verified by ideas (by mind-identity), but by sense, because the very base of medicine is in association with our basic senses. Otherwise we are simply attempting to alter history to suit a mental idea. This has nothing to do with the simplicity of Objective seeing in the Classical Taoism of the Tao Te ching.

Qing Nang Jing/ 青襄経 –Green Satchel Classic by Huang Shi Gong (200BC approx) is also a very important work, though we cannot consider it part of the "Inner Medicine Cannon" as with the Huainanzi. This is the foundational Classic of Feng Shui understanding, the equivalent to the Nei Jing, for practitioners of Feng Shui. It is vitally important that we realize the connection between these two expressions: the outer and expanded expression of Feng Shui is the same as the inner and body focused ideology of Classical medicine. There are enormous overlaps in this work. One could consider the Feng Shui practitioner to be the doctor of the out and expanded world and the inner medicine practitioner as the doctor of the inner world. While they both need to express themselves in different ways, it is important to note the nature of the spectrum of their study. Although I will not refer directly to this book, the concepts associated with the Nei Jing that link to Feng Shui, such as correspondence to the expanded ideas, such as constitution of the spirit and further relations to the origins of Chinese Stems and Branches / 4 pillars astrology, this book is the foundation for those involved in Classical Feng Shui. This practice is very well documented, much more so than Classical medicine, by teachers such as Joey Yap et al. The "Green Satchel Classic" is the oldest and most basic rendition of the fundamentals of all the schools for Feng Shui arts and so is the root for all of them. The text is sometimes hard to understand but one can immediately see the connection to chapters of the Nei Jing, which may well have been associated with each other, at the time of writing. This gives relevance to Classical Feng Shui as a combined unity with

Classical medicine, which we will discuss later. This is a referral point to those who are interested in the basis for the use of the understanding of the expanded views of heaven and earth alongside books which have more astronomical aspects, like the Huainanzi.

The Yellow Emperor's Inner Cannon: Fundamental Questions and The Spiritual Pivot
Huang Di Nei Jing/ Koteidaikei 黄帝内経: Su Wen / So Mon 素問, Ling Shu/ Rei Su 靈枢

These two works are made up of 81 chapters each (minus the lost scrolls which consist of chapters 66 to 75 from the Su Wen). This is important for several reasons. First because 9 x 9 = 81, revealing the numerological value of post-Heaven's involvement in their creation, 9 being the change of cycles of Heaven. Second, 9 x 9 forms a square, indicating the association with pre-Heaven Wu (see next chapter). Another significance of there being 81 chapters is that within these works of the early Han (100 BC), one can see the shift that took place in society. By this time, shamanic forms of treatment had been outlawed, and instead there was a systemisation based on the Taoist principles of the I Ching and Tao Te Ching, within the guise of socialised Confucianism. This meant that old ways of individual healers, who may have been (but often were not) intuitive healers but often were tricksters (as is the situation today!), were making way for a unified principle of medicine, which people could learn, embody and intuitively apply. The way of the Tao provided this unification. The exercises and understanding of the sages now had a format for mental discourse and a description in words which intuition had provided.

Within the Nei Jing Su Wen are key fundamental questions (and answers) about Nature, the physiological mechanics of energy, the nature of qi in the environment, and its effect on health and dis-ease. It also contains the application of acupuncture, massage, and herbs to the body, to allow energy to flow and so enable dis-ease to pass. The Nei Jing Ling Shu deepens this study, especially of acupuncture and moxibustion, and further explains the philosophy of a dynamic force pervading the body and all things, through the yinyang of Tao.

These works are naturalism in its fundamental form and the most important of the classics for studying medicine. These must be studied, in some form, by anyone involved in mentally clarifying Oriental Medicine.

The Yellow Emperor's Classic is considered to be the oldest classic that has been referenced in other books and that we still have copies of, today. However, the oldest book is actually considered to be Shen Nong Ben Cao Jing/ 本草綱目 (Herbal Classic of Shennong). This book was a Materia Medica of herbs, which were categorised but it did not have the yinyang understanding of The Yellow Emperor's Classic in as detailed a way. Since it is the oldest and originated in the Pre-Classical Period, along with the I Ching, Shen Nong Ben Cao Jing, it is worth mentioning. The reason I have not included it as a book, here, is because the ancient manuscript is lost. We have only books related by name, from well into the Post-Han Dynasty times; this classic may or may not be close to the original, so this work cannot be clearly identified anymore. As with all texts from this time, we only know what has been archeologically discovered or passed down through generations—scraps put together by people like Wang Bing. We must then realise that the information we are getting has probably and sometimes clearly, been altered and distorted, in order to uphold particular ideas at particular times. We can do our utmost to clarify what is ancient and unified from what is added on and specific/ stylistic, in order to uncover the base principles. Often this is best done by a consensus of those deeply involved in medicine over the years, which is how the original text would have probably developed. Nothing is certain, which is why, using what there is, we have to ascertain it ourselves, connecting to the patients and to the words of the classics, using our senses to guide us rather than having total reliance on theory. This is the natural way.

The Classic of Difficult Issues: Nan Jing/ Nangyo/ 難経

This, the third work of the medical cannon, comprises footnotes to the Nei Jing, just as the Nei Jing provides footnotes for the I Ching. In the Nan Jing, there is a deepening and broadening of pathways of understanding, developed from their source in the Nei Jing, with the writers asking questions of the Nei Jing, analysing, understanding, and developing it further. One could consider the Nan Jing as both a post-script and a commentary on key aspects of the Nei Jing. This provides a better understanding of the way of the body and how to apply the principle of post-Heaven in the form of the five phases, which are the principles most associated with the medical classics.

Readers gain the ability to see the simple within the complex. Since the Nan Jing was written in AD 100 or so, it is a more neo-Confucian work, less arbitrary, more ordered, and more attempting to create a sense of order. In other words, slowly, the time of greatness of the Tao was waning in China, but not yet.

Treatise on Cold Damage and Miscellaneous Disease: Shang Han Za Bing Lun, Later Split into the- Treaties on Cold Damage: Shang Han Lun/ Shokanron 傷寒論 and the Synopsis of Prescriptions from the Golden Chamber: Jinkui Laolue Fang Lun/ Kinkiyoryaku/ 金匱要略

This work (later split into two), was written around AD 150, by a seemingly neo-Taoist scholar, Zhang Zhong Jing/ Cho Chu Kei/ 張仲景. This work is closer to the Nei Jing in its naturalism, less ordered and less structured than the Nan Jing and the later Mai Jing. However, all are footnotes to the Nei Jing. Here, herbal medicine is more closely examined, and the application and understanding of the pathological process, based in the Nei Jing, is expanded further, offering a methodology of herbal formulae within the Oriental tradition.

Understanding of the flavours of the medicines used and how one can apply these to the body are topics only briefly touched on in the Nei Jing. In the Shang Han Za Bing Lun, however, these topics are developed deeply and lucidly, through a master who has understood the heart of the Nei Jing and the adaptation of the Taoist Way in using herbal preparations to a more refined level.

Pathological process is carefully charted, and one sees how the exterior body can be attacked and how the interior body and its relative strength towards the exterior pathogenic factor, is always at the root of pathology. Also we see how deficiency of energy on the interior also derives dis-ease without an exterior pernicious involvement. Hence, interior and exterior factors are described in these works. Broadly speaking, the Shang Han Lun deals with the exterior, and the Jinkui Laolue Fang Lun deals with the interior. Seen to be the masterwork of herbalism, this and the Mai Jing, written around the same time, remain the last exemplars of the truly great classics of Oriental Medicine.

The Classic of the Pulse: Mai Jing/ Myakukyo/ 脈経

The Mai Jing (AD 215–282) was written just after the Nan Jing. In format it is a similar work, but it focuses on the diagnosis of the pulses of the body, especially a subtle understanding of the wrist pulse. Through looking at one part, one can identify the whole within it; this is the main characteristic of this work. Again, the pulse classic is footnotes to the Nei Jing, Nan Jing, and Shang Han Lun, offering a methodology to learn diagnosis in a profound way.

iv) Viewing the Classics and the Practitioner

After these works, there are many others, of course. However, we must be aware that the history of Oriental Medicine did not always adhere to the main principles in these works. The "golden period" of understanding came and went, reflected in the "golden period" of Taoist understanding. In its wake, much illusionary thinking and lack of clarity occurred, resulting in many revivalist periods within this medical tradition, when people reverted to the classics. (See Appendix 1 for contextual history.)

These works represent something important—a single whole. It is easy to believe—if one views the works in a fragmented way—that they do not connect. This is not at all the case. If one can see the context or background, in each chapter of each of these works, one begins to understand that different views of a single whole are being expressed. Again, these are works different from today, and they require you, the reader, to engage and be part of the understanding, to allow it to rub away at one's current ideas. A study of these works opens mind-identity, rather than attempting to fill it. If viewed in this way, one can appreciate a unity throughout these works, as was the unity of the philosophy to which they are connected and of the principles of the Tao. So there is no fragmentary aspect to these works. If they are viewed as a whole, one can truly see what they mean. Otherwise, one misses the point. This is what has occurred in medicine today. Much of the deeper aspects of Oriental Medicine have been lost, as the origins were ignored. Instead, an attempt was made to piece it all together from fragments. As in a practice such as Taiji Chuan, if the body does not move as a whole, getting one isolated limb to move correctly will be impossible.

The way to view these works is much like a pyramid of understanding:

(fig.1.1)

```
                        ┌──────────────┐
                        │ The Classics of │
                        │ Internal      │
                        │ Medicine:     │
                        └──────────────┘
```

The Classics of Internal Medicine:

The Mai Jing

The works of Zhang ZhongJing S.H.Z.B.L and J.K.L.L.F.L

The Nan Jing

The Moxibustion Classic and Nei Jing

- -

The I Ching and Taoist works of Lao Tsu, Chuang Tzu, and others

Other Classics of the Tao and external Feng Shui

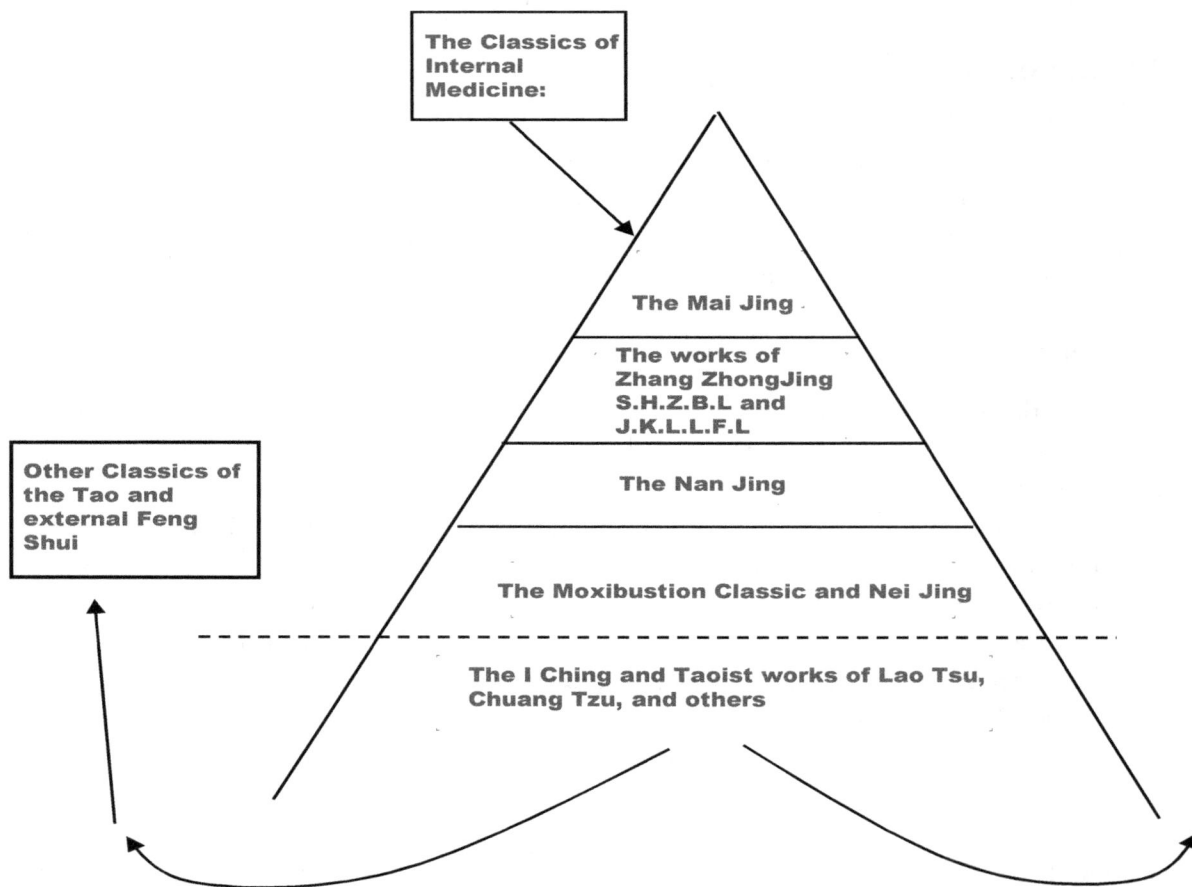

Everything above is founded on what is below it. This gives us a picture of how we can study these classics and what is important in terms of what is fundamental and what is a further elaboration of basic principles.

If we now look at the various methodologies of Oriental Medicine and represent the practitioner as a triangle, the following diagram may be a good expression of how we can view things.

(fig. 1.2)

Faculties of the practitioner's energetic body:

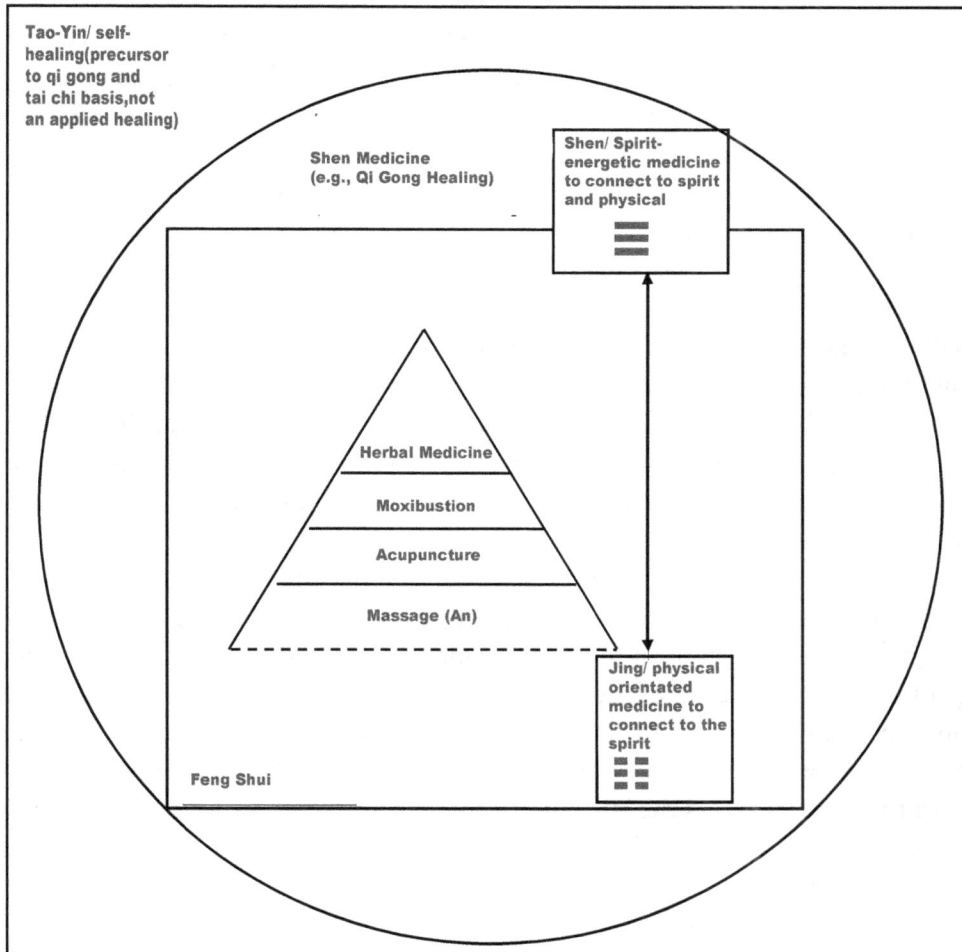

Tao-Yin/ self-healing(precursor to qi gong and tai chi basis,not an applied healing)

Shen Medicine (e.g., Qi Gong Healing)

Shen/ Spirit-energetic medicine to connect to spirit and physical

Herbal Medicine

Moxibustion

Acupuncture

Massage (An)

Jing/ physical orientated medicine to connect to the spirit

Feng Shui

If we view this diagram in the same way as the previous one, we can see the various aspects of the practitioner involved in treatment, from the more physical to the more mental, and beyond that to the purely energetic, which encapsulates the picture of the physical body (drawn as a triangle). As you can see, the classical books involved, relate to the various strata of the practitioner's interest, each building upon one another from head (top of triangle) to feet (base of triangle). Extending form the physical body is the practitioners interaction with the earth (the square) and environment, as well as heaven (the circle) and Wu the larger-outer square.

If we consider Oriental Medicine contextually, we can find all manner of different ways of healing, yet all with a united foundation. This, therefore, doesn't only consider Chinese medicine but all medicine, anywhere; the root principles are found to be the same in each culture, just as the root of dis-ease is always the same. Yinyang, in whatever form or with whatever semantic difference, is simply Oneness - that's what it means. There is a difference between the arts we classically associate with Oriental Medicine and those that draw more into the realm of the shaman-healer. I have differentiated them into those that use a physical medium to impart Stillness/healing and those which are purely an energetic medium, such as Qi gong healing method, to impart Stillness/ healing. Please note, I am only talking about healing methods not self-healing or non-applied healing (Tao-yin), which is the vital basis for the other methods but will be discussed later.

If we look from the most physical form of human-medical treatment, we start with "An"-massage (meridian-based massage) which is the traditional root of massage. Aspects of this became shiatsu in Japan and tuina in China. All these were originally united and were an expression of physical hands-on therapy, focused in energy. This forms the broad base of treatment using the physical form of the body with touch. Needles are an extension of this practice, and massage and acupuncture were often taught as one. In traditional training, after doing massage for several years, the hands and body of the practitioner developed a sense of natural intuitive touch, as opposed to the practitioner just learning the meridians and points. Mind-identity becomes quieter, so the addition of needles can be used as a conduit for energy to pass through. The needle itself has no meaning whatsoever without the stillness of the practitioner to be a conduit of the energy.

Moxa, being a herb and having a slightly separate role from the practitioner's energetic-conduit role, became a further refinement of understanding about the nature of energy points, again probably originating through some sort of physical therapy like massage-acupuncture. The practitioner needed to know the energy of the moxa, as well as of the points. It became a step towards a more mentally orientated approach, memory and knowledge supplying some of the treatment's basis, as well as intuitive touch. Lastly, herbal medicine, a far coarser strategy than direct energetic healing, could be used to supplement and further benefit the patient. This is the last refinement: true use of mind, here, to remember properties and absorb information. However, in all forms of the above training (acupuncture, moxa, massage, etc.), the body was first, always. Refining and conditioning the body and the feeling of the practitioner, through training, came before the mind was filled.

It may not have been that people learned all the forms of practices as in the pyramid, but this shows us how these practices are all variations in energetic qualities. Even the herbal doctors, who would use mind a lot, would practise some form of body-qi training. This meant they were rooted in the body, even if this was the initial learning of massage-acupuncture or exercises that would root them. It was understood that mind was secondary, that the head was a tool of the body, the top of the pyramid, not better or worse than any other part but definitely founded on the body. In this way, the energetic basis of diagnosis and the connectedness of all the processes of medicine were understood, before herbs or mind were applied.

The distinction I have made here, between what I call Jing and Shen medicine forms, shows the spectrum from the more physical to the more energetic expressions of the same understanding of Oneness. It also explains the plethora of art forms in Chinese culture: those that deal with the interior and those that deal with the exterior, and the expression that, in fact, they are all one. The different variants of Chinese Medicine are expressed by different types of practitioners, who will move towards one skill or another, based on their natural spectra of expression.

The triangle diagrams show the strength of the base and the lightness of the top. This implies strength in the roots, the legs, and lightness in the head. It means, again, an open mind, not top heavy—to be an enduring structure, like a mountain. The top of the pyramid connects to the sky and to yang, while the base connects to earth. The square background is earth itself. Herbal practise is therefore of most yang quality within the yin of the earthly realm. The next stage upwards or outwards is Shen medicines (beyond the borders of the square in the diagram). From the base of the pyramid is massage, the next stage downwards is the energy of the Earth itself (the square on the diagram) which is the practise of Classical Feng Shui. This is the art of externally expressing natural Oneness and moving with Nature. An example of a true Taoist expression of Feng Shui today, is close to the permaculture movement (more on this later). The pyramid or triangle also represents the human being in the principles of the Origin and creation (see next chapter).

There is a definite form to medicine methodology. Therefore, there is always an "I/ self" present, to some degree, in this process. Enlightenment is the ending of the idea that there is such a thing as an I/self, an understanding that there is just Nature happening, no matter what is being done. This is similar to someone practising meditation, and while doing so, the person suddenly does not exist anymore and finds just natural expression. They then have no further interest in the practice. (Although they may not be practising meditation, they may be completely intoxicated with alcohol and roaming the streets after a night on the town and then they sense no-oneness…there is no place/ situation or anything more worthy than any other).

Medicine requires a structure. It is a form of practice as well as a form to practise; this structure is Confucian. However, the essence of the medicine is Taoist or the end of form. Hence, where medicine has a structure, it is more and more Confucian about practice and principle. Yet, in its essence, it is about the expression of a person who is drawn to do such work, for no other reason than this is what they are, by nature—not by self-design, ideology, faith, experience, or anything that can be thought of. It is, in a true sense, a calling, for want of a better word. This differentiation is important, for most of what we discuss in the ancient classics is Confucianism (yang), underpinned and wrapped around a core of Taoism (Wu-yin), and this mix allows Nature to be "harnessed" for society and civilisation, whereas, in truth, the core of Taoism is the end of such ideas. Tao is not teachable; Confucianism is teachable. Acting like Nature is very different from being natural. Being "natural" is viewing outwards, projecting and imitating; Nature, as it is, comes from within and is based in Oneness. The only "pre-requisite" to practise medicine is that it is as natural to those who do so, as the air they breathe, it is instinctive for them to do what they do. There is no other reason to do medicine. In a way medicine is an art, performed with some-else; there is no altruism or compensation, no reward; it is simply done for the love of it; it is no better or worse than anything else, it is not being luckier or worse off, it is simply as lucky as a tree being very good at being a tree. Very often however, when we are what we are, this is seen as misconceived by much of society/ Confucianist ideology which sees you in a role, not as an expression of Oneness.

v) Understanding Oneness in Medicine

Further to the explanations above, I can offer an image from Jungian psychology explored and expanded by the great acupuncturist and teacher Ikeda Masakazu, who uses this diagram in almost all his lectures, it is the fundamental basis of everything to be understood in Classical medicine:

(fig. 1.3)

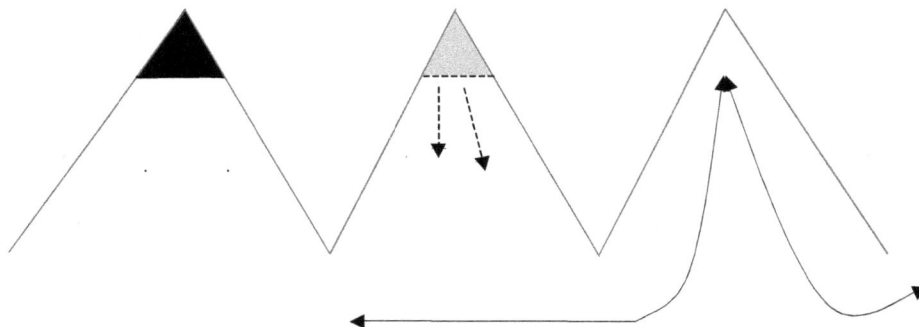

The three triangles represent three people, just like the previous triangles, which showed different faculties of the practitioner's energy. (In most cases in this book, unless stipulated, the triangle represents a human being.) Notice that the base connects them all. The first has all his or her energy in the upper part, is blocked off from the base, and cannot feel his or her feet, so to speak, and therefore fails to connect to its source. This we call dis-ease. In the second person, there is a broken line between mind and body. This person can connect to source a little, and so is in transition. The third person realises he/she does not exist as this separate being and is connected from above to below, and vice versa. This is a picture of health, where there is no fragmentation through the system. In the last two triangles; the dotted line triangle versus the clear triangle, this differentiates between mental clarity only held within the dotted line (still belonging to a seeming separate individual, still as with the first triangle, part of mind-identity), versus connection between mental clarity and the reality of collective unconsciousness or super-consciousness, the thing that connects everything in the universe (represented as the clear triangle above). This is what I have described in this text as contextualised mind, or the end of individuated "self". Again, the idea of unity permeates all these explanations; whatever the pattern just in the head of the triangle, Oneness underpins. The important point is that in the expression and the unlearning-training of the practitioner, to bring him/her to his/her senses, we build the format of learning in a particular way, so there is always a connection to the feet and less of a focusing on the head. As a result, there is always a connection to feeling and less of a focus on the mental faculty. Because it is so easy for yang to rise and be present in the head, it is important to get the energies of yin and yang to mix together. in order to create balance within the human system.

Over manifestation of mind-identity is not only dis-ease itself, but also can lead to a reliance on theory and techniques, rather than the transcending of them, and simply feeling and being intuitive, which is healing, and also living. Hence, when energy accumulates in the upper part and never moves, this creates blockage, and in the changing universe, prolonged blockage is not possible. Change will eventually occur, but this may be change that breaks a static form, which is often destructive, as opposed to the change that happens when a person goes with the flow of Nature/ life. In Oriental Medicine, we trigger the flow of natural change, rather than mind-identified movement. Both in fact are natural, but to go with Nature is not to obstruct it, to let go of "self" by itself, rather than becoming static forms that are broken by it. "Learning" in this way, is often about undoing, rather than doing and holding on. Notice too, this is the process of healing.

vi) The limits of medicine
(fig. 1.4)

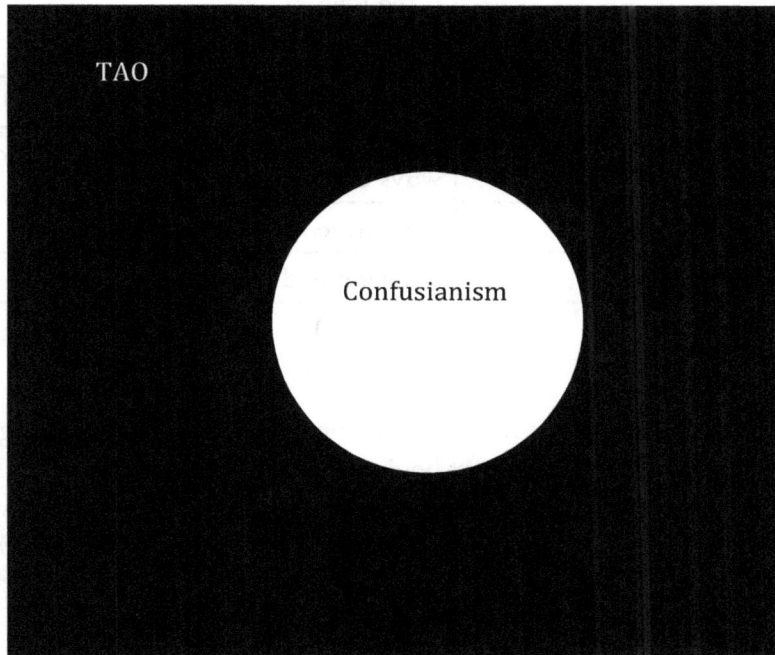

The above shows the black backing of "Tao" underpinning the foreground whiteness of Confucianism, which we will look into now:

Taoism is the broad perspective and Confucianism (please see Appendix 1 for a fuller historical perspective of this differentiation) is the narrow and the narrow is within the broad. The below table shows very important correspondences:-

(Table 1.1)

Tao (Non-dual)	Confucianism (Dualism)
Cannot be learned, is Instinct	Can be learned and is cognitive, is knowledge
Is liberation/ enlightenment or CURE	Is mind-identity/ disease
Is unknowable	Is known
There is no hierarchy	There is student and teacher, patient and practitioner
Is healing	Is the idea of healing
Is clarity	Is the idea of clarity
Is Oneness	Is the idea of Oneness
Is unwritten	Is written in Classic books
Pertains to the yin and background	Pertains to the yang and foreground
Pertains to the pre-historic and animal nature and to the indigenous peoples	Pertains to the modern world and modern ways of fragmentary thinking
Pertains to the bodyspirit	Pertains to the mind
Anarchy and the present moment	Tradition and convention

When then we talk of the Tao being the root of medicine it would seem interesting that everything we read and look at including this book and all text books and the Classics themselves all associate therefore with Confucian design, so they are actually devoid of the true healing itself...so what is the point of them? What is the point of a teacher, a practitioner or the Classic texts? What can they offer:-

(Table 1.2)

Tao	Bridging	Confucianism
Cannot be learned, is Instinct	The Classics of the Tao and Teachers/ practitioners/ people who deeply acknowledge the yin.	Can be learned and is cognitive, is knowledge
Is liberation/ enlightenment or CURE		Is mind-identity/ disease
Is unknowable		Is known
There is no hierarchy		There is student and teacher, patient and practitioner
Is healing		Is the idea of healing
Is clarity		Is the idea of clarity
Is Oneness		Is the idea of Oneness
Is unwritten		Is written in Classic books
Pertains to the yin and background		Pertains to the yang and foreground
Pertains to the pre-historic and animal nature and to the indigenous peoples		Pertains to the modern world and modern ways of fragmentary thinking
Pertains to the bodyspirit		Pertains to the mind
Anarchy and the present moment		Tradition and convention

The bridging point between the two realms is really those who have interest and deeply understand the nature of the yin in healing. This comes from several sources but at utmost all that can be done is a direction back to source, a simple pointing towards the Centre, no more and no less than this as is exemplified in the Tao Te Ching and as we go up the pyramid of medical classics this thread becomes further and further narrowed and more difficult to grasp. Therefore our rooting in this text and those who base their understanding on this text is very important.

From the Tao Te Ching:-
Chapter 71
Feeling the Truth of being, is profound.
To not know this Truth yet to believe a "you" has it, is sickness
If there is realization of what sickness is, there is no sickness.
The Natural-human is not sick because there is realization of what sickness is
Therefore there is no sickness.

This is the bridging point. The realization of the nature of sickness is the cure of it, and this "stepping out of the loop" is what teachers, practitioners and books that have the fundamentals in place (connect to the yin) can point to if the strident, patient or reader is ripe for this to occur.

As healing cannot be grasped and as teaching/ books/ true understandings it is only a bridging point or a signpost to notice this very fact, and also to realise that the instinctual senses that propel us are not under anyone's control. This being the case "Tao" has seemingly two routes to it's core, either seemingly away from it (yang) or towards it (yin). The "method" that goes away from it is the Confucian way, this actually is what most situations of learning medicine in the modern age is about. To some degree it always becomes about the teacher's power over a student, the practitioner's power over a patient, the hierarchical power of a traditional "lineage" over another one, the separation and stylism of one person over another. This is basically in essence the same way of thinking as big business, modern politics, legal systems and all the things that were rocketing forwards at the time of Confucius all over the world. In this situation if taken to its limit this drives us to a state of egoic madness which becomes at its peak very brittle and hard, when it does so it breaks open and at this point healing occurs, the limit is reached and as a result by going very strongly away from the Tao one is broken open through yang reaching a limit, this often comes with an actual death, the situation of global catastrophe and decay of the modern world is what this is about.

The other way is the route which moves toward the Centre and back to nature. In many ways this is the end of a "method" of doing something and more a deeper and deeper trust of natural processes living through a person. This is the "route" of "dissolving" associated with medicine and the Tao at the bridging point as expressed above, it is what medicine at core is all about. However we must be sure to understand that this way is not better or worse than the "method" of the yang. Some people actually need to take things to the limit and for it to crack open. This is the nature of nature, others naturally move towards the Centre. Interestingly those who seemingly go away from life's Centre (this is actually impossible, it is an illusion but it seems real: there is constant Peace behind the illusion of non-peace), may push so hard and break open completely into a liberation or realization/ clarity or enlightenment, where as those who are moving towards centre gradually may do so for the whole of their life and never break open into clarity, so the bridging route is not at all better than the moving away when it comes to liberation or total –cure.

The big difficulty with reading or looking at medicine therefore is always that it isn't for everyone. It is really only for those who are interested in it, for those who pertain to the yin really, who are interested in this, for anyone else whether they are involved in what they might call "medicine" or not it will always inevitably revert to a striving or a seeking process which may as well be expressed in the commercial world for it is of the same ilk. For the student and the patient and very often the practitioner, it is often a very confusing process as to what medicine actually is, because it can vary so much, one practitioner will have an absolute direction and form he/she is following and a lineage and a tradition and a title and a pride and a following and a way of thinking which is all about a big egoic "building up", just like a sky scraper. The patients or student in this environment will learn the same thing and pass this message on to others perpetuating the Confucian cycle of ancestor worship, and more; dictatorial ideology which emanates from imposed hierarchy always. It will always inevitably be about the egoic power of the practitioner involved however he/she tries to cover it up with what he/ she "should" be and what he/ she "aught" to be like. There is much pretence in this way which there needs to be because this is the yang, and the yang has to reach a peak of this before it breaks open into Tao. This however is no place for a person who pertains to the yin and who is looking in this direction, for this person the movement into this kind of transactional or hierarchical system will grate on the senses as the world of policies and commerce grates on the nerves. For the yin, a natural pertaining to the medicine is the key and this will draw one closer and closer to peace in an often slow and gradual way. Whereas the yang's breakthrough is like a lightning strike bursting the world open, the yin is like a slow breeze and gradual opening or dissolving of the "self" as letting go takes time. No way is better, but this book and its content pertain to the bridging point and the movement towards the Tao. This is also the origin of natural medicine. As such it is key we understand that seeking this will come to no avail, the Classic books become utterly useless and the Buddha is in the end burned, there is nothing sacred here because all is sacred in the deepest sense, the bridging point is the situation of when instead of the eyes being focused and fixed gradually they become blurred and one no longer sees the borders around things people or events. All is Oneness. This book is written for those moving in this way, which is not all people.

> *It may strike like lightning, or it may approach like the dawn: there is a source of Light that makes it possible to be aware of what we perceive, and all we do in this experiment is take a look at it. All we perceive in our life are objects, but the Light is the Seer, the Subject.* (Kersschot J. (2003), p.48)

The nature of the liberated-state is always in the background ever present, the Tao is not something that can be separated off from but as we peak into adulthood, often we are also at the height of our ideology about "self" and the possibility of HOW this dies is something that is associated with the nature of the person and the route nature takes them in life, there is no-choice here at all, as there is no-self, just dis-ease/ suffering and the end of it at a point of ripeness:-

(fig. 1.5)

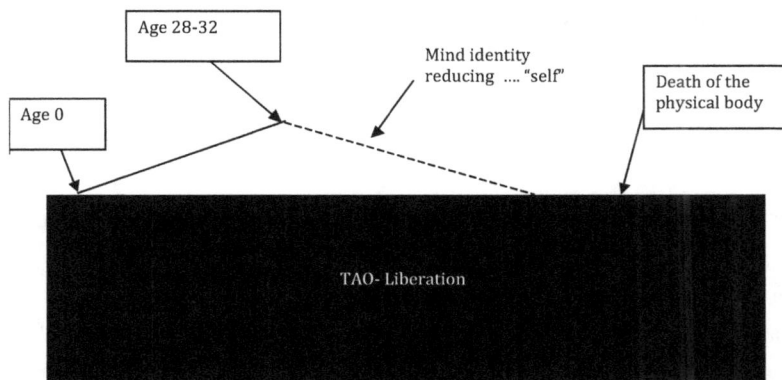

25

There is a transition in practice of medicine often:-
3. Mind-identity + Form (Water method -yang)
2. Mind-identity + No-Form (Fire method – yin (bridging)
1. Contextualised mind + No-Form (Tao-Wu)

To start with, on the Confucian side of things, there is a clearly delineated from and a mind-identified plan of action held by a teacher with "experience" and understanding in such matters which raises his platform above others. This is often a cool and direct approach, it is hard and pushy, regimented in a way one could call associated with some of the Zen sects like Rinzi. This method will break "you" open into clarity, it is the yang within the yin of medicine and as a result actually draws out of medicine and into other arenas such as status and the material world.

The second expression is the bridging point above. It is a state of mind-identity but it is warm and open, it is the nature of the female and it associates with a way akin to Rumi and the poets of unconditional-love, it is less defined more empathetic, more emotion and less thought and so often this group is viewed as being "too flaky" by the "water" stylists. However in many ways this is a point that moves towards and has innate understanding of the root of medicine greater than the yang, who are very heavily head and mentally organized although they will often suggest they are not and are in fact free, this pertains to the above Tao Te Ching statement in chapter 71 where it says "To not know this Truth yet to believe a "you" has it, is sickness" this is the way of it however. Interestingly Qi Bo/ Gi Haku/ 岐伯, the court physician of the Yellow Emperors Classic is both a Confucian and a Taoist, but he is at root Taoist, so he pertains to this expression the bridging point between the natural-person and the idea of medicine.

The Natural-person and the Tao are unknowable, but it is at this point that all roles and ideas of being a teacher, practitioner and "self" that needs upholding dies and there is a freedom. At this point the animal-instinctual nature is most prevalent and there is a natural pertaining towards moving to natural environments and away from expressions which are pretentious. It is a deep letting go and curative of the dis-ease of sufferance, allowing life to be "As It Is". Here whatever role is given is given by others not by the person. It is by the mere fact of being that he/she is healing, for there is no contention and as such the requirement for a medical approach goes, one might call this similar to Christ in "healings" for which Christ had no intention at all. This is living without intention, it moves towards a utopian expression in humankind no matter how it comes about , lightening or dawn, the result is the same, freedom. And something that cannot be sought after:-

From Tony Parsons:-

> What is sought remains hidden from the seeker by already being everything. It is so obvious and simple that the grasping of it obscures it. Never found, never knowable, being is the consummate absence that is beyond measure.
>
> Looking for being is believing it is lost. Has anything been lost, or is it simply that the looking keeps it away? Does the beloved always dance constantly just beyond our serious focus?

The very intention to seek for a mythical treasure within life inevitably obscures the reality that life is already the treasure.

By seeking the myth it dreams it can attain, the seeker effectively avoids that which it most fears … its absence.

Liberation is like a fuse that suddenly blows, and all the little lights go out and there is only light. (Parsons , 2003)

1.2 In the Beginning

i) Original Oneness

In order to understand the ancient way of thinking, we need to form understanding from its original concepts. We must look at the symbols and images associated with the beginning of life, the universe, and everything.

In the ancient Chinese understanding, when we say in the beginning, we mean the beginning before yin and yang energy expressions (see below). It is an unimaginable place of Emptiness or Void, but it is also the place that we are right now looking out of, the "original face". It is called Wu or nothingness, diagrammatically expressed:

(fig. 1.6)

Or:

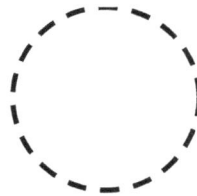

Let us look at the Chinese character for Wu and see its deep expression:

(fig. 1.7)

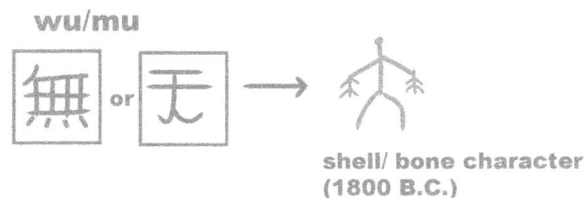

Above, I have shown the ancient form of the character, which is from engraved characters on the shells and animals bones used for fire divination practices and also writings (around 2000 BC).

Scriptures at that time were considered sacred and heavenly entities. The picture here shows a person holding something in both hands; or, it is possible that it is a man standing between two trees. Possibly, both are true. The connotation is of something being hidden—hiding between trees, one is concealed or not present and still. It also has another connotation. The two objects held in the hands could be ox tails, and the person could be a dancer—or a Wu (this wu is: 巫 – different character to the above but there is cross-over meaning) or shaman. If we conflate these images, we get the understanding of something that is secret or hidden and associated with shamans or heavenly dancers.

The meaning of Wu is nothingness, Void, absence, lack of existence (notice that this is also the basis for the Sanskrit original expression of: 0, "zero"). However, it also alludes to secrets held by the Wu shamans. In fact, it is the shamans who understand that creation comes from nothing. That is the key to their dances and to their practices of medicine and healing. So this may be why no-thing and shaman have the same root. Wu (who were usually women) were shamans who lived the art of nothingness, which is called Tao, or the Way. This is described also as Wu wei or the connection to Stillness within being. Let's now look at the characters for Wu Wei and Tao:-

(fig. 1.8)

no-thing/emptyness/void/not

animal (probably elephant)

a hand- leading or touching

wu wei

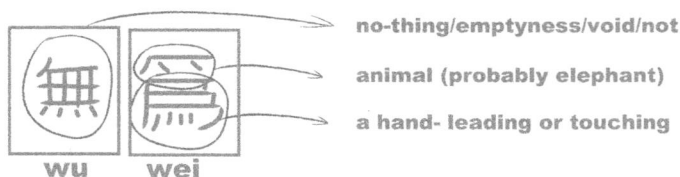

Wu Wei shows again the characture for Wu, then the second characture is Wei with 2 aspects to it, an animal and a hand holding it; leading or doing something. The animal in question is likely to be an elephant. When we take the two aspects together, we see the impossibility of a hand holding onto an elephant, similar to a person being taken on a walk by their dog! This is to express that really the leading of the hand is not really leading but being lead, it is an expression of nature leading, if you will, or a process that cannot be lead, or non-leading, or natural action, where intention is no longer involved, non-mind-identified action.

The next character is Tao:-

(fig. 1.9)

a man's head

a foot walking along a road

tao

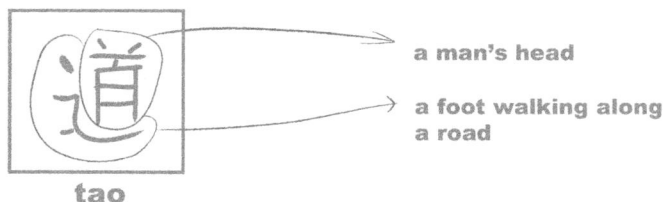

Here we see a foot walking along a road as one radical, and a man's head, as the other. This combination suggests a road being walked along by a man. The foot is making the direction, the head follows the foot; there is no goal in sight. There is no direction to this walking, the path, in a sense, is to no-where. Hence, overall, the image is one of direction, based on the movement of the feet to an unknown destination. This is the meaning of Tao; the feet know the way to walk, the way "home" so to speak, the head follows and in fact there is no where to get to.

Wu, stillness, is the Origin of change and the Origin of everything; because it expresses something that is pre-language, it is closer to a felt sense, than something words can express. This is God, Unity, Love, the Unmanifested. The name does not matter. It is the felt presence of timelessness. Why did change occur? What changed this peace or emptiness before the concept of peace or anything else existed? These questions are the most perplexing of all and the ones that are perhaps the original mental question: "Why life?" The answer is within the unity of all creation, and the question, in fact, doesn't have an answer, or is answered by the Silence of what IS. In many of the ancient texts of numerous cultures, the answer is simply "because". Why? because no-thing truly gets close to the answer, in words. That which created the question wants an answer, but if the answer is unanswerable, then this unanswerable-ness is the answer; it is an acknowledgment of the unknowable. The greatest mystery of all is the Origin of all things and why creation happened at all. It didn't have to happen! In fact, the illusion that is time and space is only real in as much as dreams and memories are real. So the question has no answer; it is looking from a fragmentary perspective, not from the Background context or Centre of things. If we could really look at this question of what is revealed, we would see that Stillness pervades everything. Therefore, there is no true beginning or end point, just the perception of change. Perception of change is actually a mask of Stillness. Hence the question is unanswered, allowing acceptance of what is/Stillness to enter.

As soon as change in the Stillness occurred, polarity began. From a place where Stillness means no-thing, for there is nothing to be still, occurs something with change. Polarity is created. The first aspect to be born from Void is yang, or the principle of Heaven. This is light, or what they call the Big Bang in modern physics. Light shows that, relative to itself, the Void that enveloped it was yin. So we gave the Origin a name, and we can call it Wu. It is the mother—yin—although it is truly before yin and yang were created. It is only because it is relative to yang that we can say that Wu pertains more to the category of yin than of yang, because of its nature of being the mother of yang and being still in relation to yang. So it becomes a yin concept. Now we have two aspects, the envelope-mother—Wu, and the universal bubble—yang. The Divine Mother is found in all ancient cultures. The Origin is always a pre-feminine entity. This, too, includes the Virgin Mary in Christianity, who is representative of the perfect Origin, the yin Wu, that which is nothing but yet from which life is born.

Note that this means that anything within the yang universe that is still, is still only in relation to that which is moving around it. True Stillness is found in the absolute Stillness of the Void that is Wu, which permeates both yin and yang and is a background to them. Note that yin and yang are not the illusion; they represent change. The illusion is to believe that they represent movement. Movement is an illusion of mind-identity, whereas change simply is. I am having to use the word change very precariously here, because generally, when we consider change, we consider it to be moving rather than a series of snapshots of Stillness. Change here means snapshots of Stillness, not movement (this will be explained further later).

(fig. 1.10)

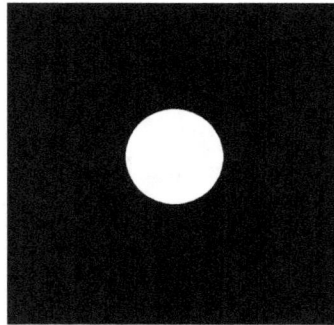

This can be considered like the beginning of a film. First there is total darkness. Then when the projector is turned on and there is light, the film starts (yinyang). This expression is a vital image for our understanding of our true Nature.

It is important to have an understanding beyond the verbalization of Wu. In Taoist expression, the truth of essential being is Wu wei, the practice or connection to the Stillness, beyond the moving universe. It is something that allows one to connect to the eternal Nature of reality. Beyond the change or enveloping the change that is the universe (yang, or Heaven, and yin, or Earth), there is this Void or Stillness. This is the source, the womb of the universe. Meditation practices often "attempt" to reconnect one to this Origin by quieting mind-identity in order for us to perceive the unity that is enveloped by Stillness. Because everything is enveloped by the Stillness of Void or Wu within the moving universe, we can feel a connection to this when we are very quiet or when there is less change. We can sometimes feel it even in the midst of turmoil and huge change, when Stillness within becomes more obvious in contrast—for it is always present. This is when we sense the divinity of being. Silence reminds us of our unity of Origin. All meditative arts, therefore, engender Stillness, to allow us to become aware of and to accept the unity of everything, giving us perspective of the change within the silence and Stillness, and freedom from the mental belief in the physicality of forms as separate from each other. Therefore, we view the eventual transmutation of these forms to seeming nothingness. If we recognise that we are, in Origin, forms of the Oneness of nothingness, we can let go of the mental perceptions that hinder our here-and-now being in reality.

Whether or not this occurs in a "person's" lifetime or not does not matter one bit, as far as the universe is concerned. It will just transmute and transform and return to total Void whether or not we can perceive. Separation is not separate; it is only the illusion of separation, smoke or clouds in the clear sky.

From within Wu, the universe is created. From here, we can move to understand that within the yang of the bubble of Heaven's light was yin. This yin, however, is a product of yang. It is yin-driven and formulated via yang. This we can call the more condensed matter in the universe and the cooler aspects of the universe. We call this concept Earth, akin to Matter, which is in relation to the ethereal and energetic,

Heaven. Note that the energetic universe is very different from the Wu/Void, yet the Void envelops and permeates it. So it must be an expression of Void as there is nothing before Void. (Void is in fact no-thing, from which everything is born). Void is Nothingness, whereas Heaven is something in relation to Earth. However, if Earth is formed from Heaven, and Heaven from Void, then everything is a form of Nothingness. Nothingness is neither created nor destroyed, but is eternal. Hence, in the illusion we are constantly in, we believe that the universe is in movement and the Stillness is outside the movement. This is the separation of God and being. If we consider that they are all one—just expressions of One—then there is no illusion.

(fig. 1.11)

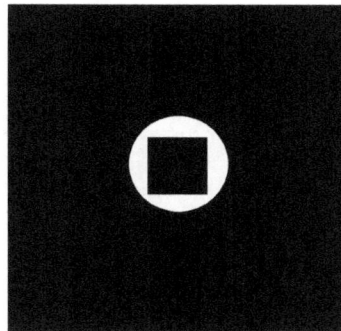

The yinyang of Earth and Heaven are the changes to the Void that we are, as much part of it as the Void itself. Hence we change, but we are still. Earthly yin energy forms the balance between the yang energy in the universe, showing the first universal principle emerging. This is called the principle of two. Wu can be said to be the Origin or the State of Unity, Original Oneness. As we go on to consider the "difference" between yin and yang, we must be very clear that yinyang is not duality, nor is anything related to ancient Oriental philosophy. Yinyang is always based in Oneness; that is why we write yinyang, rather than yin and yang separately. Yinyang is therefore always non-dualism, as is all of Oriental understanding. A circle is always a circle, no matter how many times you cut it. In the following part of the chapter, we will look into many different types of formations of this circle, representing different energy vibrations of the same origin. It is always non-dual; this is why we must look for the context—the broad view—or we can be caught in seeming opposites, seeming differences, and not see the fundamental sameness; this is called not seeing the wood for the trees!

ii) The Principle of Two, of Pre-Heaven and Post-Heaven

As explained earlier, Void envelops Heaven, and Heaven envelops Earth. Pre-Heaven means the situation before Heaven and Earth energies blend to create change. Post-Heaven means when change is born and Heaven and Earth energies meld.

The principle of two is about the two forces in the universe of yin and yang. According to ancient philosophy, there are two descriptions of these two principles. One describes them in the form of just before they started to change, so perhaps this can be described as: just after yang is formed, it forms its equal and opposite—yin. At this point, time has not been introduced. This is called pre-Heaven, and the principles of yin and yang have taken form but not changed. This can be symbolized as follows:

(fig. 1.12)

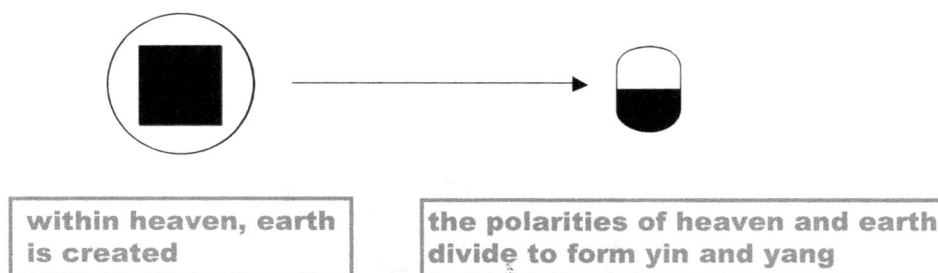

within heaven, earth is created

the polarities of heaven and earth divide to form yin and yang

When change begins, everything mixes into each other and creates myriad other changes. This we call post-Heaven, or change-Heaven. Huge shifts occur, and yin and yang displace each other to create change.

Movement, Time, and Stillness

Movement is time, time is Still, and therefore movement is Stillness: Time is the biggest illusion we face as human beings because it grew up, so to speak, with mind-identity. In fact, time is mind-identity. Space is nothing without time. Therefore, the second illusion to fall, is space. How can this be said with such certainty? Physicists today, through their partial Newtonian view of the universe, look at the phenomenon of time and how in fact it is a relative concept. There are holes in space-time, which are black holes. A black hole is so dense that it has become Wu again. We only consider time to be external to the black hole, but why? Surely it is just that there is perceived time, or as Einstein put it, "relative" time. If time is relative, it is surely a tool for humans to explain, something that they cannot really know, rather than a phenomenon in the universe existing beyond the human world. In fact, time does not exist other than as a human concept. Time movement is a concept, which relates to two points, when something begins and when something ends. Or one could say: "That bird flew; therefore, there is time". In fact, this is not the case. Mind-identity believes that the bird has gone from point A to point B, but the bird has always just been in the present moment, which actually is Stillness. This is a difficult concept to explain, because it necessitates taking away the mental image of before and after, both of which are viewed in the past storage facility of mind. Mind-identity is when that becomes a reality for us and we call it "space-time". If one is looking at the bird as it flies, with the notion that has moved it from place A, then we are projecting place A on it the whole time it is moving to get to place B. What actually occurred was in the moment. If there is no time, there cannot be any movement. If there is no movement, how is it possible that there can be a universe that comes into being or ceases to be? In fact, how can there be any such concept of the universe? If there is no difference essentially between Wu and post-Heaven, what is the universe we are part of? Note that when we try to capture movement on a camera or a video camera, movement is always divided into snapshots all playing together—many single moments.

Distance too can be examined from the first-person perspective of looking outwards from where you are now. You see words in front of you, but I'm considering the point from which the words are being looked at. If one looks from this central position (your Centre) outwards to whatever object, no matter how far away it seems, we find that all so-called "distances" are radial from the central position we stand in. Therefore, right at this moment, the whole world is appearing at our Centre. You might say " ok so what if I put my hand out and touch the book in front of me…one can do this but one can't know the distance between the book and the finger radically, one has to come to the side and measure it, hence our actual sense and reception of distance is never what we think it is. It is in fact, always an abstraction, something we can do by finding the distance between two points external to ourselves but from oneself to the object, our distance from something, this remains a mystery to the perception. One can heighten this by blurring the vision like a baby's vision; here, we see all objects as a blur of colour, and our belief in distance is gone! What is left is a realisation of all the colours happening within us. Or, one could say that we are all the colours, or the capacity for all the colours; there is no distance and therefore no time. This clear emptiness at our Centre can be called a head-less place, meaning, the place you are looking out of now, which does not have a face in the way! What you're looking out of now, a faceless, open, visual field which, if we take only empirical evidence, has no eyes, nose, ears, or hair. What you are looking out of now is just an empty clarity. So, on present evidence, one doesn't look out from a head that we see, but in fact looks out of an empty space. From this place, any measurement to an external object is always the same distance. Looked at from this perspective, the star is just as far away as the table in front of us, because like a spider's web, all of life is viewed from the centre of the web. However, going out into the world and directing ourselves outwards towards the three dimensions of space and time—rather than the one-dimensional singularity of the "original face"— we find that it takes a much "longer time" to get to the table than to the nearest star. This is an expression of Wittgenstein's expression "the subject is not in the world". The "world" is three-dimensional and is a human-world where I am considered to myself as a second or third person, a mentally imagined description of things. Time and space are common beliefs in this world, as well as good and bad, and right and wrong. From the centre, however—from the First Person not of the world, living from the singularity of the First Person, which is the same First Person that the universe is, the First Person—from here, there is no time and no distance, and everything is happening and nothing is happening.

The best phrase I have to describe this is "transformation of Stillness". Stillness changes and transforms, which is not to say that it moves from one place to another. Hence, Stillness changes, another snapshot of a moment. At the level of the most fundamental particles of the universe, it has been observed that these energy vortices occur, then disappear, coming in and out of existence like a film reel projection or cartoon flip-book where the image changes. All of reality therefore is coming in and out of existence all the time, life and death unified, formed and re-formed from the canvas of Oneness. Oneness in every nano-second of so-called time. Change is not necessarily related to time, but in the sense that I am using it, it is almost like a photograph, a still-change. Beyond this, the timelessness of being is the only way to understand this. As far as we are concerned, when looking at the difference between pre-Heaven and post-Heaven, we are simply looking at the pre-change, which is pre-Heaven, and the post-change which is post-Heaven. Notice I am not using the word changing, because this indicates movement. As we saw, the I Ching is best translated as the Book of Change. This is because change does not indicate the mental contraction of time. It is memory and mind-identity which consider change as changing. This is actually a Still universe, a quiet universe, but it seems not so to our senses when associated with mind-identity. The individual, David , Anne, John etc.. is a character or actor in the dream that Oneness is having. Reality is Oneness and Stillness and the realization that it was just an act. The idea of physical forms as separate entities and the idea of the world of time and space, are no more than just this—ideas, or illusions. All these go together. In the contextualised mind, post-Heaven existence, we are aware of change, but also of no time and therefore no space, no beginning and no end—just Wu. Wu changes to post-Heaven. There is nothing else. One cannot delve much deeper into this without knowing timeless being, which renders all these explorations totally laughable, literally. Neither this work nor others will take you closer to this. Everything I'm writing here is really about cognition of another possibility, rather than what is constantly overlooked in everyday moment-to-moment experience.

From here on, when I use the word change, you will need to understand that it is associated with the root Stillness of all things and is just a snapshot of events. Please appreciate this as an opportunity to change your perception of time, which is fundamental to mind-identity and so to illusion. Also, throughout this work, I will attempt not to use the words movement or space or time in the context of the broad perspectives and centre that we are talking about, because these are the words used by mind-identity to actualize itself. Notice that mind-identity is not a personal term either; it is a collective expression as there is Oneness. Not using these words, and/or using words in a different way makes this work more difficult to understand from one perspective, but develops interest in another way; it should show you where mental understanding stops and where feeling must begin. As I have mentioned, this work is mainly about yinyang, which is the moving universe, the tip of the iceberg, so to speak, and the phenomena within it. However, we must constantly be in context and relation to the basis of this, which is Stillness, or else the content of everything that is said here will just be an add-on for mind-identity.

In our current situation of being in the universe of change, permeated and enveloped by Stillness, the changing aspect is constantly referred to as yin and yang (post-Heaven) and symbolized (although this common symbol was created after the Classical period) as, Taiji, meaning "Supreme-Ultimate/ Absolute". Many will know this for Taiji Chuan or "Supreme Ultimate Fist/ Boxing".

Let us now look at the characters of yin and yang to understand a little more from the ancients' own pictures:

The character for Yin (Chinese) / In (Japanese) is as follows:

(fig. 2.14)

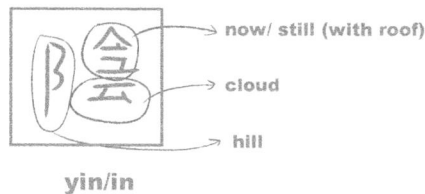

now/ still (with roof)

cloud

hill

yin/in

The character is in three parts. The hill on the left, together with the top right, is the expression for now, or perhaps Stillness. The lower right is a cloud. Together, the picture that is formed is a cloud hanging over a hill. This is key. Our understanding is of several things: The cloud has a still roof, so to speak, which means this is a gloomy scene. In fact, the cloud casts a shadow over the hill and covers it in gloom and darkness and importantly, dampness. This is our picture of yin. Notice how this has association with the Tao Te Ching's description of Wu (or the Tao, which is the Way of knowing Wu) being indistinct and shadowy. Wu, and as we will see later, the character of deficiency, have close ties to yin.

The Character of Yang (Chinese) / Yo (Japanese) is as follows:

(fig. 2.15)

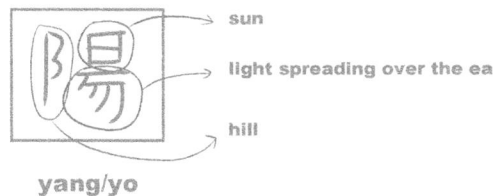

sun

light spreading over the ea

hill

yang/yo

The character is in two overall parts. The left side again is a picture of a hill. This is important because it is the same hill as in yin, but with a different right side so we can see that something different is occurring to the hill. This time, the right side shows the dawn sun rising out of the ground and spreading its light everywhere. The lower aspect of the right side could show shards of light. Hence it shows the hill in full light. Yin is always placed first, before yang, to give continuation to the notion of Wu to Heaven, Stillness to change, yin to yang. So the light indicates the cold of the yin moving away so that light can shine through. Hence the characters have a connection.

So what are yin and yang? Why do we need this mental concept? For it is a mental concept. The symbol itself is an image relating to reality, but one must remember that yin and yang are nothing new to our bodies and our feeling. This knowledge is to cure mind-identity of belief and to allow the clarity of objective (contextualised mind) viewing to be possible.

(Table 1.3)

Yin	Yang
Earth	Heaven
Female (archetype—fire)	Male (archetype—water)
Soft (but dense)	Hard (but energetic)
Is changed slowly	Fast to change
Dense	Diffused
Calm/Stillness*	Change
Silence	Sound
Dark	Light
Accumulating and drawing inwards, centripetal-cooling	Expanding outwardly , yang in yang (or internally contracting both; yang in yin) with direction and heat, centrifugal
Broad-view (from inner stillness)	Narrow-view/ focus (from looking outward)
Flexible	Inflexible
Deep	Superficial
Inside	Outside
Cold	Heat
Receptive	Penetrating
Below	Above
Absorbing in	Irradiating/giving off
Physical	Energetic
Jing (closer to Wu, but it is still within yinyang—jingshen, hence form must be let go of to become Wu)-	Shen (also Mind, and the add-on of mind-identity which can be thought of as a dis-ease of the Shen, associated more with yang than yin)

*Total Stillness without change is the pre-Heaven universe which envelops and is part of the post-Heaven-yinyang, but pre-heaven we can say is at the beginning of possible change so is itself backed by total Emptiness-Wu. Wu is like a background to potential or actual change. However, if we consider that yin is "closer", in association to Wu, than yang, then we can see that, when we look at yinyang it is within Wu.

[Note: Please note from the outset that I don't refer to yin as being contractive. Contractive implies that it has direction and "wants-to" contract. All forms of contraction that are intended are yang. When yin is seen deeply it is realised that the so called "contraction" it is often associated with, is actually a return to, the state of stillness. This is quite different from many interpretations. This is a centripetal quality which may look like contraction but this is a masculine interpretation. The yin draws to accumulation as its base

36

expression; expansion is the archetypical yang expression, but contraction is also part of its expression; it is the yang within the yin. The yin in all cases (yin in yin or yin in yang) moves to stillness.]

(fig. 1.16)

… it is the yang that is moving the yin, which is merely displaced by the movement of yang. This is a vital point that will be discussed at length later, for we must understand that yin and Wu are the root of yang; yang is the surface of the experience of the universe. Please note that the seed of yin in the yang and the seed of yang in the yin expresses that the origin of yin is found behind the yang, and the origin of yang is found from the yin. It also means that at the peak of yang, for example, seasonally, in the summer, this is associated with yin, whereas in the winter, this is associated with yang. Or, midnight is yang, whereas midday is yin. This seems contrary, but what is meant is that these points are the origin of the next phase. So at the climax of a phase, we consider this to be dominated by the opposite energy. This crossover is vital in understanding yinyang and its derivatives, in five-phase or other modalities.

The above explains the main sensory-based associations with the dynamic interchange of yinyang energy. These forces are drawn together, so they attract each other, then also displace each other, and have the effect of balancing each other out. They transform or change into each other. So what are they doing? "They are" actually Oneness undergoing change. Before the occurrence of change, there was static unity; now there is unity in change. The difference is the change itself. The occurrence of change makes yin and yang blend like a dance. When they eventually change back to unchanged, they go back to the Stillness from whence they came, to what they always were. This is called going home in Taoism. It is important to note that this is not something that humans can cause to happen or drive along through mental intention. We must know the impermanence of the changes of yinyang and the unity and permanency of that which permeates yinyang, which is Wu. Therefore, all aspects of the universe can be said to have an innate, unified knowledge of when to do what. We do not need to think about this or be something that attempts to force a change. One way or another, we simply need to acknowledge what we already are, right now—to be the tips of the iceberg. Intention therefore is really very personal, Natural-intention, or simply Nature, however, is the whole of Oneness understanding that it is moving together, Nature doesn't want anything; it just is as it is, hence this is the way humans are, under the surface pretence of society and ideology.

In the Tao Te Ching, the principle of yinyang is called often The Ten-Thousand or Myriad Things. This means the many formations that are derived from Wu, which permeates everything. Wu-yinyang therefore describes that, from Wu, the surface forms of yinyang are created. Yin is closer to Wu by inference of its nature (still, silent, etc.) but is not the same concept, because yinyang is within the world of form and Wu is underpinning it.

iii) The Principle of Three (Post-Heaven)

In between a moving Heaven and stiller Earth, humans grow. Human beings are considered, within this philosophy, as a representative for life on Earth. Life in this sense is yang formed from yin, similar to the Origin of the universe. This starts the process of understanding that whatever occurs in the macrocosm also occurs and is reflected in the microcosm. This expression follows throughout all things. Life on Earth we can call the third principle, above being Heaven, below being Earth, and between being humans representing all life on Earth, in myriad forms. In relation to Earth or dense matter and rocks, humans are yang. And so one can say that humans relate to Heaven, and planet Earth relates to Earth energy. This is why we can say that the overall process of life is moving and yang, whereas death is yin. However, it all depends on the relative aspects of the system being assessed through yinyang.

Remember that when applying the mental faculty to the notion of yin and yang, one must be clear on the relation. So if one is examining trees, one can relate trees to anything at all, but be sure to note what that component is. One cannot say a tree is yin by itself or a car is yang on its own. Yin is always in relation to its counterpart yang, and yang to its counterpart yin, or else there is disunity in the concept. This is the principle of two. Note, too, that the principle of two governs everything after it. The principle of Wu, Oneness, is the base. Hence the principle of three dictates that humans, relative to the Earth, are yang, as shown in the diagram below. The square is associated with the Earth but also the Origin. It is linked to the Origin but is still in relation to yang and so to the yin of the universe of change. This seems to pertain closer to the Origin than the yang of the universe. This is actually an illusion because the same thing permeates it all, but Stillness of the yin (relative to the yang) in the universe allows us to perceive this. Also, the square represents form and structure. Heaven is circular because it is about change and has no shape. Interestingly the circular form is often associated with the feminine and one would have thought linear expressions like a square is more masculine. However, it is really all a circle, which is the expression of everything, or a single vortex, but when representing different aspects of this, some delineation needs to be made. Also note that the circle is an attempt to represent shapelessness, and the perfect square of Wu is actually described as having no borders or no corners, in the Tao Te Ching, chapter 41. The Origin could be something other than a black square. In fact it could be pure white, but essentially it is the absence of anything knowable, and black signifies this. Square signifies connection to the yin of the universe, so it is for this reason that the Origin is expressed in this way. The triangle represents man as a representation of being between Heaven and Earth, having form but also being yang and of yang, again relating to the famous expression "all under Heaven". Perhaps this is why there is a father-creator figure in numerous religions, one who is associated with Heaven and the sky. This permeates almost all cultures. It is true that He is related to the Heavens, and He fills the Heavens, but as we can see, He is created by the Origin and She, the mother of yang, the mother of God. It is all one. Interestingly, in Taoism, it is the adherence to the Yin and the mother in the background; in Confucianism it is the adherence to Heaven and the father. This is why Tao envelopes Confucian ideology but the Confucian ideas cannot do the same for the Tao, because the Confucian is an expression of the "child" of the world, the yang.

The image of the triangle is the image of fire—the flame of life rising upwards and ever climbing— which is the Origin of mind-identity and the whole of the disillusion of human beings, as they moved from the ground to standing upright, reaching upwards constantly and moving towards extinguishing themselves, as they detach from the base of Earth with which they are one. This is the human condition, so to speak, the simple rising of yang-fire. This triangle, if seen in three dimensions, would be a square-based pyramid, whereas the square of Earth would be flat. The square based-pyramid has five points representing the five phases of humans.

(fig. 1.17)

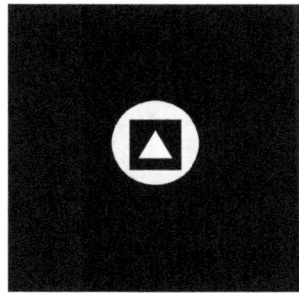

iv) The Principle of Four and Eight (and sixty-four) of Pre-Heaven

To represent pre-Heaven, or the forming of direction, we use a sequence of diagrams. From within pre-Heaven, the next expression drawn from the two separate Heaven and Earth idea, before mixing, is four; these are expressed as the four directions, each represented by broken and full lines in pairs. The broken lines represent yin, and the full lines represent yang. These describe the quality of yin in relation to yang, and yang in relation to yin; yin and yang have not yet mixed, so this means that pre-Heaven expressions of yinyang lines means something different from the post-Heaven representation:

(fig. 1.18)

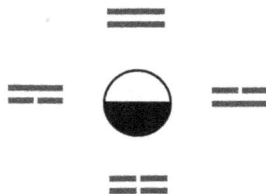

This represents the square of the Origin Wu and the earthly energy of the universe. The centre of the four does not exist until change comes about, for after change (post-Heaven), it requires a centre from which to revolve. From here, eight directions are formed from the four described; the eight directions of a square are represented by the eight trigrams, three lines to represent each direction. From here, 64 hexagraphic expressions depict 64 directions of the square. The 64 hexagrams drive, therefore, a more detailed understanding of the context or shape of all aspects of the universe. These 64 hexagrams can be rearranged to fit a post-Heaven circular formation (circular meaning mix of yinyang), but they are usually drawn in a square formation, as they originated in the pre-change universe permeated by Stillness, or prior to the mix of yinyang.

(fig. 1.19)

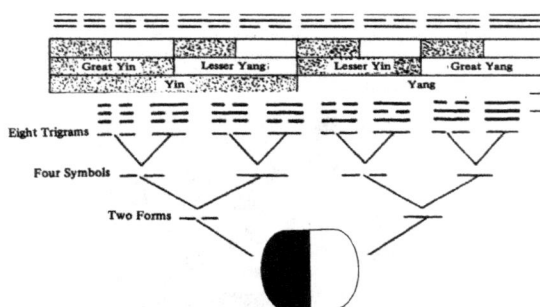

It is vitally important that we recognize that at each stage of an expression, one doesn't mix the expression of each viewpoint. So when looking at the four symbols, we must stay using the ideology of 4, and with the trigrams of 8, we must use only the trigrams to explain phenomena. In circular formation the 8 trigrams form what is known as the Fu Xi arrangement, associated with both Fu Xi and the divinatory arts, as well as King Wen and the formation of the I Ching (although King Wen is also attributed to using the 9 energies to form the 5 phases (see below plus appendix 1)). This is the Fu Xi arrangement, not that there is no central point to this arrangement; this represents the background of the universe before change: -

(fig. 1.20)

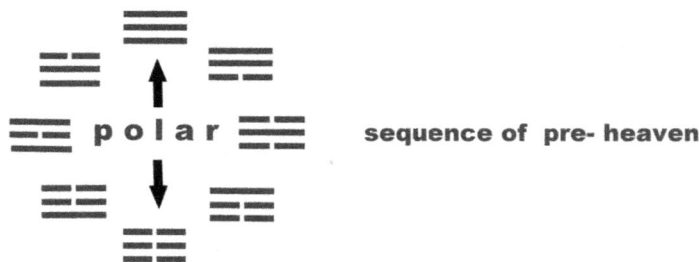

The basis of this is the 8 is polar and directional without central movement and activity. When this moves into the 64 hexagrams of the I Ching, this is no longer the territory of trigrams; the hexagram needs to be seen as one unit in relation to the other 64. This is very commonly misunderstood, and when the 8 trigrams are used to explain the 64 hexagrams, a great deal of confusion results. In this entire book, concentric circles are the key. When you view at one level, you must keep your view and orientation at that level to understand phenomena being expressed; if not, you will get very confused, very quickly. This is particularly important in later chapters, where we look at combining different qualities of expression. Sometimes it is best to differentiate phenomena in order to see them in context, seeing everything as concentric rings of expression, at particular viewing depths/ vibrations to the Origin. This allows us to use yinyang in all its various augmentations, the 64 hexagrams perhaps being the most complex and also the most fundamental due to their pre-Heaven origin. The I Ching and the 64 hexagrams date back to well before the ideology of Taoism and are imbedded in the root, instinctive expression of the understanding of phenomena. It is the foundation of all the expressions in Oriental Medicine.

Note that a yang, or solid line, is often described as the number 9, being the number associated with Heaven and yang or an odd number. The "broken", or yin line is called 6 and is seen as being square and solid but with a vessel quality to hold yin Emptiness or Wu. So, in fact, the yin is not broken at all, but full of yin; this is what is often not understood. Everything here is divisible by 3, and 1/3 of the yin line is empty. All of this represents constant relation to the principle of three. The I Ching holds much that is hidden and requires much connection, to realize its value.

For Taoism, the prime hexagram of the I Ching is Kun/ Kon/坤, which is all the yin lines. For the Confucians, it is Qian/ Kan/坎, all the yang lines. The Tao underpins and this is why the Taoist focuses always on the yin. The ways of humans and of the human mind is focused on, by the Confucians, who separate things out, hence their focus in Heaven and yang and life, the Empire and the masculine dominance. As such, hexagram 12, Pi/否 or Stasis/ Stillness, is the foundation of individual line placements for the Taoists, for the Confucians it is hexagram 63, Ji Ji/ 既濟 or Order.

(fig. 1.21)

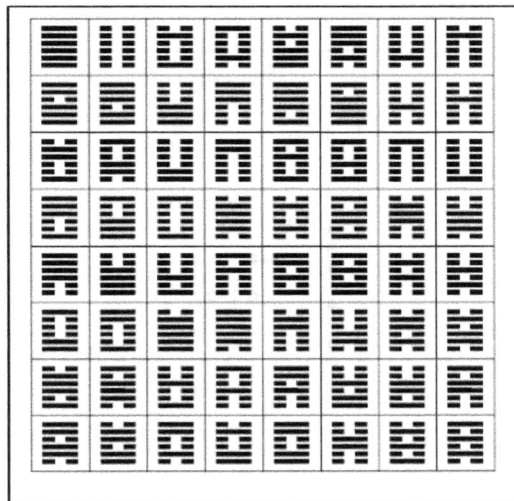

v) The Principle of Pre-Heaven, 10 Celestial Stems

The principle of 10 in pre-Heaven relates to the creation of the post-Heaven energies. Five relates to humans (see below), but humans or life could only occur after the universe was in motion, so we can say that this stage is the pre-arrangement of the energies, that formed the energies within the universe, prior to change—or perhaps as they are just beginning to change. They perhaps could be described as the rim edge of the post-Heaven universe:

(fig. 1.22)

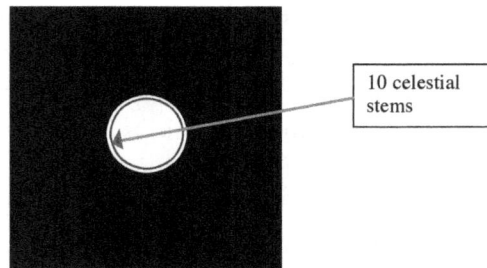

The Origin of this pre-Heaven 10 energy is expressed on the Yellow River map, or the He-to map, connected to the most ancient aspects of Chinese philosophy. Here it shows the energies of five being firstly related to the pre-Heaven four directions. It shows the sequence of creation of these energetic bodies as they start their change and yin and yang start to blend. (This formation is implicated in Su Wen, chapter 4.)

The following is the Yellow River map (see appendix for historical context):

(fig. 1.23)

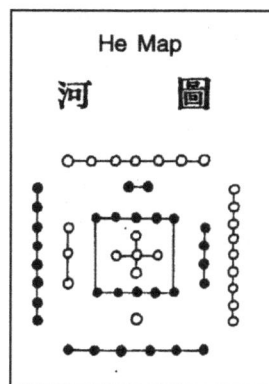

The map shows a sequence inside to outside:

(Table 1.4)

Creation number	Second-time creation
1	6
2	7
3	8
4	9
5	10

The 10 stems from the He-map (above):-

(fig. 1.24)

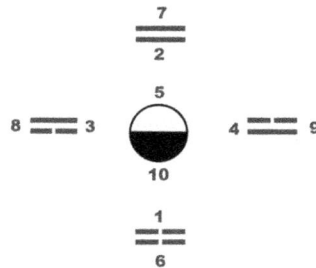

The symbol associated within the 10-stems above is called Wuji, (rather than Taiji). Wu, as we saw before, means No-thingness or before yinyang change. The Wu is the circle within the centre of the symbol. Ji, here, means ultimate or absolute. Hence together the meaning is "No-thing Ultimate". This is the point just before the mixing of yinyang in change (Taiji). Hence the sequence goes: Wu, Wuji, Taiji, or Void, Void-ultimate, Supreme-Ultimate.

This sequence is the creation of the principle of ten. However, when applying the 10 sequences into post-Heaven, (remember that the principle of 10 is related to the initiation of post-Heaven, not actually post-Heaven itself), to make it suitable for, and in relation to, human life, it is often associated with post-Heaven five sequences, called the five phases. However, we must be sure to note that 10 is between pre- and post-Heaven and therefore adheres to neither and to both. It can be associated with the five-phase principle; this forms the cycle of 10 stems used in the calendrical system with the 12 branches. Kabbalistic and Arabic numerology use 10 and 12 combinations to form the sacred 22 symbols. Hence, 22 is a universally associated numerology and is associated with Heaven and Earth energies. [Note:in describing the principle numbers, we are looking at the roots of universal numerological understanding from Greek to Egyptian to Judeo-Arabic to Chinese; it is all one. We are not including, however, associations with the linguistics associated with these numbers. For example, 7 in Kabbalah is important because the word seven has a meaning of fullness, or completion; there is relevance for this however (please see Appendix 2). (Although, 7 does have associations with Indian understanding also and always with the spirit expression, 9 and 7 therefore have some relation to one another. 7 is a focus in Chinese expressions but in the star constellations 7 is the number of stars in the main constellation, the Big-

Dipper, and also it associates with the number 28 (see below) . It is associated with the yin quality within the yang (odd-number= yang); in a sense it is both deeply yin, yet energetic superficially/ shimmering). Or, in Japanese, the number 4 is associated with a linguistic meaning of "death"- shi / 四. These are part of individual cultures, the "Tower of Babel", so to speak. We are looking beyond this to the numerology behind the language so we can see, without words and so without as much mind-identity. This too was the Pythagorean notion, to end belief, or to reach pure reality through number. This also is the Taoist expression, as number is a description of something symbolical. that is not biased by ideology; it is pure and objective. So number and mathematics is in a sense the universal language of the Ancients.]

Note: the secondary five aspects born from the original five aspects are yang in relation to the originals. These 10 expressions are known as the 10 celestial stems:

(fig. 1.25)

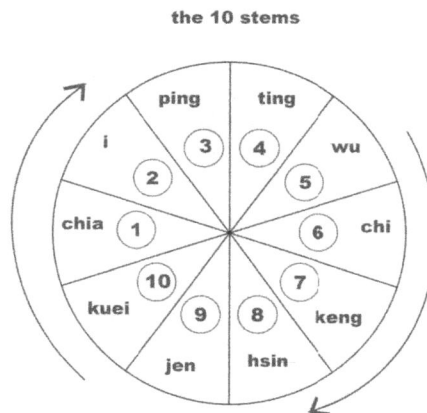

the 10 stems

vi) The Principle of Nine: The Universe in Change (Post-Heaven)

Whereas one can see the principles of two and four and eight and sixty-four, all relate to the pre-change or pre-Heaven sequence, when the 10 celestial stems meld to form post-Heaven, the energy of nine is formed. As soon as this takes place, yin and yang merge, forming post-Heaven. Hence the principle of post-Heaven two (yinyang) relates immediately to three (as man is formed) and to nine, and from here, five can be explained. We will look at this now.

Numbers of the pre-Heaven are:
(2), 4, 8, 10, 12, 28, 50, 60, 64

Numbers of the post-Heaven are:
(2), 3, 5, 7, 9, 11, 25, 81
(The above are all the main associations with all the derivations of number-principles in the main classical texts.)

The number two is interchangeable. As we can see, even numbers represent pre-Heaven, and are seen to be static and relate to yin. Hence, we say that even numbers are yin. Odd numbers, therefore, have more to do with change and represent yang. They relate to the post-Heaven.

Going back to nine energies: When the sequence of trigrams after the universe incorporates change, and then change begins, this melds the triagraphic sequence to create the following, but please note that there is always a centre to the 9 energies, whereas the 8 triagram has no centre. It is empty and still, the 9 are like a wheel:

(fig. 1.26/ fig. 1.27)

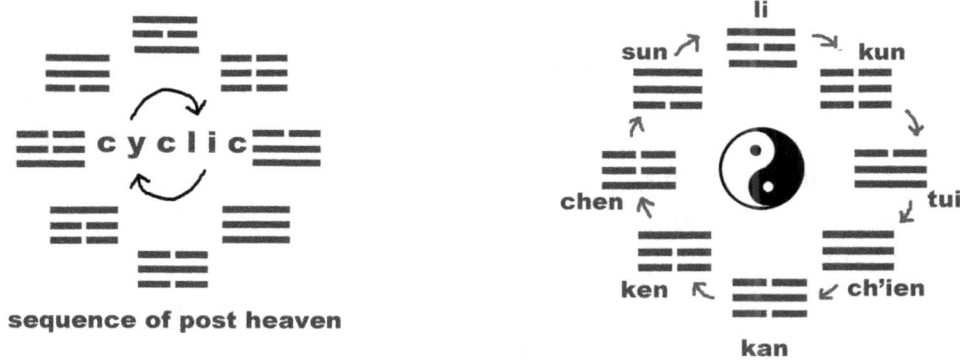

sequence of post heaven

Here we can see the triagraphic explanation of the Luo Su Yellow River Map (see appendix for historical context). This explains the moving universe and the beginning of the understanding of the nine phases of change. This is represented here by the numbers of the magic square of Oriental philosophy:

(fig. 1.28

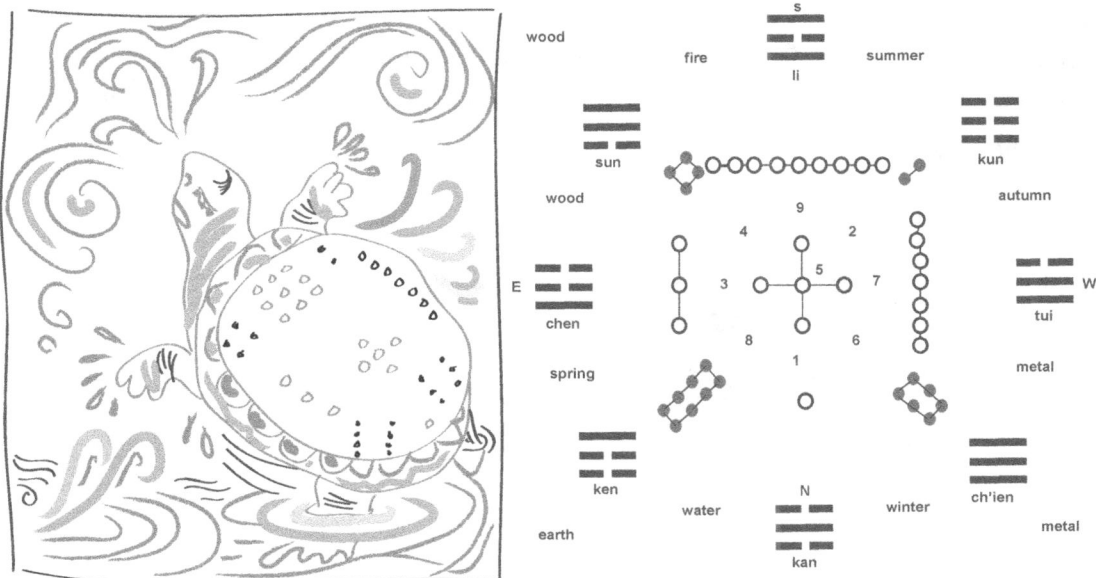

(Table 1.5)

4	9	2
3	5	7
8	1	6

The "magic" square is a symbol of the sequence of change in the moving universe. All lines horizontally or vertically and across the diagonal add up to fifteen, but the "magic" aspect is mathematically and energetically deeper than simply this. It is represented in all the ancient worlds, and even similarly today on the touch pad of any telephone; this square is deeply embedded in a natural symbology of life. Its relation to the various pre-heavenly arranged directions of the universe gives us a notion of change, and also of the quality and feel of change. This goes back to the merged yin and yang symbol, which is simply applied to the nine numbers, directions, and trigrams in the following all-encompassing diagram, which is perhaps the most important to understand in all of Oriental philosophy:

(fig. 1.29)

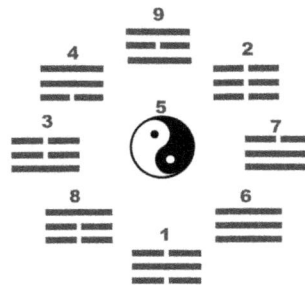

One can extend the scope of the principle of nine by laying it over the pre-Heaven directions, to give understanding of phasic Nature in relation to direction. Also, phasic seasonal change can be expressed if we consider the Earth to be the central number, five. The same applies to directions, because direction is meaningless for humans, except for orientation on Earth, with ourselves as the centre.

(fig. 1.30/ fig. 1.31)

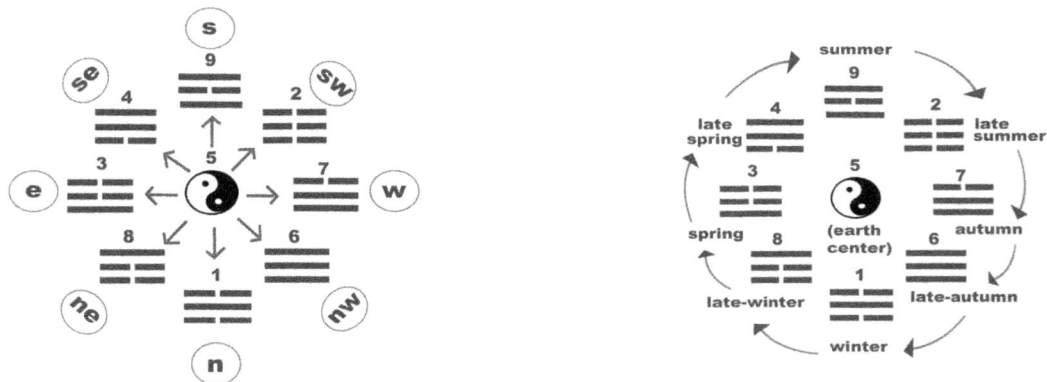

Lastly, the sequences of numbers one to nine, within the magic square, is the change of the universal flux. The energies of the universe expanding and accumulating follow the sequence 1 to 9 and 9 to 1. One represents earthly energy (1 to 9) within Heaven, and represents heavenly yang energy (9 to 1) within the Heaven. These sequences are like "DNA" strands of the universe, so to speak; they are a helix of expansion and accumulation that mixes and ties the post-Heaven universe together, and can be charted over time. To know this is to understand the nine sequences, in relation to heavenly and earthly phenomena. However, consider nine as a heavenly principle, even though application can be made to Earth. Generally the 9 energies are known as the 9 Palaces or 9 regions of the sky (see Ling Shu chapters 77-79) or as the 9 Stars (see below) but essentially they are associated with light and yang energy, this is their most important association. (9 x9 = 81 which is a common number, representing 9 in the classic texts.)

(fig. 1.32/ fig. 1.33)

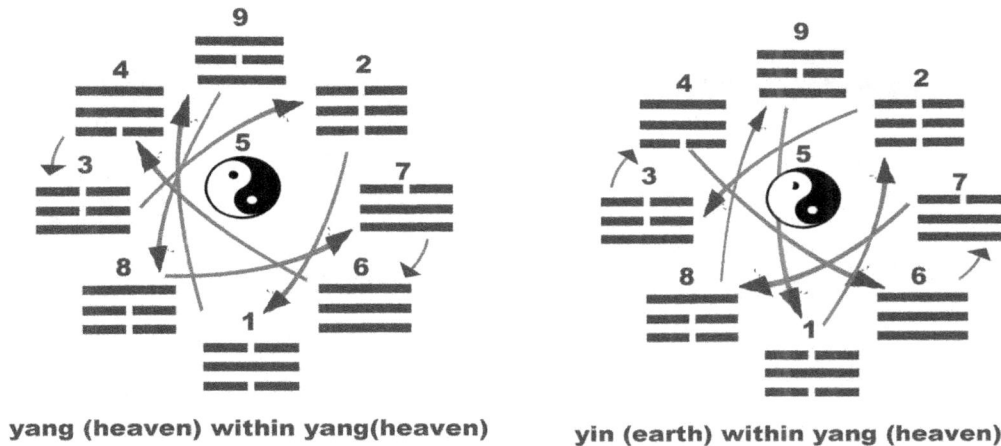

yang (heaven) within yang(heaven)   yin (earth) within yang (heaven)

Universal Change and the Nine Energy Sequences

The universe is in constant change, and the energies of Earth within Heaven and of Heaven within Heaven are constantly changing towards expansion and accumulation. The Heaven energy sequence starts with fire and ends with water. This shows an accumulation process. So numbers go from nine to one. The Earth, or yin within the Heaven, is propelled by the opposite force, which counterbalances the yang. This energy goes towards expanding, and so the numbers for yin change from one to nine. When the accumulative, or cooling, force overcomes the expansive force, which is especially when yang expands to the greatest degree, it cools and goes to its opposite, of cold yin. The opposite is true of the yin; when it comes together and accumulates inwards, it forms massive density and heat, which is the yang within the yin. This creates the implosion of what physicists call the Big Crunch. The expansion resultant from this is called the Big Bang. The flux of the whole universe is therefore likely to have a double helix of Big Bang, Big Crunch, Big Bang, echoing into infinity. During a period of expansion, there is always a restraining force of accumulation that eventually overcomes it, and then there is a restraining force of expansion to overcome this. The question of constant change from this expansion and accumulation will perhaps come to an end, for the same reason that it started—simply because. Each time there is a Big Crunch, the universe can go back to the Origin. However, the Big Bang occurring spontaneously from the Big Crunch is not an inevitable outcome:

(fig. 1.34)

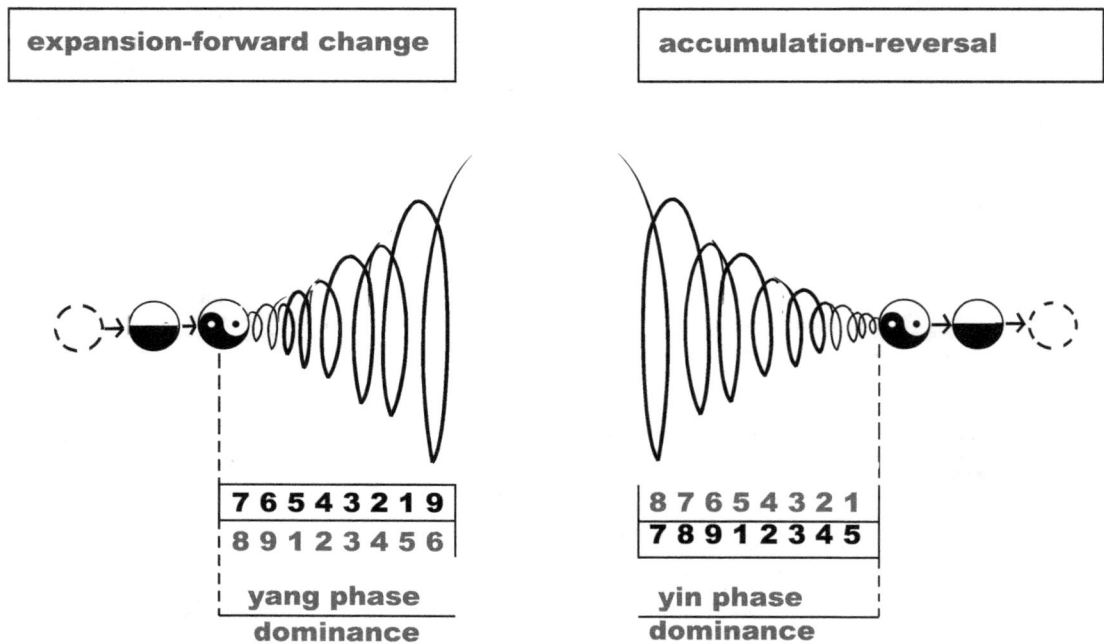

expansion-forward change	accumulation-reversal

7 6 5 4 3 2 1 9	8 7 6 5 4 3 2 1
8 9 1 2 3 4 5 6	7 8 9 1 2 3 4 5
yang phase	yin phase
dominance	dominance

Perhaps black holes hold the answer to the next possible expansion, (or not), of the universe. Black holes are part of the mother-yin or the Origin pre-Heaven/pre-change universe. Essentially they are drawing back the energy of the universe to a Stillness; the universe is not a separated aspect from the mother-yin but within it. Hence these holes are drawing back to the Origin. Therefore, it is probable that the Big Crunch will unite and so end all change. The change of the yin phase, when everything in the universe is cooling and comes back to a singularity again, can be considered to be the same notion as going home within Taoist understanding, going back to Void.

This idea is shared in the numerous other ancient understandings such as the Vedas of India. The interesting aspect is that we can see, at the moment, change in itself, going in a particular direction—I could say "time" here, but as we know, time does not exist. So changes are going in a particular direction; we get older, not younger, things break if you drop them to the floor, etc. When the Big Crunch or "inhalation" of the universe occurs, change will occur opposite to this and so the universe will be drawn inwards. All previous events will flow opposite to how they are at present; everything will be a mirror of how it is now. This return "home" is joyous in its wondrous nature, as we see total balance come to all events that ever existed. This is how we can never know, in a sense, how many times this process has occurred. The infinite nature of the in-breath and out-breath of the Oneness of the universe is always a mystery to the individual, and known to the Central Self, the singularity of Stillness within seeming change. It may be too that the universe just keeps expanding and cooling, as it is now, and time never reverses and collapses back. If continual expansion keeps going, then the universe will return to the mother yin in this way, everything cooling to a point where everything stops moving. This again would be the end of change. (See Appendix 2- page 783)

The expanding and restraining forces, when combined, create the spiral image found in all of life and in all aspects of Nature. The yinyang symbol called the Taiji symbol has this effect. The yang principle is expansive, and heat rises. However, the yang principle's direction of change in its force of expansion is to be driven downward towards the yin, in the extreme expression of yang. The yin is accumulative, and cooling occurs downwards. However, the force of the direction of change of the yin is to rise up towards the yang. This occurs in the direction of the yin and yang meridians of the whole body. This is the expression of the opposite force within the yin or yang. This opposite aspect drives the energy around and simply means that when the force reaches its maximal point, it changes, yielding to its opposite. So even when there is light, there is darkness, and in the darkest times, there is light.

The nine energies are associated energetically (i.e., have energetic resonance with) the seven stars of the big dipper between the yang Polaris star and the yin Vega star. These make up the nine stars, which have different coloured light emanations relative to each other. These were charted by the ancients from intuitive sense, and it is how their origin is usually considered. Hence the nine energies were related astrologically, and these stars form the supreme yang and supreme yin between (in the seven stars of the Dipper) the yinyang movements. Hence this represents the universe—yin energy and yang energy, mixing. It is, however, incorrect to assume that the stars were the origin of this understanding. It was in fact the intuition, and stillness of minds of the ancients, combined with the astronomy they viewed, that revealed the nature of the cycle of nine as the broad movement of yinyang within the universe.

(fig. 1.35)

All numerological methodology today in the West, probably originating in Egypt and Babylonia in from about 10000 BC (related later to Pythagorean mathematics), has a connection to the cycles of nine as the fundamental base (please see Appendix 2 for correlations). These concepts are universal in human ancient history as essential precepts.

vii) The Principle of 11

The number 11 represents the 11 meridians of the body and the combination of the 10 and 12 energies. This principle is a yang number, as it is an odd number, and the meridians are the yang aspect of the system. Eleven is actually associated with 22, which is 10 and 12, so half is 11. We will touch more on this later.

viii) the Principle of 6

The number 6 is almost always associated with the number 12 in the classics. In the Classics, 6 relates always to the 6 Fu organs or the body , the 6 climates or weathers and the 6 groupings of meridians called the divisions or warps, which are in relation to the depth of energy of the body meridians. The 6 always is associated, therefore, with those expressions that are "outer" and a movement towards a connection with the cycles of 12 of earth. These expressions, are always drawn into the 5 –phases expression in the Classics because 5 is the foundation of Human senses. Hence these are not the main issues to be focused on. (This will be refuted by those of the 6-qi school theory of the body, but this is not a Han Dynasty model and as such is not part of the Classical Canon. Please see Appendix 2 discussion on page 800 for more detail)

ix) The Principle of 12

The sequence of 12 relates to the change of the Earth and the solar term change in relation to Earth as central importance. All aspects of earthly existence are related to the sequence of 12. This is the basis for the correlation of the Earth and human sequence in the Chinese calendrical system combining the 10 celestial stems, which act as a pre-heavenly reference point, with the change of the earthly energy—12. This creates cycles of 60 (see diagram that follows). The 12 sections associated with the Earth are called the 12 earthly branches. The 10 stems in the calendar act as heavenly stems, but this is of the pre-Heaven He To basis (see 10-stems section above), which means that it is really the 12 branches that rule this expression. This is because they are within the post-Heaven expression, so the stems add definition and embellishment to the branches, but not the other way around. The calendar is much more about the cycles of 12 of Earth than a true picture of Heaven, because 10 does not move: It is still associated with the structuring of the universe before the post-Heaven arrives. Nine is the true nature of post-Heaven energy.

We must consider the difference between the astrology of Western origin and the energetic cosmology (rather than astrology) of the Ancient East. In Chinese concepts of the exterior, actual astral bodies are not looked at as specifically fundamental as in modern Western astrology, but instead, are appreciated as phases of change of the energy over areas of the sky in relation to Earth, involving aspects between the physical bodies, as well as the physical bodies themselves. Chinese concepts look more at the overall picture, rather than specific points in space. This creates less reliance on the physical and is based in the overall energetics. It is an attempt to see the Oneness of something. Modern Western astrology is all about physical bodies, and this is where there is too narrow a focus. However, there is a number sequence of 12 involved, and so this is an Earth-related sequence, which ties in with Oriental understanding and the ancient origins of all ancient astrological studies, as well.

Astrology in the modern Western world is very different from the ancient Greek connection (Hellenistic astrology), which would have been closer to the energetic associations beyond specific astral bodies with which the ancient Chinese system still is aligned. This is clear too from the fact that the outer planets, commonly used in astrology today, were not known by the ancient Greeks, as they could not be observed with the naked eye. The great misunderstanding is that astrology has undergone improvements since ancient times. Actually, it has undergone supposed "advances in accuracy", which means that it has become more specific and detailed, yet the whole principle of astrology is about the cosmos it is within and the energetic associations connected to it—the general or contextual, rather than the specific.

The number 12 is a major link to most other calendrical philosophies and methodologies of divination and astrological understanding. This is not only the 12 signs of the zodiac, seen in Vedic astrology, it also connects to an Egyptian understanding of the 78 cards of the Tarot, derived from the number 12, and Kabbalistic understanding is associated with the number 12 as well. The Tarot derives much of its meaning from numerology because the 78 cards of the Tarot are equal to the sum of the first 12 numbers (1+2+3+4+5+6+7+8+9+10+11+12=78). Twelve is repeatedly used to represent a complete cycle of events and experiences: the 12 months of the year, astrology's 12 signs, the 12 Disciples of Christ. All associations to cycles of 12, whatever their combination, are about an association to the Earth energy. A principle of 12 can be expressed in many ways, but the principle is basic to what is being observed.

The 12 branches cycle:-

(fig. 1.36)

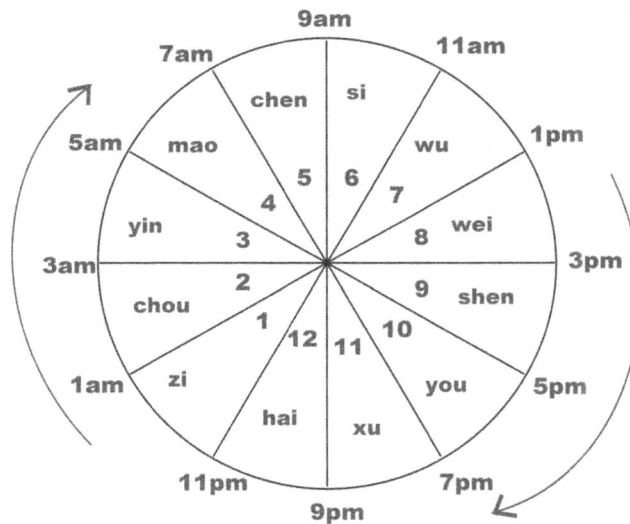

The 60 celestial stem-branch relations, combination of 10 and 12 principles:

(fig. 1.37)

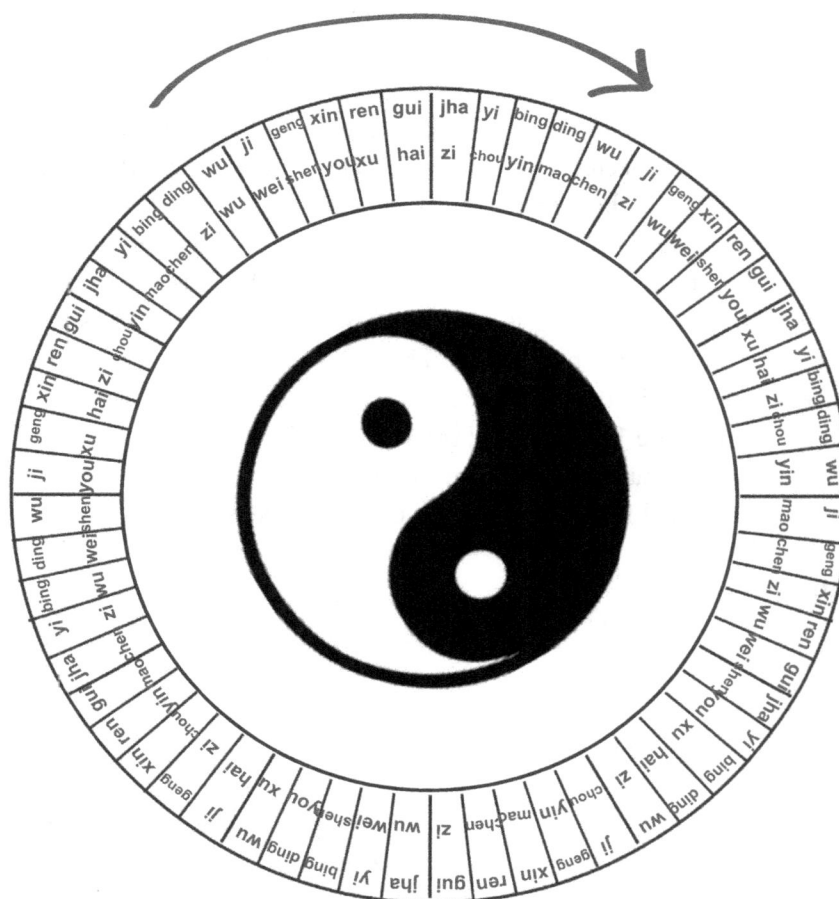

From the 12 branches: 12 x 2 is 24 and this relates to the 24 seasonal markers which are also associated with the Chinese calendar and are still used all over the East Asian regions to understand the nature of the seasonal changes each year.

x) The principle of 28 constellations

The principle of 28 is a yin number (even), so this makes it associated with Earthly phenomena. However there are 28 constellations of the night sky. The 28 star constellations are split up into 4 x 7: 7 in each of the four directions from the centre point of the earth, with the Polaris star as an anchor point. It is interesting that really this is charting the Heavenly expression but what is being charted is not portions of the sky itself but actually the content of the sky, the stars. The stars are yang, points of light, yet they are viewed from the point of view of earth; hence the 28 constellations are a combination of 4(yin) and 7 (yang). The 4 however dominates here because this is a structural diagram of earthly origin; it is a positioning and mapping out of something rather than being energetic and in a sense non-specific, or not associated with "bodies" or any "thing" but more the energetic region, such as the 9 energies (see below). This expression is not used as much in association with medicine but the 28 constellations are expressed in the Ling Shu, chapter 15, and are loosely associated with the body as an expression of Unity of macrocosmic to microcosmic. The constellations are expressed in relation to the meridians of the body interestingly, being between the energetic/ spirit and more physical (see later sections to understand this more deeply). 7 is very much associated with the female and the yin overall; although it is a yang (odd) number, it is the most yin of the yang numbers (associating with metal and P'o/ corporeal energy (see later sections). This again indicates a movement towards the yin overall, or perhaps we could say extreme yin reverts to yang. There are also 7 emotions mentioned in the Classics (chapter 8 of the Ling Shu) overall, though always this is rendered to 5, to make it useful clinically, however the 7 is an important association towards the numerological isolations of the 9 energies and spirit.

7 also is very important in ancient numerology of the ancient Babylonian/ Pythagorean and Greek expression; this does connect to the 9 energies (please see appendix 2).

xi) The principle Post-Heaven Five: The Principle of Humans

This is the principle through which the entire universe is seen, because it is humans that have constructed the mental images, above, in order to understand our place in the scheme of Nature. It is important to note that all of the above are mental creations, expressing how the ancients sensed human existence, in relation to the vastness of the universe and all phenomena. They aid mind but are only an image. Sense dominates all mental notions, and therefore, although these notions are a road map, they never give us the full feeling of experience. Please be constantly aware of this. Humans are deeply connected to the number five and sequential nature of the five energy. We have two arms, two legs, and a head—five exterior appendages. We have five deep organs (from which the perception of the world, in five phases, comes), and all things relate to five within a human. Hence, often humans use five to describe the universe around them. This is why Chinese philosophy became totally enveloped in five as the main principle:

(fig. 1.38)

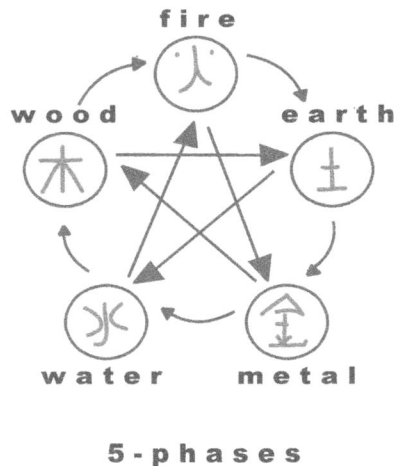

5-phases

The five-phase principle is the one that relates and originates from the inner organs of the depth of human beings (Zang organ), right out into the exterior universe, and provides us with an understanding of ourselves, or the connection with ourselves throughout all aspects of the universe—from the breadth of the 10 stems and nine energies, to the Earth's 12 branches and all living things around us. The five phases is the last, as it is the most recent arrival, with that of life. Heaven and Earth formed before life and humans. Human five is a derivative of nine. It is associated with the yang more than the yin energy of Earth, although the female is yin within this yang. (Five and 5x5 = 25 are asserted many times in the Su Wen and Ling Shu texts and are really one and the same expression, 5 x 10 =50 and is the association with 5 and the 10 stems, creating the yin number 50, interestingly 5 x 12 =60 and this is association of 5 and the 12 branches forming the yin number 60; the stems and branch calendrical system therefore is constantly dominated by yin numbers therefore relating to Earth).

These are very brief explanations of the number cycles associated with Oriental philosophy. I call them principles because they are more than just number cycles. By looking at this diagram of concentric circles, what we can see is that these represent resonant cycles of change, almost like resonant frequencies of energy. These energies are often related to humans (five phase) changes, in order to relate to them in human terms. This simply applies the yinyang of man to all things. This was the process by which Oriental philosophy was unfolded and understood. Oneness could be called the resonant frequency where all the rings move as one.

(fig. 1.39)

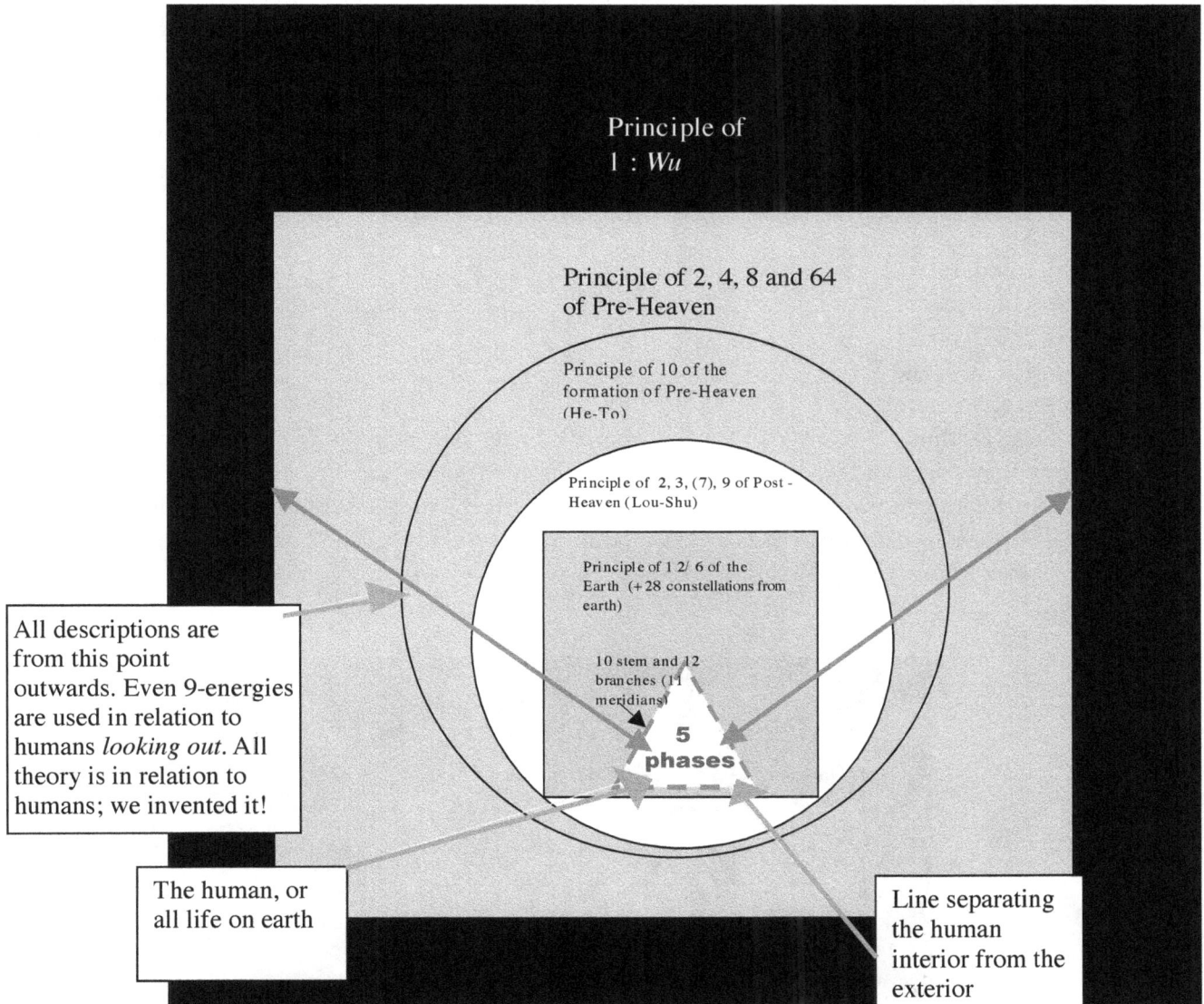

Principle of
1 : *Wu*

Principle of 2, 4, 8 and 64
of Pre-Heaven

Principle of 10 of the
formation of Pre-Heaven
(He-To)

Principle of 2, 3, (7), 9 of Post -
Heaven (Lou-Shu)

Principle of 1 2/ 6 of the
Earth (+ 28 constellations from
earth)

10 stem and 12
branches (11
meridians)

5
phases

All descriptions are
from this point
outwards. Even 9-energies
are used in relation to
humans *looking out.* All
theory is in relation to
humans; we invented it!

The human, or
all life on earth

Line separating
the human
interior from the
exterior

The above is recorded in the Tao Te Ching, the last lines of chapter 25:

Life/Humankind is formed from Earth and follows its way
Earth is born from Heaven and follows its way
Heaven is born from Void and follows it
All of this is Naturalness and is Nature itself.

Notice that in the above diagram, first is Wu, second are the principles of the pre-Heaven, and then post-Heaven, and then Earth's change, and only last do we have the formation of the human being. This follows the track of the evolution of the universe from the beginning. The Wu permeates everything. Then governance is derived from the preparation of the Big Bang/principle of ten, and then, as we go inside the change of the universe itself, lastly humans. Humans are part of the expression of the myriad things, all derived from the Stillness of Wu.

The above diagram is also a statement of the following:

1 implies 1
2 implies 1 and 2
3 implies 1, 2, and 3
4 implies 1, 2, 3, and 4
5 implies 1, 2, 3, 4, and 5
(And as such, it continues …)

So everything is really part of the original circle, the "original face". This "original face" is everything that exists, and when we look out from our own centre and also know the in-looking at the place we are looking out from, we recognise the same in the microcosmic as is in the macrocosmic.

The previous diagram gives us the view from outside of Heaven and Earth—in fact, the view from outside of everything. Compare this to the diagram that follows:

(fig. 1.40)

This is Douglas Harding's image (See bibliography:-Internet Reference 3) of exactly the same thing, but from the microcosmic perspective of the individual's experience. From our perspective as humans, we are headless visually, and if we look from this perspective, emptiness is behind us and is in the transparency of the actual visual space we look through. As we move outwards, we get the world of form and society and nation and world and universe—and again nothingness at the macrocosmic limit. What this shows us is that there is nothingness at the centre of the "onion" of ourselves, but also nothingness as we travel to the outer reaches of space. It is in fact a constant nothingness. Even the forms in between are transient and are not really made of substance; they only hold the appearance of substance. Therefore the microcosmic picture is exactly the same as the macrocosmic: inner-space and outer-space are one and the same—with the world of form as part of it and between the nothingness. Douglas Harding also points out that there really is no difference between microcosm and macrocosm. It is all a matter of perspective. We are a microcosm of the universe and yet have it entirely within us; this is the paradox of truth. It also makes clear that observation (what one could call "pure" science) and fragmentation (modern science) are very different. Observation of the microscopic is one, but then the separation out of what you see to be separate bits (often used for some intention or purpose, to bolster a pre-arranged ideology), is separation, but one can separate out at any level from the microcosmic to the macrocosmic; also one can see the unity at any level. This is the huge difference between modern scientific rationalism and Ancient energetic understanding.

"I am" in its true sense, means all of what is, which is very different from someone saying, "I am this body, this mind". What it really means is that I am the entire universe, the universal body; it is recognition of that which one truly Is rather than the image or external appearance. Harding expresses this brilliantly simply, by saying, "I am not what I look like!" The point he often makes is that, dependent on the observation point, one can go from the outer-space view of "You" to the galaxy view, the solar system view, the planet view, the country view, the city view, the person view, the cellular view, the atomic view, the subatomic view, and then the space between the subatomic material or what we could call inner-space. These are all views of the same Oneness, the same "I am". The key issue we find is being caught at what seems like the human level and believing this is all "I am" means. Harding expresses the following in consideration of this from the modern perspective of a simple work environment (from The Hierarchy of Heaven and Earth (1952), p.147–148):

> "Information from below and instructions from above somehow find their way to my desk, and have there to be reconciled: that is my whole concern. How they arrive; how my instructions are carried out; the mode of working, and indeed the very existence of my innumerable subordinates [for example cells of the body]– these are matters, which I seldom consider. But when I do so, I conclude that each official, whatever his grade, is in a position like mine; and that his duties are, first to receive and co-ordinate and pass up the data presented from below; second, to receive and apply and pass down his superior's instructions …. In a sense it is they, and not I, who inhabit my cubicle on the middle storey, as I inhabit theirs."

This can be represented as this simple diagram, each level being a subordinate or superior to the one above or below it, each of a different order, a primitive fractal arrangement:

(fig. 1.41)

Harding's point is that seeing the human level rather than the higher levels of the cosmos, or the lower levels of the intra-cellular, is actually in itself a realisation of Oneness, acceptance of what is happening right now. From inside (below) and from outside (above) we are formed, and we are a vessel of energetic movement, what Tony Parsons calls an "explosion of life". It becomes very difficult to point out what is "me" in this (being both inside and outside,... so there is no inside and outside in reality), because the me is not only part of everything else but also the centre—"I am, that I am".

If we consider emphasis on when we close our eyes, rather than open them, the night rather than the day, the time asleep rather than the time awake, the spaces in between the frames of a film rather than the film itself, the infinitesimally small moments when a subatomic particle is out of existence rather than the instant later when it comes into existence, anti-matter rather than matter—if we look towards the place where we are looking out of, rather than what we are looking at, one knows the enveloping nature of God or Oneness within, as well as its manifestation without. However, we are addicted to the manifestations, and we forget about the origin. In the 2000 film by Edward Yang "Yi Yi", the small boy, who is the main character, takes photographs of the backs of people's heads, as his pastime. On being asked why, by his father, his answer is, "Because people only can see half the truth".

For a human categorising everything in relation to himself, everything must be related to post-Heaven five to understand its change in human terms, because humans have a "five-dimensional" base perspective, so to speak, the 5-phases. It is important to note that numerous people over history have looked at one or another of these sequences and decided that the only important one was x, y, or z. This, as we can see, gives only a partial view. With the original principle of Wu or unity, there can only be a real understanding by accepting all as one and seeing how all things relate. This image of concentric circles has been used timelessly by the ancients to describe this unity. In fact, the original symbol of yinyang merging in post-Heaven was designed with this in mind:

(fig. 1.42)

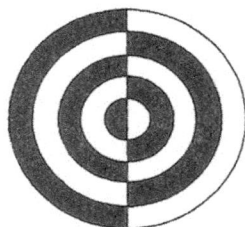

The exterior means that there can be an interior. Without an interior, there cannot be an exterior. All aspects of the universe are one and are inseparable from each other. However, each concentric ring needs to understand its unity to the whole, or to the rings that encircle it. The interior of a human life is made up of variations in the five physical constitutions and nine spiritual energies and is augmented by the stems and branches of its meridians, which are connections to earthly energy. This formulates the human being as a tip of the iceberg of Wu or conscious awareness. If this constitution is in relative change and in communication with the exterior, then all is well. If this moves against the exterior and does not adapt well, or the exterior is too imposing for the interior to be able to survive, then the interior always gives way to the dominance of the unity that it is part of and inhabits. We call this process death-rebirth. It is the balance and communication of the relationship of interior cycles and environment, relative to the exterior environment that makes for health. Health and life also incorporate death. However, because one can see the context of life and death and does not hold to them, all is well and there is no suffering. Dis-ease is the fragmentation that does not see this, and so is the origin of suffering. Suffering therefore is not being able to see the Wu that envelops all phenomena and all things and therefore unifies, whereas yinyang, looked at from within its whirl, can seem fragmented to mind-identity, that is fixated on its whirl, rather than rooted in Stillness.

In the next chapters, we will be describing the constitution of a person. Constitution means the way a person's energy is in body and spirit, in relation to the natural order of things. Illness derives from fragmentation of the mind-identity, of the Wu-body-spirit. In these cases (which applies to almost all humans), a person's illness or behaviour may split away from what one would expect from his or her bare constitutional expression or spirit constitution joined with the body constitution. This of course is due to the fact that mind-identity has trapped his or her being from authentically being itself, often due to past events replayed endlessly within the person. This happens to all of us, to some extent or another, which is why these studies are useful. It allows us an objective reflection of perhaps what we could describe as our original Nature, or the colour or flavour of it. This can lead us to sense ourselves, to let go of mind-identity to arrive at a clarity of feeling, through acceptance. It is not an attempt to label but to experiment with what one believes one is, and the actuality of what one is in complete unity and relation to the whole. There is no other reason for this study; it simply helps us to discover what we already know about who we are, but are not able to access, because we have trapped ourselves from Ourself. This fragmentation causes most dis-ease patterns. Be careful, therefore, to see clearly this study from the second part of this book, in this series, about internal medicine. In this book, we look at the Constitutional Pattern (CP) in the medicine, and we look at the Dis-ease Pattern (DP). These can intertwine at some point, but can be entirely different. In medicine, one always goes to treat the DP in the context of whatever the CP is.

In this next section we start to look at the constitution of both Jing or body and Shen or spirit. We have to realize that mind is a function of spirit within the body and so mind-identity is the pathology of this combining, or at very least, a process or expression derived from the union of bodyspirit in humans. This aspect will be the discussion of the aetiology and pathology sections of this work. In these next parts, while there may be mention of mind-identity and an awareness that this is part of the growing human and its expression, constitution is especially an idealistic model from which we can look at what we might call

'Ideal human nature', almost like the expression of Qi Bo. In this section we see the sage-expression within every human and I will just touch on the points within the constitution that can be the root of expression of a particular type of mind-identity, be it mental or emotional. In the aetiology and pathology sections of the text there will be a focus on the mind-identity as the basis of pathology, or suffering manifesting within the ideal expression. Life and death of the bodyspirit are inevitable and also totally in harmony with all of life. The resistance to this is called mind-identity and this therefore, is the process of sufferance, which is dis-ease, only found in humans. This is the basis from which medicine is most deeply recognised. Now however, we will not look at pathology or medicine but just look at the ideal and the Eden of possible expressions of the human in bodyspirit combinations with nothing in the way of this foundational view.

Part 2

The Constitution

2.1 Differentiating the Physical Constitution

Let us begin by defining the terms Jing (Sei in Japanese) and Shen (Shin in Japanese), as used in the classical material and as they will be used in this book.

Jingshen is written as one word, as is yinyang. In the term yinyang, yin is placed first, showing that the mother of the universe is yin and first. However, as we will see, Heaven's energy tends to be primary, ruling (yang) energy within the post-(partum)-Heaven universe, and so directs the yin (Earth) within it.

The character for Jing has three ideographs within it. By looking at the structure of the character, we can derive a closer meaning. Jing is most often translated as essence.
The character has these three parts:

(fig. 2.1)

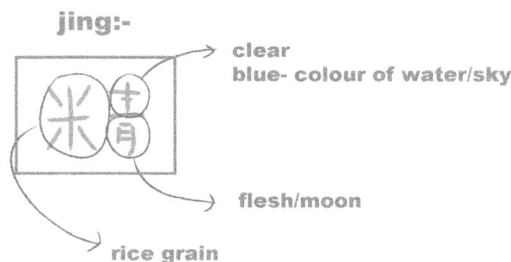

jing:-

clear
blue- colour of water/sky

flesh/moon

rice grain

If we draw these together, a grain of rice and flesh, both of which relate to physical energy. This is why we relate Jing to the energy of the Earth. Therefore, it is more yin in quality, in opposition to Heaven and the yang. Earth energy has actual form; it is something you can hold and has mass, for mass is condensed energy rather than stasis. Of course nothing is static but in change. Hence Jing relates to the construction or density of the body. The quality of Jing is our physical constitution and the energy that quantifies our nature through our relative size and the power of our bodies, the shape of our bodies, and our ancestral heritage, compared to one another. These are all attributed to Jing. This could be said to be close to the fragmented image of DNA within modern Western models.

Male essence Jing and female Jing are sperm and egg respectively—again, in a modern Western, fragmented approach. If we see the sexual connection as a joining of spirit and body, Heaven and Earth, then this has a fuller meaning. In addition, Jing means constitution, which is not just the sexual fluid that represents the hereditary chain to reproduce the same body structure, but also is the structure itself, the quality of the structure, as an overall unit of measure. This is Jing. We can appreciate Jing in different ways: macrocosmically and microcosmically, even within the person. Jing is prenatal or can be described as a pre-Heaven energy, meaning before birth. This concept of Jing macrocosmically conjoins with the image of the universe before its birth, which microcosmically relates to the foetus in-utero. So we can say several things about Jing: it is of the Earth; it is of the body; it is physical and visible; it is refined and pure; it is dense and yin. Therefore, Jing relates mostly to the four and twelve principles of the Earth (yin energy). Jing's relation to the kidney yin energy (please see later in this section for more details) also links it with the ancestral energy and the brain. When Jing combines with Shen, or heat, there is a movement of both fluid-yin-Jing and heat-yang-Shen together. This constitutes heated Jing, which is what we consider to be a live person.

[Please note that Jing is the most yin within the body energetics. However, outside the body energetics, we can say it is related to Po. Po is the corporeal quality of spirit and will be discussed later. The energy of Jing in relation to Po is that Jing is cool within the energetics of the body. Po, however, could be said to be the aspect of the exterior within the interior of the body energetics, it is like Wu in relation to yin as discussed above, yin being within the post-heaven (life) expression. It is purely the dead weight of the body and not within the range of energy that supports life within the human yinyang spectrum. It is similar to a stone or a piece of metal within the body, so to speak. It is a spectrum, so when the alive body-spirit of human energy ends, another spectrum takes off—the body in death-decay. This is another expression of why the words life and death are truly meaningless.]

The brain is simply an extension of the bone marrow, which is dominated by the kidneys Jing. It is not only powered by the kidneys Jing, but also is the store of ancestral memory of physical experiences of the human being since the dawn of time. The brain exists very much as one with the physical body. However, we can say that the formation of the impetus to think (mind) concerns the spirit. The brain records experiences; it is the physical substrate for recording like a CD or tape. These are not just the experiences of a single person but of the whole history of life, and prior to this, of the whole universe, bedded into the expression of the physicality of the body.

Memory, one can say, is stored within the brain-body, like a recorder of every experience. The spirit is that which ignites and animates the body, and the mind is part of this, whether identified with itself (dis-ease) or not (health state).

Mind-identity/dis-ease is the distortion of this memory to form illusions like day-dreaming, cinematic projections onto the world as it is. This is like smoke rising from fire. The illusions, which are always of separation, divide things and splits things off, labelling, and categorising. These functions are not a problem, but if drawn into them, we form an identity with the idea of separation. As a result, we are in the grip of fear—fear of the "death" of this idea. This is what we call the "fear of death", but it is actually only an anxiety about the idea of death that we hold onto.

Since the ignition of this capacity within human evolution (probably around 100-40,000 years ago), ideas of separation have been memorised in the tissues—or Jing—of the body and passed on as an ancestral message. Hence, this agitation is also our heritage. In deep meditational-relaxation states, it is said that "past lives" can be remembered, which of course must be true and is not surprising, as held within the body tissue is dense energetic memory. These memories however are not owned by an individual but are part of the "super-conscious" or mind-identity of humanity's experience of mind, hence the idea of individuated soul moving from body to body is a total ideology of mind-identified egoic expression and focused on keeping the individuated state alive at all costs, because if not, the mind-identity dies into an unknowable place where it cannot be an individual any longer. These ideas (e.g. Karmic retributions) run throughout often the ancient world also but are not part of the deep source material in any culture. At the heart there is Oneness in all of the expressions of fundamental truth. History is history; it has no relation to the present moment other than a re-looking at the history, which is in the end useless in the present moment situation.

Memory is not the problem; it is really what is occurring now that is key. What is now is all there is. Spirit, in its activation of these memories, occurs now, in the moment. In fact, even the illusions it conjures up are arriving right now. What occurs in liberation, or contextualised mind (state of natural health/freedom), is that the identity of a person being some-thing is realised not to be true. The shift in perception contextualises mind to the perspective of the first person Singular, which is the Emptiness from which we look out at the world, the background and centre of what is going on in the foreground. This differentiates memory from abstraction of mind-identity. Abstraction of mind-identity is an add-on; it is not the whole picture. From context, one can see the whole picture, and therefore mind-identity is just smoke, not absolute structure. However, with so much memory, one could be lost in mind-identity for many years, even a lifetime. It may come and almost go, like a dance. This is not a problem. It is how it is, and very often it is infused with tremendous suffering. Nothing can be done about this, only undone by "opening" to what is, not what could, should, or might be. This happens as it can. Once the self starts to dissolve, the memories, which came from emotional reactions, are realised to be happening to No-one, hence they are just as they are. They lose their power and fall away like smoke, easily and immediately. There is often a dance between the two places of mind-identity and context mind-identity till mind-identity gives way.

Jing, or the physical form, is not a true state of static form, but a changing form. It is only an illusion of mind-identity that sees this body as being "my" body or that body as being your body. In fact, it is cells of the same body, or fingers of the same hand. The ancestral Jing has memory connection with everything since the original birth of the universe, so a past life is really the past life of all humans, not of "a" human. We can't own a past, just as we can't own a body—this one in front of me or any other.

As an overall principle, the Jing relates to before-birth energy, or prenatal energy, and to collection and gathering in order to manifest a form-like entity, which is impermanent.

Please note that the spirit energy is different. It has no past collected and is the actual life of a person. It has no memory and occurs only now. Past, therefore, can only be read in the moment. Spirit is not accumulated; unlike Jing, it is instantaneously now.

Shen has two parts to its character:

(fig. 2.2)

shen:-

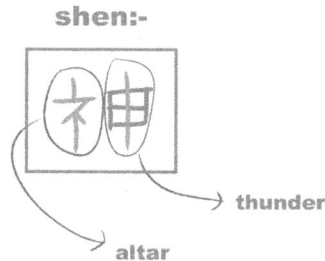

→ thunder

→ altar

Shen is made up of the image of an altar and thunder. "The altar of thunder" is very important. Thunder is the energy not of the Earth, like grains of rice or flesh, which has material substance and is of below and ancestral. It is the energy of Heaven, the sky, and the altar, representing an offering, perhaps a place in the body, which holds Heaven's offering or is a conduit for this. So what can this mean? This represents the spirit, the spark of life given from above (or all around … post-Heaven). It is heavenly; it has no memory and is occurring now, like thunder's effect. It is not heavy but light. It is awesomely important and powerful, as it is an altar, and a temple is where the altar should be, in the Heart. Shen relates to the principle of the universe, nine. Spirit has no memory of the past. It is totally present and always in now time. In relation to Jing, therefore, Jing holds memory but is not activated until the spirit is involved. Hence Jing, too, is in the present. Mind is formed when Jing and Shen meet, and then often illusion sets to form mind-identity. Shen is the post-Heaven activating, changing spark of life that gives life to the body. Shen animates the body Jing.

This animation process can connect to the Wu, which permeates or underpins both Jing and Shen, where mind-identity is seen for what it is, another manifestation. Or it can formulate the distortion of mind-identity, as the changing will investigate/activate the memory function of the body Jing and identify with it as being separate. Hence it is mind-identity that is a distortion of what the spirit does in action on the body. The body is still; the spirit is activity; both are present until the mental-processing movement of the spirit activates the system in a way that warps. Therefore, mental process is a dis-ease of spirit relation and therefore of yang relation rather than yin.

The spirit activates the body, giving it movement. The body is pre-Heaven/natal, and the spirit is post-Heaven/natal. However, the spirit also activates the body movement, and this causes mind-identity, which is driven by the spirit. So spirit is life, but also it is the yang that derives mind-identity. Mind-identity is like a resistor wedged between body and spirit, and it prevents their union from being effective or feeling complete.

Mind is an aspect of the body physically, but run by the spirit. Hence the spirit's now-ness must be present (with a connection to Wu) for the body. If mind-identity is resisting, this cannot occur effectively. The spirit is one with the body. Its nature is now-ness viewed through the limitation of the bodily form but aware of the presence of Stillness behind all changing forms. Hence the spirit within the body is an aspect of the wholeness of Wu—a tip of the iceberg. To dissolve the identification with mind requires acknowledgment of the unity of body-spirit without mind-identity's resistance and attempt to fragment it (an impossibility for self will). Mind-identity is a relatively "new" manifestation in humanity, and it has not reached full expansion yet. This is where much the suffering lies, in the willing of it to dissolve. This is very important.

Yinyang or jingshen (body-spirit) are platforms of constant change and should not be made rigid when applying to relationships, form, experiences, or anything else. If mind-identity looks at them as they seem to be, they essentially contain a great illusion— permanent absolutes to be adhered to. In fact, they are impermanent and based on Stillness. It is the unification of the still base that is permanent, not the foreground changes. Hence if one hooks on to the fantasy that something within yinyang is real and therefore is going to stay the way it is, this creates immense suffering in mind-identity, for it all passes away, including the seeming individual. Therefore, we have to go deeper into the Wu that pervades yinyang. This means to look inwards at the emptiness and to be anchored in it once more. As children, this is easier because mind-identity has less ability to abstract; the body may have memory (even of trauma pre-birth or otherwise), but it is generally accepted and the present is primary. However, as we grow into adulthood, we are taught (as part of the function of mind-identity of society) what is right and wrong, good and bad. The memories are activated as "past lessons" or "future (past-based) plans". Here we live in another world cut off from what is actually going on now. The return to now is a dis-identification with the memories of ourselves (self-image). All images are not what we are at Centre, instead just a clear empty space where the world is happening within. Then we return to what is natural.

This shows us the difference between the body and the spirit. It is of course true that they join together within the human structure, and also of course true that wherever there is yin there is yang, so spirit energy and body energy are all one. However, it is important to differentiate before we see the clarity of their unity. Jingshen relates to the human principle of the post-Heaven five phases. Qi is associated with the movement of jingshen as one. Hence this makes up the principle of three within the body: Jing – qi – Shen. In the macrocosmic, it is Earth-human-Heaven (see previous chapter). Because qi moves the physicality, one could say that qi relates more to spirit and to mind, whereas the body vessel is too dense to be energetic. Jing energy is of the earth, but can't really be said to be the actual earth material itself but a refined essence of it within the human body, like a representative of earthly physicality but not actually the dead-weight that the body-shell is made from. Hence one could say that the Jing-qi-Shen is all a form of Heaven (Shen macrocosmic) in relation to the physical substrate or weighty density of the body's shell (we can call this Po, as will be discussed later) or Earth energy (Jing macrocosmic). Therefore, humans are more spirit-Heaven than they are body-Earth, if you like. We are moving entities, like fire. We are not still, like stone. So we have many more problems than stones! Or seem to! Remember yin and yang are always in relation to each other, so one has to identify the parameters of their use.

(Please note again that I am using jingshen in terms of yinyang. Although Jing is associated with the yin and the physical, and Shen with the yang and spirit as overall macrocosmic terms, they can also be used in the microcosmic. Jing can be used to describe the energy of the kidney's yin fluids, and Shen can be used to describe the energy of the heart's expression. The microcosmic are not relevant above, as we are at the moment looking at a broad view. Shen macrocosmic means all the spirit energy as an overall picture here; Jing macrocosmic means all the physical energy. Later we will discuss the various microcosmic aspects of Shen and the various aspects of Jing.)

i) Conception and the Formation of the Jing-body

At the beginning of the universe, the Void gave birth to Heaven, or yang, and from the fact that there was yang, yin was formed. Thus started the universe of change. We have differentiated the Shen body and the Jing body. Here we explain how the different energies—the Jing—of the woman and the man come together to create the physical quality and constitutional quality of the vessel for which the spirit energy colours the existence of life. Conception is the first aspect. The formation of the body as an entity is expressed in chapter 2 of Book 9 of the Mai Jing. In this chapter, the pregnant woman is said to form a body or form flesh within her. This is different from the idea of spirit. Direct reference is made to the body. Activation of this body comes at a later stage, with the first breath igniting the spirit, or visa versa, and forming the seemingly separate individual of the baby. In utero, there is no separation with the mother.

The Male Jing and Its Origin

The male Jing energy is the energy of the constitution. This energy is replicated and expressed in the yin energy within the left kidney. This is where the Jing or essence is stored. However, it is expressed throughout the system and is literally the power and energetic strength of the constitution. The greater the quantity of the Jing, the stronger, more dense and larger the physical body can be. The Jing, from a microcosmic perspective, is only the kidney essence, including the semen. More broadly, it literally is the constitutional strength. Either way, it is a far more physical substrate than the spirit-Shen.

The male essence will accumulate in the testicles, which are considered to be the exterior kidneys—and they actually look like the kidneys, as do the ears, which are another expression of the kidneys on the exterior. The penis, however, is dominated by the liver and is connected with liver energy. The Wood/liver organ energy is full of heat and blood; hence the penis is a tendon and is actually connected to the base of the spine, which especially expresses where the energy of the semen originates, almost literally from the bone energy and marrow.

The seminal fluid is a cold fluid; the sperm themselves are hot. There is a combination of the energetic properties of saltiness and bitterness in the emission. Hence, when there is ejaculation, the male body loses cooling energy and starts to heat up, but also loses heat and cools down and becomes more tired. Male body energy as an overall entity is seen as the Qian trigram (Heaven) – pure yang before sex moving to Li / Ri/ 離 (fire) or energy after sex, Li is a female expression, Qian is yang. The male body is yang and heated, but the essence-Jing of male body energy is fire within water. Yang energy is in relation to two main trigrams, pure yang, or Qian. Pure yang is the representation of the yang principle in action. Male Jing essence is said to be circulated and conserved through abdominal breathing into the space where women have ovaries. Here, male body energy is enhanced, and the power of the yang aspect within the yin, or the potency of the sperm, is encouraged. This is part of all practises of the inner arts of reconnection to natural movement, such as Taiji Chuan and Qi gong.

Pregnancy, too, is a more physical, Jing process. The woman must eat more and create more blood. She is creating a physical baby, a blood form. In men, meditative practise and special exercise can expand the space, whereas the ovaries are in women to create energetic power to be used to enhance health. In

Taoist practise, this recycling of the energy and building of the lower abdomen is seen as the "energetic baby" or "immortal foetus". This explains why practises such as these were associated with men almost exclusively. This builds the yang Qian (the trigram for water, see next chapter). Commonly practices that focus on this region are about longevity and ideology about prolonging life and sexual power. Sexual energy is simply the exuberance of an animal, the extra that can be used for reproduction. Sexual output is really, if felt rather than thought out, simply neither too little nor too much just as it is with animals in nature, it is self naturally regulating if one lets go of ideals and fear. This easily turns into a male dominated compounding of self-manipulation rather than natural health. No animal attempts to live longer than its natural expression. These practices very often are huge red-herrings of so called "spiritual growth" but actually enhance the egoic sense. All that these practices can do if they are understood is to help mind-identity dissolve and move towards natural movement and away from form and self-improvement ideals which are not real and not Taoist but Confucian ethics.

The space where the yang energy of the Ming Men/ Mei Mon can be infused in the lower abdomen is also the place where yin (Jing) can be supplemented (not replaced) via the bitter flavour from food. The male can form the energetic "embryo" within his abdomen, which is the power of energy created from Emptiness/hollowness, condensing yang energy. If the energy of the masculine is not anchored, then it rises up to the head; the yang of the core of Qian can therefore move into the upper body and head. This has the potential for madness or mind-identity domination. In women, the body (and specifically the vagina) is represented by the Kun trigram of Earth, the opposite of Qian. The Earth is yielding and a hollow space within it for the female body to take the yang of the male. This of course is the womb and vagina and the place from which life comes. This is the archetypical expression of the female body in health. Her body is hollow and yin, hidden, whereas his contains light. This is how yin and yang merge. The hollowness of the female body energy makes for a more receptive energy and one that has no yang internally, so she requires direction from the masculine body principle. The emptiness of the inner yin means that there is nothing to rise into the head of a woman and drive her to mind-identity/ madness. Hence she is calm, and without specific direction until it is given from the masculine body. Please note this is all about body male and female, not about particularity and complexities of the spirit-Shen involved at this stage. When the woman is pregnant she holds yang, which is associated with the Kan/ Ken/ 乾 trigram and is associated with water, Kan is actually a yang principle. Martial arts and "spiritual" practises are often dominated by men. It is not only that men have been overlords of this territory, but also that the female body is innately something that men have difficulty with, and this is why men need further training to anchor their energy and sink the mental energy. The Tao Te Ching is constantly advising that men need look no further than the female principle to understand reality.

The Female Jing and Its Origin

The female Jing is related again to the constitution in a broader view. More specifically, it is the eggs she carries and the ovaries that are supplied by the kidney energy, specifically of the right kidney. The egg contains an active energy that is yang. The movement of the egg through the fallopian tubes and into the uterus and its initial expulsion from the ovary are all yang processes. Again, the connection with the spine, of the uterus and ovaries, is important. Less obvious energetically, the ovaries via the uterus originate from bone-energy/marrow. The female energy as an overall entity is seen as the kun trigram. The empty lines represent the empty quality of the vagina and womb and the requirement of the yang principle. Pure yin energy associates with the Kun Earth energy. Pregnancy in women is a physical manifestation: she grows a vessel full of powerful physical energy to be manifested. A process of the mother's energy feeding the child occurs. The anchoring of the woman's basal (right-kidney/ Ming Men) energy through the interior nature of her genitals makes, innately, for a calmer, more anchored mental pattern.

ii) Sex and Conception

For men especially, sex is a yang process, and so yang is used up during sex, as well as yin, as a secondary. If the yang is released also, this helps the body not get too overheated after sex. For women, sex uses up yang, because nothing is released at this stage. Sex therefore is a yang/male-dominated activity. If we think of the monthly period (or post partum) for women and the expulsion of sperm for men, as cyclical patterns, we are quite accurate. The period after sex is an important time for men. The days after the monthly period are an important time for women. During these phases, the man and woman are essentially empty. They have lost yin and yang energy and are basically "smaller" in energetic weight, so to speak. The turgidity they felt has passed. It is in fact a small death. This time can be very helpful for men and women because the body is more fragile, and so the heat of mind-identity, is not so full and exuberant. The person is forced to take things slower and become more still. This allows them to let go of identification with "self" more easily. Of course, this is not a recommendation to destroy ones sexual reserves. It is simply important to see that these cycles exist in men and women. It is vital that one respects and allows these cycles to occur, rather than restrict, resist, or control any aspect of these cycles. One should literally follow the flow of them.

Monasticism and abstinence from sex to enhance training must be understood to be ascetic and are unnatural unless deeply needed by the body—which means the person is tired, aged, or sick. Withholding of sexual energy can be just as damaging as constant, addictive release. It is often a waste of energy to attempt to control sexuality, but better to accept it and know its movement, for it is, in itself, the cycle of life. One must ask oneself why, other than to follow an egoistic path, one would abstain from such a process. When sex becomes as normal as eating or drinking, this is an accurate sense of it.

To know the emptiness of post-sex for men (talking about body expression masculinity only) is to realise the futility of the ego and how easily the body becomes weakened. This weakness is good to experience, especially for men, because it can deeply make the man realise the lack of power to control and to dominate, for his seed is passed on. This allows him to know the truth of life—that all things live and die, and that ultimate power is not found in control, but in acceptance. Women have no choice other than to follow their cycle, which is to their great advantage. After sex, the way a woman allows her partner to be accepted in his vulnerability is actually an entry point for men to experience trueness of love beyond fear of annihilation. The woman knows this innately if she is in touch with her deep sense. The body of the female is always both a "partner" and a "mother" to the male body. The male body is always a "partner" and a "child" to the female. "Friendship" is involved in all of this as uniting Oneness.

The penetration of the penis into the vagina is the mixing of the yin and yang energies. It is as basic as this really and the ideas of sperm and egg are really Westernised fragmentation of the idea of Essence of Jing of male and female which is an overall expression of male and female energies, not specifics. The power of these energies is the power of union and of conception, the lack of potency is the lack of this fusing energy of conception. This is all. Ideas we have beyond this actually make out thinking fragmented involved in the microscopic-materialistic rather than the energetic. The heated masculine penis and the cooler or open/empty female vagina represent the completion of a cycle. The sperm and egg are known to be a representation of what is visibly seen. Hence, in classical understanding, we simply see the vagina as being emptiness or yin and the penis as hard with heat, also within (this is the sperm). The sperm is salty energetic heat, and the vaginal mucosa is also salty energetic heat. The hollow space of the vagina is yin, and the penis itself is yang with heat, but has hard solidness. This creates the phenomenon of unity of yin and yang, which makes life possible. Note that the Tai chi symbol indicates this. The seed of yang in the yin is water meaning pregnancy; the seed means that this region is controlled by water. The seed of yin in the yang of fire meaning the male after sex; that this region is dominated by yin. Hence we have the male/female balance.

The classics say that pregnancy is actually a 10-month process of 30 days per month, so birth occurs sometime in the tenth month. Interestingly, 10 relates to the pre-heavenly formation of the universe in the He-to map, showing again the unity of microcosm and macrocosm. The expression of nourishment of the foetus is mother to child and via the mother's meridians. Notice that it is explicitly expressed in the second chapter of the tenth book of the Mai Jing that it is the meridians, rather than the organs. From this, we can extrapolate what is explained in the following section.

iii) The Process of Pregnancy

The process of pregnancy is governed by the pre-Heaven connection and the 10 stems concept. Post-Heaven is the actual emergence of the foetus from the womb. However, the pre-Heaven (womb itself-empty-hollow-still/Wu) and pre-Heaven formation stage (associated with the placenta and 10 stems) is expressed as the growing foetus in the womb. This can be represented in the following diagram:

(fig. 2.3)

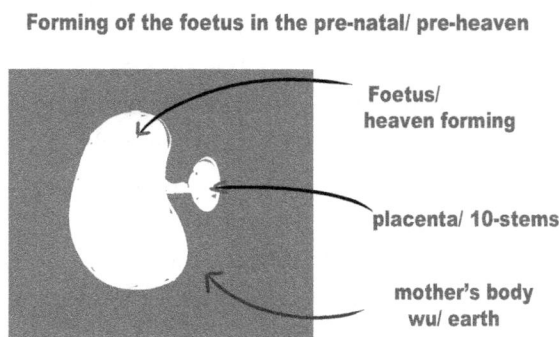

Forming of the foetus in the pre-natal/ pre-heaven

Foetus/
heaven forming

placenta/ 10-stems

mother's body
wu/ earth

Notice that the placenta is associated with an extension of the liver organ energy within the body—in other words, a part of the uterine wall, which is part of the liver system organ. This of course is initial yang and connects mother to foetus. Note that this is in relation to primary principles and is the 10 stems of the post-Heaven formation stage. The foetus is the growing yang, ready only at birth, and the outer darkness is the yin of the female womb relating to Wu and pre-Heaven and Earth, due to the fact that Earth is within the post-Heaven change of the universe. Notice that we explain the Jing body's formation before we look into spirit because yin/Wu is primary. Although Earth is Stillness within Heaven, birth always arrives from Stillness, so it is the mother-yin that we constantly look to for the Origin. Although the spirit then activates and moves the body, it is the form developed from the yin first, before the vessel is ready to be irradiated by the energy of the light of day—birth.

Within society and general culture today, there is no mystery concerning any part of the process of the body. It's all been scanned, logged and charted, so you know, mentally, what is going on. Women have ultrasound scans, so they can see the foetus and check on its status. Everything has to be scanned and rescanned in order to know things are going well. The process of probing, wether into the dark-yin of space, the dark yin of the earth or the dark yin of the female is a highly masculine process. This, in mind-identity turns into desperate seeking, outright rape, or the ideology of "pioneering" change. In some areas of life there is some, very limited use for this type of direction, but only when the emphasis is in conjunction with or for the yin. Some things are left best with mystery intact. The most profound understanding of all in fact is to accept mystery and not to understand it, as it is unfathomable. Pregnancy for women is very much a return to this sense; instinct. This is a mind-identity-based/ masculine energy dominating society, which has no interest in the natural movement of things. Ancient ways of understanding view the woman as she is, in the moment. Pregnancy is when her mass and size become larger. She develops a greater physical form, and this is representative of the form of Origin. She becomes Origin. Hence, as a very wise woman explained to me, "A woman must be in touch with the invisible during pregnancy". This means that she must feel that which she cannot see or does not know with mind-identity. Also, she must lose her old ideas of herself and adapt to life. The universe is living through her, as it always was but now it seems more obvious as it has physical presence also, and during the pre-birth/ pre-Heaven stage, she becomes deeply in touch with Wu if she lets go of mind-identity. In pregnancy the woman is both yin and yang, she is female yet holding the yang within her, the seed, this is the Kan trigram and represents this overall expression. Mind-identity is associated with the pushes and pulls of civilisation, but she must become simple and introverted. She does not need to know what is occurring within her, for the knowing is not in mind-identity but far deeper, in the stillness that is present within. When words could not describe the process of pregnancy, and mind-identity was not active, birthing was easy. In the Bible, when Adam and Eve are cast from Eden, God says: "In pain shall you bear children". The pain is not about physical pain, but the knowledge of pain or mind-identity of pain; this means suffering. And so there is now great trepidation and fear associated with childbirth: Will there be pain? Should I have a Caesarean? And so on. If we are born into the world without intuitive faith in being present, there is little hope that we will recognise this, and so the message of mind-identity and separate-self is passed on from mother to baby, as are all other aspects of expression.

71

The more still the mother is during this time, the greater the power and the quality of energy of her offspring. It is as simple as that.

iv) Growth During Pregnancy

The most in depth energetics of the process of foetal growth and development we have from the Han-dynasty Classics comes from Chapter 2 of book 9 of the Mai-Jing or Pulse Classic (See Shou-Zhong Y.,2002). Herein there is an explanation for the foetus's energy drawing from different regions of the mother's body during its growth and so we get an idea of development.

The sequence as explained in the Mai-Jing is as follows:-

Month 1: Liver meridian
Month 2: Gallbladder meridian

Month 3: Pericardium meridian
Month 4: Triple-warmer meridian

Month 5: Spleen meridian
Month 6: Stomach meridian

Month 7: Lung meridian
Month 8: Large Intestine meridian

Month 9: Kidney meridian
Month 10 (baby is born within this month): Bladder meridian

The other meridian of the Small Intestine (and also the heart-meridian although this is very likely to actually mean the pericardium meridian, see page 295 for further explanation) is used most strongly after birth in the production of breast milk.

What we can see above is the nature of the meridians being drawn from by the foetus. What this implies is also the organs the meridians are associated with in the mother: this importantly has implications to the energetic qualities involved. Ikeda Masakazu's clarity of understanding the Classics has allowed us to realize that organs and meridians have different energetic qualities. As a result this is what we can say about the nature of the energetics for the 10 month period:-

(Table 2.1)

Month	Organ Energy	Meridian Energy
1	Liver (Spring – pungent)	Liver (Autumn - sour)
2	Gallbladder (Autumn – sour)	Gallbladder (Autumn – sour)
3	Heart (Summer- salty)	Pericardium (Summer- salty)
4	Right-Kidney (Summer-salty-pungent)	Triple Burner (Summer - salty-pungent)
5	Spleen (Late-summer – sweet)	Spleen (Late-summer –sweet)
6	Stomach (Late-summer-sweet)	Stomach (Late-summer-sweet)
7	Lung (Autumn)	Lung (Spring –pungent)
8	Large Intestine (Spring-pungent)	Large Intestine (Spring – pungent)
9	Left Kidney (Winter –bitter)	Kidney (Winter –bitter)
10	Bladder (Summer – salty)	Bladder (Summer – salty)
10>	Small Intestine/ Heart (Summer – salty)	Small Intestine/ Pericardium (Summer – salty)

The above shows the energetic qualities of the organ and meridian energy influencing or being drawn from the mother to the foetus during pregnancy. As we can see the predominant expression of the sequence is through the 5-phases of growth i.e. Wood tonifying (tonify means to give power to, or strengthen/ nourish), Fire tonifying Earth, and so on, as would follow the 5 phase cycle of creation (please see later sections for further detail on 5-phase energetics). However notably the periods of month 1-2 and 7-8 are complex. This is because they are associated with the wood and metal energies, we will see in later sections that there is a lot of overlapping of these energies. However months 1 and 2 seem to have an association more strongly with autumn and 7-8 more strongly with spring as an overall expression. One needs to take account of this in the following explanations of the various stages of growth through pregnancy (see below). Also there is only the cool energy of water applied in month 9, this is because month 9 is the end month for growth really and so the end of the sequence, also the kidney is the only true coolant energy of the mother's body, generally humans have more heat than anything else which is why overall this process is hotter rather than cooler. Note too that the tenth month is the month that during which there will be birth and after this time the small Intestine and heart organ and small intestine meridian and pericardium meridian (the meridian of the heart organ) are used to provide energy thought the breast milk.

The reason for this is that the foetus's energy is drawing from the mother. These are the regions where the mother's organ energy will be drained at the same time that the meridian energy creates the same meridian energy within the foetus. It is the aspect that the foetus desires. Hence the flavours of the meridians are drawn off in specific order in accordance with the foetus's growing needs. Notice that the Zang organ comes before the energetic-relative fu organ, because the Zang is the interior. Then the exterior (fu) forms afterwards, again showing the importance of the five-Zang as the roots of the body.

Month 1

The first few days and weeks of pregnancy are the most important and also the most powerful. Initially, exponential suction occurs within the body, powered by the liver meridian. The liver meridian is sour in flavour and contains much of the nourishment energy of the yin quality that is needed just before a massive change occurs. This energy builds up in the form of what might be called hormonal load to the system in fragmented Western terms, which is a physical expression of the response of the mother's energy, which is being internally drawn by the foetus—drawing inwards into the internal organs in order to be used for the foetus's growth. The energy of the foetus strongly draws the juices of the mother's energy inwards, and the mother responds by creating a lot of blood. The liver organ contains this blood and is also drawn from strongly at this time. The complex of pregnancy-symptoms in this month is due to the very different nature of the 2 energies being drawn from the mother's energy, both the yang-red aspect of the blood and the white-yin aspect of the blood are drawn by the foetus. It is the most intricate time of pregnancy and is the most "critical" stage to begin things with strong foundations. The mother's appetite starts to change, but with the sudden change to the environment of her inner body, her appetite will allow her only the foods it needs, and the start of cravings for certain foods may occur, especially if there is deficiency for specific things within her body. At this time, the tendon energy is prevalent in the foetus. The sour flavour that predominates in this period, is the drawing inwards of the foetus's energy, suckling the woman's body from the inside, as if the baby has energetically latched on. The woman responds by allowing her energy to be drawn.

Month 2

The gallbladder meridian is the next meridian to tonify the foetus. This, too, is a sour meridian and has the effect of doubling up on the liver and starting the increased growth period. This energy feeds the foetus's tendon energy, which is the first to be created, like a small mass of tendonous tissue. Again there occurs a further suckling of the foetus energy, with a strong suction effect from the gallbladder meridian (sour).

Month 3

The pericardium feeds this next section. This is the energy of the Ming Men of the mother, and it is the hottest time. This energy increases growth exponentially, and the digestive capacity also is increased during this phase because the energy of the pericardium is salty and makes the heart beat stronger. The baby's heartbeat will have started during the second month of growth, but now becomes strong and vibrant, and the blood vessels of the interior and heart network start to develop. The heat of this stage nourishes the foetus's energy.

Month 4

The Triple Burner / Triple Burner meridian governs this phase. It is the phase most closely connected to the Ming Men or right kidney energy of the mother. It is a powerful heat that speeds up the exterior formation, especially of blood vessels and the back of the body. The Triple Burner phase is hot and expansive and provides the extension of the mother's source energy around the body of the foetus.

Month 5

The spleen meridian dominates the fifth month. This is when the foetus puts on bulk and body mass. The fleshy fatty tissues start to increase, as do the needs of the mother to supply herself and the foetus with food. This is the most plentiful time, the late summer; everything levels out, and the massive change stage has reduced. Now growth is more gradual and smooth. Flesh or fatty tissue accumulates mainly around the forming organs and on the insides of the tissues, bulking them out. The mother's milk glands will enlarge and breast size will increase considerably at this stage.

Month 6

This concerns the stomach meridian energy. Again, it is a bountiful time where the energy of the stomach is at a peak and provides the foetus with a lot of nutrients—and the mother with a large appetite. At this time of her pregnancy, she can eat the most. She and the foetus will put on considerable weight. The baby will begin to show more extensively than before. This is when the exterior fatty tissues covering the muscle of the foetus become more substantial. Again, more milk production and breast enlargement occur at this stage.

Month 7

The lung is the next meridian to affect the growth. This is the meridian of spice, and its effect is to create skin and body hair. In addition, the mucus membranes start to become more formed. The skin for this phase is of the front or yin surfaces of the foetus.

Month 8

The large intestine is the next meridian to affect growth. The skin and body hair of the yang surfaces, as well as the gut, start to become more prevalent. The foetus and mother excrete a lot of unwanted material after the gorging of the last two months on the way to an easy birth. Toxic substances are cleared from the system, and the body becomes more solid and firm as the foetus moves towards coming out of the womb.

Month 9

The end of the pregnancy is nigh. The bones need to be solidified before the end, so the ninth month involves strengthening the bones and bone marrow. This is the part of the pregnancy that most taxes the mother's Jing energy, because the bone marrow is pure yin essence and this is drawn by the foetus, from the mother, to top up the reserves of energy in its body. The mother may look fully pregnant by this stage and ready for birth. The energy she loses is cold, bitter energy, so she will become very hot at this time, also there is overall a lot more blood and heat around in the body so this too will make her overly heated.

Month 10

The baby is born sometime during this month, so it is the last month of pregnancy. The bladder meridian not only again draws the kidney yang of the mother, to double up on the TB (triple burner) energy from before, but also the bladder gives protective covering and power to the defensive wei qi on the foetus's surface, ready for the colder exterior environment. This energy feeding to the foetus helps the mother to loosen and open the bladder meridian, and this opens the uterus for birth. The mother's body is fatigued by this stage, but the bladder energy electrifies the system for a final push to allow for new life to emerge.

Note that the small intestine and heart meridians are not included. There are a number of issues with inclusion of the heart meridian (which we can consider to be the pericardium meridian as the use of "hand shao yin/ Shoin/ 少陰" in the Classics is energetically rooted within the hand Jue-yin), which we will discuss later. The small intestine is said to be involved in post-partum milk regulation and production, so is not part of the 10 months. One could perhaps consider the small intestine meridian to be connected to the bladder meridian, both being aspects of kidney yang energy or the Tai Yang/ Taiyo/ 太陽 meridian. Also, the development of the small intestine comes with the input of the mother's milk and the beginning of the digestive process, so this is why there is relation to it, tonifying the baby's system after birth. The small intestine therefore connects most strongly with the spleen and the digestion in this process, making refined blood from the hottest and most refined area of the mother's digestive system, then sending this qi up to the heart and chest which, instead of blood, is further transformed to milk during pregnancy. The small intestine meridian/organ and heart organ connection are in this case about the use of blood.

The issue of breastfeeding infants and for what length of time is only a contemporary issue. It is natural that the baby will attach to the breast and natural that the infant will diversify into foods and leave the breast. It is not something the mother needs to think about, but rather something the mother can watch occurring. With the development of the child's first teeth at around six months, this also expresses that the digestive capacity has reached a certain level of development that can move onto solid foods, so Nature is showing a way. Hence, between six months and a year is a natural period to move on to solid food. If pre-Heaven is strong, then the breast will not be needed as much as when the pre-Heaven energy is weakened—for example by premature birth, Caesarean birth, and other traumatic occurrences during birthing that can greatly weaken the child and require more nurturing of the refined energy of the breast milk. (Women will be blood-deficient throughout breast feeding. The milk will draw from their reserves of blood, sometimes making them quite tired and also not sexually interested for quite awhile after birth. Her energy is concentrated on her child. Having children in succession, without recovery of the energy, drains a woman's energy. We will consider this in later sections).

Please note that the inputs of the liver, gallbladder, and kidney meridians at the first, second, and ninth months, consecutively, are vital aspects of the process. At these times, the foetus's body receives vital energy from the mother that it will use its whole life and will be difficult to reform within its body. The essences that pass to the foetus during this time are the pre-Heaven essences of the sour and bitter flavours. Although the other energies can be supplemented easily through the digestive system after birth, these energies are the ones that cool, condense, and draw inwards—and thus are filled up only once before the "spring is left to uncoil", so to speak.

As an overall process, the prenatal period is sour-bitter, accumulative and gathering, whereas everything after the first breath is essentially pungent, salty, and sweet—which means yang and expansive. (We will discuss the flavour relations later.) This is important because pre-Heaven is always associated with Stillness/Origin, which pertains to the yin and permeates everything, whereas post-Heaven is associated with change and yang.

(Note: Any woman who is pregnant needs to create a lot of blood. Hence most foods, treatments, and herbal nutrients are used at this time to strengthen the blood. Pregnancy is not an illness, and so practitioners, the mother herself, and her family aid her through her pregnancy to obtain blood-creating conditions of warmth and heat that will supply the needs of the foetus and mother throughout pregnancy. The foetus is a blood formation. Therefore, the mother is always deficient in blood. It is rare that any other treatment principle is used during these 10 months.)

v) Birth

The baby should be born when it is ready - a time when the mother is as relaxed as possible. The uterus of the mother will open. This process is that of the bladder meridian activating her back and sacrum - allowing the uterus to warm up and expand. This opens the uterus at the front of this' area, and then stretches to allow the baby out. The process of the baby being pushed through the uterus is akin to the birth from darkness of the yang from the yin. Hence, this pushing process activates the baby's exterior body and stimulates blood and energy flow to the skin to be prepared for the exterior world. Without it, it is too fast a process, and the baby is not ready to emerge. There is a great misunderstanding involved in treatment , of any form being required for birth and the "induction" of labour. The main point being that labour and the process of birth is entirely natural and nature knows the process. It is only when the woman is out of touch with her sense that "induction" of labour is required which really means simply, as with all issues of health it is a process of dissolving and relaxation. Very often especially if it is the mothers first pregnancy she will be caught within a tenseness and fear about the process of birthing often due to projection from what other people have told her or the general tension of the world around her in modern society.

Birthing naturally occurs when there is no "due-date" in mind and the mother is simply allowed to be what she is and allowed to have an environment that is warm and comfortable and simple and still to give birth in. As in nature the birthing position that best allows the woman to feel most comfortable is the best, this does not necessarily mean water birth but it is also almost never lying on one's back! The process of childbirth and pregnancy can be a situation where the woman is strongly directed to sense her body, what she needs, what she wants to eat, how she wants to sit, when she wants to move and exercise and stretch, all of these come into the picture of the birthing process and women , if they listen to the instinct rather than what it says in a book or what friends tell them, can know the natural way instinctively. This is true of all of life but it is particularly focused in pregnancy because just the life process of sleeping, eating, digesting, defecation, urination, sex and conception and pregnancy all these are "animal" expressions which are seen as requiring an "agent" or practitioner involved to "guide" the process. Nature has no guidance and it is really simply the woman acknowledging her own innate sense of herself and allowing life to move through. This is simply about letting go of the mind-identity at the least when it's time to give birth and this often happens due to the intensity of the experience, but ideally all the way though the pregnancy process and life for that matter.

Pregnancy is no different, no more important than any other part of the continuum of existence it seems a great occurrence for the human within the world of "good and "bad" set up by mind-identity but in essence all things are "sacred" those who can and can't have children are all One. Otherwise we are saying the spring is "better " than the Autumn. For humans we pertain to the yang so we associate with Birth more than with death but this association is one of the issues that prevent the allowance of things to be as they are. For women before birth it is important for them to keep the body moving and walking regularly. This keeps the uterine region and the back of the legs and hips circulating, having a small amount of alcohol can also help to relax tension and bring about the natural reflexes that respond to birthing. It is all entirely about relaxation. The opposite of this is the cold and clinical methods of western hospital birthing where the mother is seen as a "problem" to be fixed and the birth is a "dis-ease". The midwife of ancient times (associated with the doulas of today) are similar aids to the process of natural relaxation for the mother providing a calm and soothing environment often away from the more yang approaches of "observation" and "probing" which occurs often throughout pregnancy - causing constant tension for the mother. The simple allowance of a calm environment is something that aids the natural spontaneity of birth and means that contractions reach a natural ripeness and birth can be over in a matter of moments, the woman readying herself for the expression of contraction, much like an orgasmic contraction in fact, which can be experienced if the woman is fully relaxed, rather than in tension, nervous contracted and tired. The suffering of the mind-identity is the suffering of childbirth. If far when this goes child-birth is not painful and "Eden" is seen behind the mind-made world, as it always was. (Ina May Gaskin, is very much worth investigating as a woman who has understanding of the ancient clarity of natural birthing, she must be considered in part of the Classical lineage of understanding, see annotated bibliography).

Tension during pregnancy causes all sorts of abnormalities of timing and positioning of the baby. The mothers' relaxation and realization that the process is truly out of their hands, which for some women is a great difficulty because they are so used to feeling in "control" of "their" lives, for them pregnancy is at last a time to relax and so becomes a highlight of their existence, the rest of the time there is turmoil! There is an unravelling process going on of spirallic movement within the uterus. This only can unravel fully and clearly and without mishaps when there is an allowance rather than a resistance to change. Therapy at this late stage of pregnancy, before birth, simply allows the woman to relax and this will also follow suit with the foetus, which will engage into the right position. It's all about letting go of ideas, and going with the flow of natural vibration, just as with any life process, it is absolutely all the same. Everything that occurs to the mother is occurring to the child.

The moment the baby is out of the womb, he or she must take the first breath. This first breath is the most vital aspect. From pre-Heaven energy, the baby must move to the post-Heaven energy of air and milk and the world's energies. The first in breath is instigated by the baby's own Ming Men beginning to start in response to being outside of the darkness of the womb into the light. The breath is drawn down, anchored into the lower abdomen by the left kidneys energy. The first out breath propels the baby's lungs and own Ming Men system, which had previously been supplied by its mother. At this moment, the baby is an entity in its own right and is coloured by the spiritual energy of the universe. This is now seemingly a separate person and as such associations with the universe at the time of birth is often very carefully looked at by those involved in Feng Shui, as a guide to understanding the nature of the individual expression, although he/ she as we will see, is not an individual at all but a finger of the hand of nature. (This will be explored in the next chapter). The placenta should come away easily from the uterus, as if a stone is plucked from a ripe plum. The placenta should be ripe if the baby has been born at the right time. The umbilical cord is cut when it stops pulsing. As the Ming Men energy of the mother powers this, the nutrients of the placenta must be allowed to pass into the baby for as long as possible, as this is precious pre-Heaven energy supplied directly to the baby's internal system. Breach position and other abnormalities of pregnancy are very much associated with the mother's natural sense of herself during pregnancy and the adjustment she makes to what she feels rather than attempting to adjust to the exterior.

All things affect the development of the child, from the moment of conception right through the time of pregnancy and birth. Conception should never happen, for example, during drunkenness or violence, although often women have remained deeply Still even in the most horrific of situations, proving that no matter what force is applied nature always envelopes with Oneness. The ideal conception takes place with caring and compassion, and this should continue in feeling throughout the pregnancy. Violence and strong language, pungent food, and aggressive behaviour are kept away from the woman so the child is tonified to the greatest extent. Notice how the woman embodies stillness, especially during pregnancy and birth. Women innately are able to be closer to the sense of Oneness and Stillness because they embody this and are tied to their bodies far more than men, whose energy can dissociate into mind-identity too easily. Women's bodies are yin in relation to those of men; hence women are more still, physically, than men and more easily able to connect to Wu. The mother is one with the foetus at this stage, and also to her exterior. There is total unity, and the mother therefore, if she listens carefully to herself, will not tend towards behaviour patterns that induce excitement or over-expression. She should also avoid becoming too cold or too hot. She needs to be centred and still, much like the Earth energy itself, which moves only gradually and nourishes all of life.

vi) Growth

The growth process is described in chapter 1 of Su Wen. Please note that the female child energy is related to the number seven and growth cycles of seven, and the male with the number eight and growth cycles of eight. This is important because the principle of the nine spiritual energies seven relates to dui, or the girl, and eight relates to gen, a boy. Again, the governance of "time" or moment to moment change, is associated with Heaven, whereas the actual growth is a formation of Earth or physical energy. The kidney yang is associated with the spirit and the changing of cycles. Hence the spirit is the timing of events in the now.

In chapter 1 of Su Wen, Qi Bo, the Taoist nature-man, answers a question about the physiological growth of the of human:

(For Women's body)

> *For a woman, her kidney energy becomes prosperous when she is seven, as kidney determines the condition of the bone, and the teeth are the surplus of bone, her milk teeth fall out and the permanent teeth emerge when her kidney energy is prosperous; as hair is the extension of blood and blood is transformed from the kidney essence, her hair will grow when the kidney is prosperous.*

> *Her reproductive cycle- kidney based energy appears at the age of fourteen (7 x 2). At this time, her Ren channel begins to function and her Chong channel becomes prosperous and her menstruation begins to appear. As all her physiological conditions become mature, she can be pregnant and bear a child.*

> *The growth of the kidney energy reaches the normal status of an adult by the age of twenty-one (7 x 3), her wisdom teeth have come through by this stage and her teeth are completely developed.*

> *By the age of twenty-eight (7 x 4), her vital energy and blood become substantial, her extremities become strong, the development of the tissues and hair of her whole body are flourishing. In this stage, her body is in the most strong condition.*

> *(based-on Wu N. L and Wu A.Q, (1997)).*

(For Men's body)

> *For a man, his kidney energy becomes prosperous by the age of eight. By then, his hair develops and his permanent teeth emerge.*

> *His kidney energy becomes prosperous by the age of sixteen (8 x 2). He is filled with vital energy and is able to let out sperm. If he engages in sexual intercourse with a woman, he may make her pregnant.*

> *By the age of twenty-four (8 x 3), his kidney energy is well developed to reach the state of an adult. By the time his extremities are strong, his wisdom teeth have grown up, and all his teeth are completely developed.*

> *By the age of thirty-two (8 x 4), his whole body has developed to its best condition possible, and his extremities and muscles are very strong.*

> *(based-on Wu N. L and Wu A.Q, (1997)).*

Birth through to 10 Years (Spring/Wood Phase)

The human body, constructed from the essence of male and female, now has access to the post-Heaven world. Growth begins, and the main area that is weak is the digestive system. This is vulnerable to attack from exterior dis-ease. Often, children will develop coughs and colds, but their immune system is very powerful and will clear these problems immediately, especially if the mother has breastfed the child. The breast milk is actually a form of blood transfusion to the child. One drop of the mother's milk is equivalent to fifty drops of her blood, it is said. So the milk continues the process of supplementation, but through the coarser, less refined route of the digestive system.

Up to the age of 10, the child may have digestive issues but can usually recover from them quite quickly. After this time, growth becomes much stronger during pre-puberty, and the digestive system becomes stronger and better developed.

Between birth and the age of five or six, the child is absorbing much from the exterior. The output energetics are significant, but the full strength and construction has not unfolded. Children have a powerful ability to absorb and learn mentally at this stage because development is more in the physical structure than mind, so the mental processing is not yet in place. Children simply absorb osmotically, the way adults wish they could still learn. Because of learned ways of behaviour in "civilised" society, we forget how to learn—and with it, how to be in the now. With that, we forget or abandon our childhood understanding of being natural as well. However, memory at this time is perfect, and the way children learn about their exterior will often colour what is to come. This is similar to the way the foundations of a house deliver structure and therefore instability and rigidity or stability and flexibility, as well as longevity. Physical activities and learning through the physical body and sensory experience therefore is the prime focus at this stage, if one over stimulates the mind too early this will jump the gun and the child will almost forget about the body in some cases becomes only focused in the mental and the simulating of the mind. This makes for mind-identity to become much more of an easy strong hold later. Hence ancient practises of parenting, especially when applied during these formative years, are likely to acknowledge this and help the young energy develop its true, authentic sense and be clear in the moment. This is perhaps the ideal of this short but vital phase of life.

10 through to 32 Years: Men (Summer/Fire Phase, and Earth Phase to the Beginning of the Metal Phase)

During this period, the body of the male develops and becomes much more yang. Exterior energy develops, becoming harder. Before this time, the male child usually looks quite scrawny or underdeveloped, but this is simply the energy on the inside. When the energy moves to the exterior, the testicles drop down and the penis becomes larger. All aspects of the exterior of the male body structure become larger; the energy simply moves from inside to outside. By the age of 32, the male body has grown to the greatest extent it can and is at the peak or plateau of human existence.

[A note: During this period, teenage angst is a common phenomenon in contemporary Western culture. However, this should be the spring and summer of life. The reason for the discrepancy, of course, is the unnatural control placed on teenagers, which their energy naturally pushes against, with violent results either internally or externally. Teenage angst is, in fact, perhaps one of the last symptoms of rejecting society and mind-identity that created it. However, rather than teenagers being criticized, society needs to be looked at deeply, or perhaps we need to look within to see how the ideas of separation-society are all about a mind-identified character with a name "me" or "you", always an identity being formed. The teenager is often attempting to find individualism because this is what society/ mind-identity demands but also is the natural energy of spring expansion pushing to the surface and becoming free of constraint. It is however lonely in individualism and a huge questioning of existential concerns dominates because this energy is not considered beautiful and wanted but rejected and stiffened off. In Ancient ways the expression of the total energy of the body delivers a sense of freedom, passion exploration and excitement, a celebration of living for the teenagers, not a anxiety driven idea that they need to become "someone" or "something" more. It is more a time of celebration than anything else and this is not accepted nor desired by an over conforming militaristic self-protective society. Watching this is that which has no name, and so no form and no society. Living from within, the outer world too has no separation.]

10 through to 28 Years: Women (Summer/Fire Phase, and Earth Phase to the Beginning of the Metal Phase)

Women develop faster than men because, unlike men, their energy does not need to unfold on the exterior. By their early teens, most women have grown to be their full size and overall body shape. The breast size increases, and the uterus function begins with the period cycle, in time with the change of the Earth and lunar calendar, tying the female system with the physical energy of the universe. The female body does not become as exterior as the male but simply can expand. It does not necessarily change as much as the masculine body, because the energy of a woman is still often on the interior, and this does not change during adulthood. The interior energy also helps women live longer because they do not use up the energy in exterior, yang expression.

Again, note that stillness is more natural to the female form, and so mind-identity is slightly less of a problem for women than it is for men. The expansiveness of the exterior and upper is the expansiveness of the head and upper body; this affects the masculine more than the feminine. The problems humans face, are largely due to male body based thought patterns, as they are more of mind-identity than women's, because mind is yang by nature. This has forcibly and brutally repressed women for many generations. However, the root of acceptance and peace is actually closer at hand for women than for men due to the Earth-bound nature of the female form.

vii) The cycle of life of the Jing-body in male and female:-

Kan can be used to express the life cycle of male energy. Men in relation to women start life yin, and then after puberty they become more yang and peak around 32. Then there is a downward movement of energy back to yin again, so if drawn, this would look like a peak of yang energy. This is the opposite of women's energy, which starts life with a yang velocity, as girls are a lot more yang then boys of a similar age, then after puberty and with the menstrual cycle, there is a propensity for women to become cooler and more yin. After this time, menopause often heats up the body and is considered by many women to be a second birth of yang. Hence women's energy follows the pattern of the Li trigram. This is also why Kan is yang and Li is yin.

(fig. 2.4)

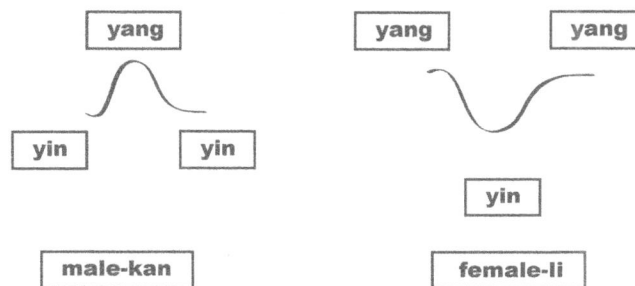

From this point on, there will be a plateau of existence (Late Summer/Earth Phase), and then the body will go through its Autumn (Metal Phase) and Winter (Water Phase) as it moves towards death.

viii) Natural death of the Jing-body Vs pathology:-

Su Wen Chapter 1 continues to express the cycles of the Autumn/ Winter phases of life, in both the male and female systems, the body simply breaks down with time and becomes more and more yin, becoming increasingly colder until death, the spirit's animation diminishing in brightness like a bulb losing its brightness.

Autumn and Winter Phases, the Life-Death Cycle

One could call the full stage of the body the fire and Earth phase, which is up to the ages of 28 for women and 32 for men. This is the completion of expansion. The autumn and winter stages therefore are the movement towards accumulation and the completing of a phase of existence. The phase has simply changed form. The body cools (autumn and winter phases) and is then cold (changing form or death), meaning that the spirit has fully dissipated from the form into the exterior; this is another way of saying that the yang ki has left the body. The spirit is not seen as an entity that moves around and goes into and out of existence as a form or shape that is separate from other things. The spirit is constantly leaving/ transforming from the body, from the first day of life, and slowly letting go. It is infused into all the expressions of the life of that person and the expression he/she has in the exterior world. Hence, like a light bulb giving off light, in the end, the bulb blows and what is left is a place where there was once more yang ki. Now this has gone; it goes down and up, the ethereal or yang spirit diffusing upwards and outwards, and the yin body (spirit) passing into the ground.

The concept of ghosts is prevalent in a lot of ancient Chinese belief systems, but classical Taoist philosophers, or neo-Taoists such as Wang Chong, AD 27–97, gave chase to the idea of ghosts. (This too is reflected in Su Wen 11, in relation to belief mind-identity being a block to treatment).

According to Wang Chong:

> *"People say that spirits are the souls of dead men. That being the case, spirits should always appear naked, for surely it is not contended that clothes have souls as well as men". (from his "Lun Heng").*

From the classical Taoist viewpoint, the concept of ghosts as being entities that are of dead spirits is impossible; ghosts and spirits were the old shamanistic language that was outdated by the Taoist five phases and yinyang. Both the ancient shamanism (rather than its complex aftermath) and Taoist understanding are one (similar could be said of Hinduism and Buddhism, Judaism and Christianity). Mind-identity, at the time of Wang Chong, had changed again (i.e., this was the revival of Taoism 500 years after its origin). Once again, after expanding outward into Confucianism from Taoism, the simplicity of neo-Taoism was an intuitive response of those connected to the lineage of ancient understanding (Tao) to once again control the ever-expanding mental state of the people at the time, showing us these patterns of expansion and the reversion to simplicity like ripples on a pool, happening worldwide, not just in China. Wang Chong was able to quickly cut through fragmented ideas. The "spirit" of all things, all aspects of life, still remains, but we would need to be here and now to experience this; then we too may describe things as the ancient shamans used to—it depends how you try to put the unexplainable into words, however the whole notion of Taoism is to get to a linguistic meaning and essential meaning beyond words and beyond belief that therefore all peoples and cultures naturally adopt, which is expressed in how no matter what the religion yinyang is never seen to be in contention, and is sometimes adopted as an expression of unity.

The process of pathological change/ pathogenesis a purely natural expression, if thought about (i.e., by mind-identity) is the most feared of all concepts. Animals do not fear death. Animals are alive untill death occurs. The thought processes around death are its danger. A candle flame does not think of how much it has left and stop burning; it burns till it is no more. This is the same with all life. Life is like fire; it burns till it is no more. We expand and burn till we reach our limits and there is no more—some slowly, some fast, dependent usually on the environment and the mental fragmentary processes.

The problem comes when we reflect on life and what will occur beyond death, which is unknowable to the alive mind-identity. Masters throughout history have told us to "die, before you die". This means to let go of attempting to know the unknowable and be here and now. Death, therefore, is actually a notion of mind-identity; mind-identity fears the death of itself and relates degradation of the body-form to impending doom, when in fact it is change of form superficially and constancy of unity behind and beneath this—Wu. Life too is impermanent and changing; therefore life turns to death, and death to life. Our perception is that death means the end; this is the illusion of looking down at our feet and believing we are separate from the Earth, or looking up at the sky and believing we are separate from the Heavens; all an illusion. The prominence of the Source, Wu is the true nature within all forms; thus using the word death has many implications. This is the true Way. It is actually the over-concern with prolonging life that is the biggest problem in the Modern West. There is no presence of mind-identity, which means Stillness or contextualised mind, to actually adjust for death and change. As with all things, change occurs, just as we grow and change form, so death is another change of form and is not truly a death/end, so there is no problem. Only mind-identity sees it as a problem because the ego/identification with mind must let go and die, which means non-existence, which cannot be THOUGHT out. Therefore, it is feared, just as all things that are unknown to a mind-identity attached person. The only problem is for mind-identity to accept contextualised mind or no-ego, for it is the expression of expanding yang within the body and it cannot imagine a point where its change cannot exist. This would annihilate all it knows. It is hard and unyielding, and so it falls hard at the end of the expression of the life cycle.

When humans spontaneously let go of mind-identity during life, death is as it is, as if taking a shower in the morning. It is truly as easy as this. However much there is a holding on is the sufferance that occurs, both during life and inevitably towards death, as often people close to death must face a lifetime's worth of holding on, only to find that it all has to be let go of anyway. This of course can be physically very painful, as chronic holding like this will become more and more dense in the tissues.

People don't die because of a dis-ease; they die because it is time for change. And again, the word death only denotes this "moment" when it has actually been a lifetime of deaths; the word really has little meaning. This is very important. One's mind-identity is one's dis-ease, and the dis-ease is the point of transition. Hence as transition occurs, it is accepted or fought-resisted. This is freedom, or sufferance respectively.

Chapter 1 of the Su Wen describes the aging process of men and women and what will occur in the transition towards death of the body. It is interesting to see how, when we have a cold or dis-ease of some kind and we recover from it after a short or long time, we often have the feeling of being re-born somewhat. This, as with all changes, occurs because we have changed during this time. The cycle of man is eight years, for women it is seven. During this period, the body will change and transform. Life and death are simply aspects, mere aspects, of this wheel.

In Su Wen, chapter 1, Qi Bo answers a question about the physiological decline of humans ...

For women:

The physique of a woman turns from prosperity to decline gradually after the age of thirty-five (5x7). So, by this time, her Yang Ming channel turns to debility, her face becomes withered and her hair begins to fall.

By the age of forty-two (6 x 7), her three yang channels (Tai Yang, Yang Ming, Shao Yang) all begin to decline. By this time, her face complexion becomes wane, and her hair begins to turn white.

After the age of 49 (7 x 7), her Ren and Chong channels are both declining; her menstruation severs as her Reproductive energy is exhausted. Her Physique turns old and feeble, and by then, she can no more conceive.
(based-on Wu N. L and Wu A.Q, (1997)).

For men:

By the age of forty (5 x 8), his kidney energy turns gradually from prosperity to decline. As a result, his hairs begin to fall and teeth begin to wither.

By the age of forty-eight (6 x 8), his kidney energy declines even more. As the kidney energy is the source of Yang energy, the Yang energy of the whole body begins to decline, due to the decline of the kidney energy. As a result, his complexion withers and his hair becomes white.

After the age of fifty-six (7 x 8), his liver energy declines in the wake of the deficiency of the kidney energy. As liver determines the condition of the tendons, the deficiency of the kidney energy will cause malnourishment to tendons, which will become rigid, and fail to act nimbly.

After the age of sixty-four (8 x 8), his reproductive energy is exhausted, his Jing and vital energy is reduced, and the kidney energy becomes weaker. As kidney determines the condition of the bone, the debility of the kidney causes weakness of the tendons and bones. Thus, at this stage, his essence and vital energy turns to the utmost decline, his teeth fall off, and every aspect of his body becomes decrepit. (based-on Wu N. L and Wu A.Q, (1997)).

"Pathos" means suffering and, suffering and pain are very different things. (Please see aetiology section) Hence, pathology is the study of suffering. And if this is true, then it is really the study of the nature of mind-identity and the traces of its effect on the tissues of the body. This is not to say that the tissues don't undergo natural aging in natural time, but then this cannot be associated with suffering, as this is natural. Here we make the key differentiation.

The above makes a very important point when considering death as pathology. Just as life has all its processes, such as pregnancy, birth, coupling, and mating, the process of death is just as important. In fact, it is vital that one considers it as of equal importance and doesn't attempt to treat the "pathology" of death. Death in itself is not pathology, or it is just as much pathology as pregnancy. It is only pathological; in fact, if it doesn't go through the processes extrapolated above and is instead cut short or held back in some way. This is pathology, so pathology and sufferance is mind-identity. However, natural death and reaching one's full potential as far as years goes, be it a hundred or not, as long as life has been fully lived, is not a pathology but a welcomed time by those who have clarity of mind-identity to let go of life as it passes, rather than all at once, at the end.

Pain and symptoms are the objective expression of change that eventually leads to death of the organism. Pathology / dis-ease, however, is the process that practitioners are interested in. The dis-ease may be cured, but the pain may still be present, though it will not be felt in at all the same manner. This is curative. Sometimes the pain will go away, and sometimes not; both situations are a cure if mind-identity/ dis-ease, is resolved. Hence, to reiterate, within this text, when we talk of DP, or dis-ease pattern, it means the pattern of pathological symptomology associated with a mental disposition that is the origin of the felt suffering. Without suffering, the body's natural ability to heal is at the highest level possible for that organism.

It is very important for a physician to differentiate the point at which a process of pathology becomes part of the process of dying. This is vital, as it gives one an understanding of the depth of the illness and when a natural process is drawing the body energy, so "cure" become an irrelevant concept and one is simply allowing a process to be smooth and regulated, rather than held on to and with great sufferance. The Buddha said, "I teach suffering and the end of suffering", meaning mind-identity and the end of mind-identity. "Pathos" means to suffer, not to die; dying and the process of death do not have to be a sufferance, unless mind-identity is involved in attempting to prevent its process. The fear and trepidation associated with the word death is due ONLY to mind-identity's belief that death = death of itself, which of course it is and which of course it cannot see beyond, to understand that life and death cycles are only of the phenomenal world and the root of this world, and all that exists in it is unified by the acceptance of Stillness, Wu, God, or whichever word is your preference.

The only true pathology begins and ends with mind-identity and our lack of awareness of the interior and our relation to the exterior. The rest is going with the flow, whether that means finding yourself pregnant and about to give birth or whether it means to be born, or whether it means to be in human form and let go of this. All is necessary, all is natural, and all is one. It is the belief that death ends something and that life and death are separate that is the big illusion that causes what the Buddhists call "sufferance" or mentally associated/created pain. (Please see the aetiology and pathology section for a full discussion of mind-identity/ dis-ease in the true sense.)

vii) The Types of Jing Body — The Origins of Humans and the Different Forms of Physical Constitution (Su Wen 12)

Now let us look at how humans are formed, quite literally, from the different regions of the planet. It is easy enough to see how different cultures have different physical attributes. This is due to the Earth. The Earth is different in its contours in various areas. It has different arrangements of heat and cold, light and dark. The Earth itself looks and feels different in all these areas, and humans are created structurally from the rocks and plants and animals of an area of land. People are the true inhabitants of a land, blending into it as if there is a seamless connection between man and land. Examples still today are the indigenous cultures found in Africa, China, India, and Australia etc.. These peoples have grown directly from their land, for thousands of years, and through their spreading and settling in different regions, they are part of these regions.

Su Wen 12 is, for me, one of the most important chapters of the whole of the Classical Medicine canon (see page 585). It is a concise and clear chapter, containing information about people of different regions and their physical attributes, their diet, the kind of treatment that is most beneficial for the constitution of that region, and region-specific susceptibility to dis-eases. Its brilliance is that it conceptualises a great many issues in understanding the methodology of Chinese Medicine. It also explores the constitution of purely Earth-related energetics, viewed through the five phases.

It has been said, in historical and archaeological records, that humans came mainly from the African regions and then spread out to the rest of the world. (D'Adamo, 2001). Energetic (genetic, in Western fragmented science) chains can be traced back to long historical connections for most peoples from Africa. As people moved away from this hot and arid region and went to colder climates and became nomadic and multicultural, their physical structure changed. This is charted in Modern Western medicine, through changing, evolving blood groups and other fragments of knowledge. However, if we look at the mixes of people, their quality of skin tone, and the hardness/softness of constitution, we can identify the quality of the physical structure and its cultural origins, and the relative strengths and weaknesses that form what are now mixed constitutions.

As always, with humans, we use the principle of five to group and understand the different constitutions. The following chart gives an idea of the energy of the five phases and their use in understanding the human body.

(Table 2.2)

	Wood	Fire	Earth	Metal	Water
Season	Spring	Summer	Late Summer	Autumn	Winter
Pictorial representation of energy effect of season	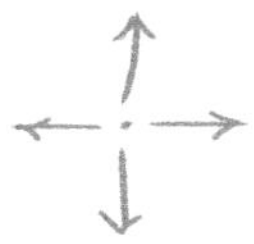	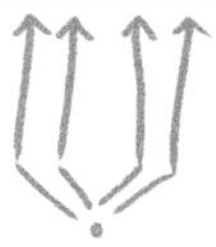	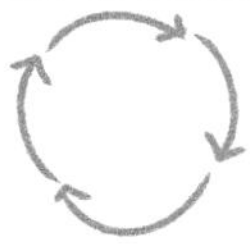	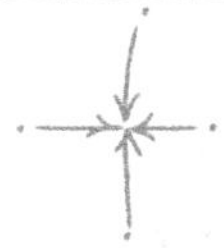	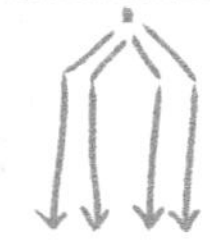
Shape	Cylindrical	Triangular-pyramid/Pointed	Square/Flat	Dome/Arch	Irregular
Time of Day	Morning	Midday	Afternoon	Early Evening	Midnight
Daily Rhythms	Male erection; female clitoral activation, period flow; bowels open exercise, empties abdomen and chest of heat; movement is expressive; eyes are dryer; focusing of eyes is better	Food consumed, peak of energy, heartfelt, expressive, sexually expressive, in-breath is most powerful	Food digested, mood is steadier, talkative, gently expressive, mouth has more saliva, lips are wetter	Cooling, draw in heat from food, wear warmer clothes, do less, calming, meditative, eyes can take in light very well, lungs breathe out deeply	Cold, silent, dark, hidden, sleep, deeply covered, heat deep within, hearing is acute
Temperature	Warming	Hot	Mild	Cool	Cold

Zang Organ	Liver (Ming Men, Right Kidney)	Ming Men, Right Kidney (Heart)	Spleen	Lung	Left Kidney
Flavour Associated to Tonify Associated Zang Organ	Pungent	Salty (and pungent)	Sweet	Sour (no lung essences exist, Gallbladder being surrogate Zang organ)	Bitter
Yin Meridian Associated	Lung	Pericardium	Spleen	Liver	Kidney
Tissue Type (Zang Ki = Tissue Ki)	Muscles (red part)	Blood Vessels	Fatty Tissue	Skin and Skin-like Tissue, Membranes and Tendon/Facia	Bone
Entry Point of Flavour	Tendons/Skin	Bones	Fatty Tissue	Muscles (and surface)	Blood Vessels
Flavour Goes Into (expressed in)	Muscles (and surface)	Blood Vessels	Fatty Tissue	Tendon/ Skin	Bone
Fu Organ-Meridian Associated Through internal-external Pairing	Large Intestine	Triple Burner/ Small Intestine	Stomach	Gallbladder	Bladder
Fu organ-meridian associated though energetic similarity	Large Intestine	Triple Burner/ Small Intestine and Bladder	Stomach	Gallbladder	-
Flavour Associated to Tonify energetic Associated Fu Organ/ Meridian	Pungent	Salty-Pungent	Sweet	Sour	-
Depth of Energy	Deep	Both, between Earth and Wood (Pericardium) and most superficial (Triple Burner/ Bladder/Small Intestine)	Below the Metal Layer	Superficial	Deepest

Spirit Quality	Holds Ethereal Soul	Houses the heavenly Life Spark/Yang	Holds the Intuition, Inclination/ Impulse to Follow the Heart	Holds the Earth-bound Soul	Holds the Wilful Action of the Heart
Spiritual Expression	Assertiveness	Joyfulness	Sense of Feeling	Consolidating	Wilful-fearless
Pathological Emotion	Anger	Anxiety	Muddled-ness	Grief-struck	Terror
Colour	Green	Red	Yellow	White	Blue/Black
Governed Region of the Body	Sides of the Body (inner and outer)	Face	4 Limbs and Abdomen	Chest	Lower Back
Sense Associated with Zang	Sight/ Image (note this expresses out—yang—as well as letting light in—yin)	Expression Organ (voice; no sense, and no exterior contact)	Taste	Smell/Touch	Hearing
Sense Organ Associated with Zang	Eyes (inner eye is yang expressive of Shen/liver organ, outer eye is yin absorptive of light Gallbladder/ liver meridians)	Expression Organ (Voice; no sense, and no exterior contact, expresses in the tongue-body)	Tongue and Mouth and Lips (mouth and tongue surface is spleen, lips are more associated with stomach)	Nose/Skin, (the open hole on nose and pores represents the lung organ, the energy of warmth is the lung meridian)	Ears
Direction	East (left side of body)	South (front of body)	Southwest and Centre (centre of body)	West (right side of body)	North (back of body)
Age of Human Life Phase	0–18	18–32	32–45	45–65	65+

Many charts similar to the above exist with far more categorisations than I mention. In the five phases, we cannot use trigrams, because these pertain only to the principles of eight and nine. Hence, we use the phases each by name. Also important to note is that the relation of the season to the organ may not be the same as that organ's associated meridian, because meridians and organs do not always act in the same way. Notably, the Metal and Wood organs and meridians balance each other out in the system, but we will look into this later.

Note on the Unity of all Aspects/Phases; The View Out Versus the View In

As always, when we are seemingly separating off one group from another, we are not actually separating. These are phases; all phases are within everything/ circular. What we are looking at is the relative balance of energy in the human body, which is an aspect of the fractal of the universal Oneness. This relates to any separation into any group that is made; all are one within a spectrum of energy, and we are viewing a section of the spectrum and finding the relative qualities within it. Green light has different shades from green moving into blue one side yellow on the other, but we can say a particular range is "green". Hence this overall bracket of green has all the other colours (to some extent) within it. Green in relation to all the other colours is a further differentiation in the macrocosmic. It is the same feel with energetics, any bracketing is not an absolute, it is a range which contains all other aspects just in a qualitative mix that allows us to call it a specific name.

The basis of understanding yinyang is the circle that holds yinyang together, and the background of Wu. This circle is constant, and this means unity, but within each aspect of that circle, it seems different if looked at from near. From afar, it looks as if it is all one, which it is. The perspective from which we can see unity is Stillness—this is Wu, the context of all things. Or, one could say it is like a tree: the branches may be of different heights and look different and be of different colours in light or in shade, but the root is the same. This is the understanding we have when looking at aspects of a five phase, nine energies, or any of the distinctions we make. Within everyone are five phases. Within everyone are nine energies, and the root is unity. This work simply allows one to see oneself as aspect of the whole, "a finger of the hand of nature" whole within itself but only as a "finger" of nature, and so to dissolve the notion of the egoistic sense of self (mind-identity). All symbology of circles within circles follows this same concept. Now we will look, with a background of this clear understanding, into the phases of Nature in the physical and more spiritual expressions. We are viewing the various expressions of the physical form (yin), which in turn encapsulates the aspects of the spirit (yang) that both have their Origin in Wu. In the deep root, all is one and begins with Stillness/Wu.

In the Tao Te Ching, this reality is expressed as "the uncarved block", as the Origin from which all the myriad creatures and forms spring (in- looking at No-Self, then looking outwards form here). Hence all categories are from the uncarved block, and the expression of a phase of form/spirit is merely an expression of this uncarved Being-ness. The carving is attachment or mind-identity, that which makes the forms of the world numbered and categorised (out-looking only , (from No-Self) forming belief in a projected self), and capable of becoming objects of attachment and identification. Our perspective must always be from Stillness so that we can observe free from differential judgements, and so that our perspective transcends time and space. This means in a sense seeing out from within or seeing in from without, in the end it becomes a two-way Seeing effect. This is explained more complexly in the work of Douglas Harding.

The Paper Tube Experiment

In order to understand fully the aspect of a part within the whole, the following experiment is helpful:

1. Get a paper tube or rolled up newspaper and peer through it as if you were looking through a telescope.

2. Focus on an object with one eye. Keep the other eye closed. This represents mind-identity's view of reality. It can see only aspects, not the whole.

3. Slowly draw the tube away from your eye, but keep looking down it at the object.

4. Now open the other eye, but keep looking at the object through the tube. This is now a representation of categorisation used in this work and in all of Chinese philosophy; it is looking at something in the context of all its surroundings. This is also the difference between Ancient Eastern and Modern Western medicine, diagnosis, and many other aspects of modern life. The point is that through viewing unity and an aspect in context, one can see the whole picture.

5. If we then get rid of the tube all together and sense the Stillness behind the whole of the scene, this is Wu, or Void, the background of everything, which envelops everything and is everything. Wu does not exist in time, but in the present. One finds that it is the place where one was viewing from that is the key to this sense of "background", there is in fact just Emptiness viewing the scene, No-One is in the way of the view, this is "I" the original "I" or the "Original face", which is the same face "we" share.

(Please see the work of Douglas Harding on using a paper tube, to further this and learn to discover looking from the first person point of view.)

viii) The Constitutions

The classics explore constitution extensively; the following are based on the classical descriptions of the Su Wen and Ling Shu.

Wood Constitution

These people come from areas ranging from the cold northern regions towards the temperate regions. Having a Wood constitution means that the Wood energy is strongest within the person. This also means that the weakest area tends to be the opposite area, the digestive system and/or the lung Metal system. Wood has a strong association with the muscles and their power, and also with the liver. So these people have strong livers and strong muscles. As a result, this will form their character, which will be vibrant and assertive, powerful and expanding.

Blood will be plentiful, and the eyes will be large. This excess of blood will create stagnation patterns within the system. The heat of the blood does not get out to the skin, and so the skin becomes hairy to prevent loss of heat and to increase circulation of qi at the surfaces. The skin colour is olive.

The lung organ is weak, making the outgoing breath poor, so the qi is not pushed out to the exterior of the body. The Wood individual may get colds if the energy of the body stops moving outwards.

This type is often related to pioneering behaviour because they have an aggressive tendency. They can see themselves as isolated units and try to take power from each other. Their temperament drives a lot of body heat upwards, and the stagnation of blood blocks it by creating a mind-identity that can be rigid, wilful, and unyielding. This mind-identity can take over easily. They are often Caucasian in constitution and in relation to the world populous. They will have hard, strong bodies and can have high blood pressure. They are especially predatory and territorial and will have a diet of meat and animal food. They have large noses, due to needing to take in large amounts of air because their lung system is less efficient.

This relates most to a Pitta type in Ayurvedic medicine.

Fire Constitution

These people are generally of the southern or hot regions. They tend to have very powerful hearts and chests, as well as upper-body strength, but the legs are generally thinner. The skin is black or darker in colour due to the kidney energy being burned out. Hence the sun's energy burns the skin to a darker shade; this blackness is bitterness, which is a coolant for the body, offering protection against the sun. The skin has some hair, but not too much.

These people do not live very long but live with much joyousness; they are vibrant and expressive. They can be irritated easily and become aggressed quickly, but less so than those of the Wood constitution. They are more expressive and dance orientated. These peoples can be highly intelligent, but in a musical and expressive way; they do not hold on to things for too long. Common groups are the Africans, the Aborigines, and the Brazilians. They often live close to the equator.

They can become anxious and are often affected by rash action due to fearfulness. There is high tension and emotion in this body structure, and it often cannot keep going for a long time. The body gets exhausted through overexertion and overheating. The nose is large and needs to vent the heat of the heart as well as gather ample supplies of air, because those with this constitution do not have the most effective lungs.

This relates most to the Vata type in Ayurvedic medicine.

Earth Constitution

These people come from temperate regions where the weather is mild and food is plentiful. Their bodies are rounded, and they have an agricultural or nomadic background. Their food involves intake of fatty foods such as dairy products, making them rounder and fleshy. Their lips and mouths are large, and they often think of food—or their lives revolve around food and family. They live in areas of mountains or valleys, and often live cattle and farming animals of some sort. They have quite naked skin of a yellow or brown coloration and are simple in nature. Examples can be people of Middle Eastern and Indian descent, or Mongolian and some Chinese. Their digestive systems are the strongest, but they have weakness of the muscles. They are generally more flexible and less harsh or aggressive. However, they can easily be followers of aggressive people, who can dominate them.

This relates most to the Kapha/Vata type in Ayurvedic medicine.

Metal Constitution

These people come mainly from the Asian region and from quite windy and cold areas of the northern hemisphere, or are Polynesian in origin. Their strength lies in their ability to breathe out and express strong defensive energy against the environment. Their lungs are strong, and their skin is totally naked. They have many more pores on their skin, making it more powerful as a guard against weather. The skin is white or creamy in colour. Their noses are small because they have efficient lungs that can take in a little air, which suffices for their system. Their eyes are small and formed to protect against the wind.

They can survive a great deal of difficulty and complain little about it. They are sensitive to strong odours and dislike heat, as their body systems work better in cooler temperatures. They eat vegetables and seafood mainly. They are very peaceful by nature but can hide great aggression and vengeance, usually for short periods—though they hold on to it for many years. They keep going for long periods and can usually see the big picture rather than being egocentric. Examples of this construction are the Japanese and those in some Chinese regions.

This relates most to the Kapha/Pitta type in Ayurvedic medicine.

Water Constitution

These people come from the northern regions close to the pole and the dark lands where there is the least overall light. Their skin is the palest of all of the types, but also thick, though not as developed as the Metal types. They have a white and very tough skin, with hardly any hair. Especially on the back and extremities, it can be leathery. They have the strongest bone construction and can weather the deepest cold. They have powerful kidney energy that keeps them warm in these temperatures.

They are generally quiet people, but also very wilful and can make a great effort for a long period of time. Generally there is little vegetation, so they rely on the sea for their food, or they eat animal food that they raise. They are generally nomadic peoples. Their weak area is generally the heart, as they have high blood pressure issues due to their powerful nature. They have the strongest constitution of all, with large ears as their main feature. They can live the longest. This is the stillest of body forms and therefore less mentally active, although with the capacity to be so, because the brain is well cooled. They are Eskimo types and Tibetan/Icelandic and Canadian Indians.

This relates most to the Kapha type in Ayurvedic medicine.

ix) Ayurvedic Constitution and Chinese Medicine Constitution

Ayurvedic medicine is perhaps the only other tradition as ancient and rooted in Stillness as Chinese Medicine, although Chinese Medicine has developed external techniques to a higher level, whereas internal medicine of the region is practised at a high level by Ayurvedic practitioners. According to some authors (Ros, 1994), there is evidence that the origin of Chinese Medicine is actually an import from India with the influx of Buddhism into China, possibly from the Qin Dynasty (259 BC–210 BC), and perhaps by the originator of the yinyang school in pre-Han China, Tsou Yen/ Zou Yan (305–240 BC). While this is possible, the expounding of the understanding of the Chinese after this time, formulated through Taoist understanding, is likely to already have been underway before this point, in association with Yu the Great and his discovery of the Luo Shu map, and King Wen's formulation of the eight post-Heaven trigrams, based on the Luo Shu and its formulation into the I Ching. It is possible for the flowering of cultures to occur at the same time; this has happened many times before and is part of the Oneness of human expression. In either case, the point is that the two systems are parallel and show this parallel deeply. Treatment principles should therefore follow exactly the same base in Chinese and Ayurvedic practices. Agreement is often found if the practitioners are of competent clarity in each tradition. If the root is the same, all forms simply express the same clarity. Ayurvedic principles and Chinese principles have the same root. They gave rise to all the other medical traditions of the Ancient East.

Chinese and Ayurvedic practises always attempt to allow the person who is in a dis-ease state to rest the mental illusionary state and come back to the natural state of the constitutional make up and its relative imbalances within itself, which can be supported by diet and practises that support the individual body structure.

The Ayurvedic doshas may be implicated in the three types of body mentioned in Ling Shu, chapter 59. Three forms of body are represented: Fat type (Kapha), Puffy type (Pitta), and Fleshy type (Vata). These 3 types are mixes of opposite energetic qualities: - Kapha is water and earth, Pitta is fire and water, and Vata is ether and air, which is metal and wood combination (see below).

The three constitutions of Ayurvedic medicine come from various combinations of the Ayurveda's five phases. In order to get a general idea about similarities of the energy types used in the ancient world, the chart that follows offers possible correlations between Chinese five phases, Indian five phases, and Greek five phases. These are seasonally based.

(Table 2.3)

Chinese (circa 500 BC)	Indian (circa 500BC)	Greek (Pythagorean, circa 500 BC)
Wood *	Ether (Space)	Ether (Space)
Fire	Fire	Fire
Earth	Earth	Earth
Metal**	Wind/ Air	Air
Water	Water	Water

* Wood and ether are synonymous, but in the relation that we recognise that wood has to do with spring and rising energy. Ether is like vapour; it is rising out of water and, like a mist, is going upwards

towards fire. The Chinese use the expression wood or the rising of sap in a tree, life and expansion. This relates to the pungent and open expression of spring.

**Metal and wind are synonymous also. As we will see, wind relates to the metal energy. Wind is cooling, and this is its primary function. The cooling effect is also movement but in a downward direction, opposite to that of ether/wood which is rising upwards. Wind, as we will see in later chapters, has very much to do with emptiness, cooling, and death and relates to the lung organ function. We will see how vitally important this correlation of cooling and wind are, not only to the five phases but also to the wind energy within the nine energies of Heaven, and overall as the basis for getting a clear picture of the metal phase. Since it is true that India is the origin of the Chinese five phases, then the relation of the cooling air of metal and the ether (ethereal, and also related to alcohol, a liver tonic) nature of wood is very important to our full comprehension of these seasonal expressions.

The Pythagorean system is amongst the oldest forms of Greek understanding. It pre-dates the Aristotelian four elements, which leave out ether and are based on a more structural (yin) expression, also connecting to the cycles of 12 in the Greek Zodiac, which relates to earthly qualities. This leaves out the more human five expressions, which is more yang, like human energy overall, and associates with life. As a result, the Aristotelian system, in its lack of inclusion of the Pythagorean five phases, begins to come out of alignment with the rest of the ancient world, and the medicine becomes more and more structural and less and less energetic. This has formulated the Modern Western mind and to some degree Newtonian ideas of physics and the world. A view based on pure structure or pure formlessness is never the whole picture. The whole encompasses both yin and yang. The Pythagorean system, however, is in confluence with the rest of the ancient world. In recent times, it has been said that Plato has links with the Pythagoreans (internet ref 1), but as this would have been heresy, he hid this in his work. This would make clear sense, as this would connect the ancient pre-Socratics to a lineage, which almost dies with Plato. Aristotle creates quite a new "modernised" (and interestingly hierarchical Confucian) view, which Plato can see, but he also knows his ancient heritage, which dies with him. This may or may not be the case, but the Greek philosophy was originally based on the five phases also, which is the key here.

x) Yin and Yang Deficiency Involved in Constitution

In Chinese Medicine, just as we can say that for a wood body type there is an area of strength, we can also say there is an area of weakness. Hence, the way we can discuss the constitution, in terms of Chinese Medicine, is by looking at the strengths and weaknesses of the various Zang organ systems involved:

Wood Type Body

This is called lung deficiency/liver excess type. This expresses that the liver organ is strong, and the lung organ (and kidneys, see page 506) are weak in comparison. It also describes that this is a yin deficiency. Yin deficiency simply means lack of yin of the system. Yin deficiency in its mild form is a state of health. This is because a person who is yin deficient is radiating from the exterior gently and also is warm and alive. Yang deficiency of any kind is a form of dis-ease or leakage of energy within the system. The person is getting colder and moving towards death rather than life. This is a "serious" (often, but not always, long-term) illness and one that needs immediate attention. When we are assessing constitution, we do not discuss the yang-deficient body structure, only the yin-deficient ones. This is because, constitutionally, all people born will prosper, whatever their body quantity of energy, if in good health and placed in an appropriate environmental/climatic situation that does not place too many demands on the physical system. The yang-deficient phase arrives when this structure has broken down for some reason and there is weakness. Hence the natural health state of the body is yin deficient, so all the types above are in health and therefore mildly yin deficient.

The expression here of a pattern of disharmony is therefore the constitutional pattern of disharmony, not the pathology of a patient. A patient who has a Wood type body may have a completely different body structure, which would mean illness. The attempt, in medicine, is to get the patient back to a place where the energy can express itself most efficiently, and this is the constitutional pattern, which we will call the CP. The above are all CPs. A DP is a dis-ease pattern, and this can be present at the time of treatment, or may have been present for much of the patient's life. Until he gets back to his or her CP, the patient will not feel fully healthy. Hence the CP patterns above are not DPs, but the patterns of well people. The patterns, however, are given the same names, so be careful when differentiating a CP from a DP pattern type (see pathology section, Section C, for DP patterns). Please note "Pathos" means to suffer in ancient Greek. So Dis-ease or sufferance/ pathology is really the main focus of the study in medicine; it is not just about the associated pain patterns but the cause of this being problematic for the person, sufferance. We will look more into this in Section C.

The organ that the phase belongs to will have the strongest energy in these patterns (again, these will all be discussed in detail in Part 2 of this work):

Wood type: Lung deficiency liver excess CP
Fire type: Kidney yin deficiency with heart heat CP
Earth type: Spleen yin deficiency CP
Metal type: Liver yin deficiency CP
Water type: Kidney yin deficiency with bladder heat CP

Notice that there is no heart deficiency as a CP. This is because the heart, especially, is the flower of the body system. The Fire energy in the body originates from below in the right kidney and culminates with the other energies in the body in the heart itself. Hence when we describe Fire, we are actually describing the kidney yang. This being the case, there can only be four possible deficiency roots: lung, spleen, liver, and kidney. A Fire type is therefore similar to a Water type. The key difference is that the heart is overly strong in the Fire type, whereas in the Water type, it is the kidney yang or bladder energy that is strong. This differentiates upper body energy and lower body energy. One would expect that winter and autumn energy would be cooler, and this is true, within the spectrum of the mildly yin-deficient CP. The Metal and Water types are cooler, but often because they are in the context of a cool environment. The human and environment are one, but the key is to see them in relation to each other. The most overheated is the Fire type, and the coolest would be the Water type. Although both have a similar CP, the body structure that goes with this pattern is quite different, and one needs to therefore see these CPs as contextual number-plates, better described by Fire type, Water type, and so on. The type is the strength area in the Zang organ. The most conflicting types are the Metal and Water types. If we look at the strength energy being of autumn in the Metal and winter in the Water, these areas are associated with the yin energy of the body.

Because humans are yang, their amount of yin/pre-Heaven energy will be the potential power of the system. Hence, those who have strong yin will have power and potential to express relative yang energy to match their basal yin power. (Yin in this sense is referring to internal pre-Heaven essence, so it is completely separate from the notion of yin of the exterior environment being colder relative to the human/body-spirit energy.) As a general rule, this means that they use it up. Hence a lot of pre-Heaven autumn energy internally (sour flavour) contributes to using up this strength, creating a yin-deficiency in the Metal energy of autumn within the body. Winter is the same: the strong yin of winter (bitter flavour) creates a strong outward use of this, forming a deficiency in the winter energy within the body. This is why the Metal type is liver yin-deficient but part of a cool environment, and the Water type is kidney yin-deficient and part of a cold environment. This complexity is why we usually consider types rather than CPs when describing the constitution. Remember, however, that a yang-deficient constitution is not really an effective way to see people.

In these descriptions, we can see that constitution very much relates to the type of climate and region of Earth from which people originate. Nowadays, so many of us are broad mixes of these constitutions. However, we will have a body construction relating to one more than the others. This is our physical constitutional type, and we often need to look at the area of deficiency to see how to stabilise our system and replenish what the strong aspect draws from the weaker areas. Remember that the body must be in unity, so if there is too much of one thing, it becomes unbalanced, static, and vulnerable to exterior change, which can break it down. This means the system stops being flexible.

These types represent the Jing body, the construction of the structural formation of the body as a particular type of physical form that is energised during life but that passes down to the ground when we die. This covering of energetic structure holds within it the soul or spirit, and this is the Shen body. The spirit-body is one within the structure and interplays with the Jing body, either in a way that matches or mismatches the physical structure it is within. The unity and acceptance of the two aspects completes the understanding of the expression of a particular human form and especially the way to his/ her good health. In the next chapter, we will look at the Shen constitution and how this is different.

Notice that the classics of Chinese Medicine deal almost exclusively with the Jing body, aiding its performance and longevity. However, there is another method of practise of Chinese Medicine.

There are two main branches of Chinese Medicine: one is through the body (yin), and the other is through the spirit (yang). Both methodologies are valid, and both are effective when there is recognition of Presence in the moment in practice. When anyone truly is "them Self" in the deeper sense, meaning Oneness here, there simply expresses like a musical note of existence, and through this will also echo the Stillness of unity behind the surface. The note that is the expression of a more physical energy will relate differently from that of a note of more spirit energy. However, there is unity in both of them. It is only on the plane of the body-spirit or yinyang that interactions occur of changing form.

And so the expression of one note will affect some people differently than others. In the ideal practitioner like Qi Bo, there will be a connection of the practitioner with Wu through the body-spirit being an instrument of Oneness. Some practitioners will use tools of form to create their expression; others will use tools/expressions of ethereal energy to do so. Those who are more yin can be more still and act as conduits of Wu/ Oneness. Those who are more yang are more expressive and form change, which returns the patient to Wu through movement to direct the patient back to Wu.

Of course, in reality, most practitioners are involved in treatments that affect both levels, because people are mixes of these energetics of yinyang. The more subtle the sensitivity, the more he or she can invoke stillness through the quality of being, the true presence of what he or she is—(or in fact is not!)—rather than through any particular action or activity. Often this will seem miraculous or awe-inspiring to those still within the mental illusion, and yet it is totally natural/simple and non-abstract to their deep sense of self and to the practitioner who can take no responsibility, for it is not about a separate sense of him/her but a recognition of unity with the patient and all things that has invoked natural shifts. Hence we can call this aspect shamanic or Shen medicine and the other aspect Jing or physical-energetic medicine. Both are important. We acknowledge both in their proper context.

2.2 The actualization/expression of the Spirit within the Physical Vessel

The Shen body is derived not from the principle of human (five) but from the principle of Heaven. This is the principle of nine, which is fundamentally the most important for describing the attributes of the spirit in relation to the movement of Heaven. Before we look into this, it is important to see at what juncture the spirit energy is seen to reach its point of actualization in the process of pregnancy and birth. Please remember we are discussing Shen-spirit here without the involvement of mind-identity, which is an extended warping of the spirit expression into a dream of imagination. The expression of spirit here is as a pure or perfect form, a natural expression.

i) Actualization/ expression of the Spirit

There are 10 months of pregnancy, which is about the celestial pre-heavenly stems, almost like the re-formation of the universe as in the He-to map, but this time within the post-Heaven universe, and physical-Jing growth forming the vessel to hold the spirit. This is a cycle within a cycle. So the organ systems, the storehouses of the spirit, are formed in relation to the five-phase principle. We looked at this process earlier; now we look at birth again and the engaging of the spirit energy within this process to form life outside the womb.

Pre-birth

The pre-birth state of the mother is vital in many respects. She represents at this time the universe before manifestation or expression of life or yang. Hence the main principle of her being embodies Wu and the absolute yin at this stage. Women at this time in their lives are at their most intuitive and sensitive

99

to the body and also the inner rather than the outer experience of their life. Many ideas have formed over the millennia of what happens during this time of pre-formation of the body-spirit union, but this is to miss the point. The mother may well have a sense of her newborn, his or her personality or expression, images of future possibilities of what this expression may be like or atunement to what some of the ancient tribes people called a "song" or vibration of a child before it is born. Simply this is the recognition more and more physically from the energetic that something is changing and that something is happening. A mother's personality will partially alter during this time due to the nature of the expression they are carrying , although it is important to recognise that what is occurring has not yet come to fruition.

While many of these insights are interesting and obviously experiences of great significance for the sense of femininity and realization of Oneness forming seeming twoness, it can also be clouded by ideas which are quite delusional. This expression happens all the time through nature, and as we say the move from the He-To to the Luo-Shu is the pre-heaven to post-heaven change, which is fundamentally different. However much there is focus on what is occurring though the gestation and birth period, it is more important to sense what isn't changing, what actually stays the same, what is anchoring this whole process.

There is a background of Oneness even through twoness seems to be occurring. The mother senses both, she can sense the expression of the to be baby and yet she is sitting in Stillness. Where as yin and yang (heaven and earth) finding expression within so she is the space for this to occur in. In this way the whole notion of the individuated soul and the ideologies around the "Incoming soul" or "child choosing the mother or mother choosing the child", doesn't come into it. These are ideas associated with distortion of the ancient ideas of Oneness and the perfection of Uncarved nature or innocence to mind-identity for the unborn child. As with everything else, the ideas the mother invokes at this early stage effect the process of her baby's mechanisms. If at this early stage in-utero she is already seeing the child as separate from her, as an individual soul, then the mechanisms of response to this idea when the child is born is more easily triggered causing the origin of mind-identity and sense-of separateness which is the basis of illusion and suffering. The mother in fact can have an instinctual-natural expression which is one that retains unity with the baby to not see the separation and to be as one with it, inside and outside the womb until the child is able to move on its "own" expression although in fact, it will not see it like this it will just be continuous sense of Oneness with life. Rather than identifying individuality and getting swept up into reiterating this and also focusing on it and making it important for the mother to be is actually a furthering of a dis-ease, it is not encouraging the unity of Oneness and as a result has massive influence on human society. If from the very root of life, the female is looking towards separation and ideas of individuated souls she effects all of the populous to see in this way. This is the origin of fear, and the ideology comes from this sense. As we will see the originating process is the very root of dis-ease so at this time more than any other there is a subtlety about meaning and expectation and direction of what is to come that is derived directly from the mother's own perception of either separation or Oneness. From Oneness is derived peace, from separation dis-ease. Those looking to attempt to find a way of looking at individuation is making the child, adult, even in-utero. This of course is an entirely unconscious process, for those that look towards individuation are not choosing to but it is simply that the world seems so much like this that there seems to be nothing else, this is how it is, and this is how the message of separation takes hold within society.

When pre-natal awareness does not look towards Oneness then from the outset human beings are lost in mists of separateness. This too associates at the end of life with the Near-death experience which is Near-death not death, death is Oneness but near death implies twoness and the ideas and wants of the "separate soul" to continue. At the point of death, just as In-utero there is Oneness behind whatever abstraction is being presented in image or sound or anything.

To keep the mystery as mystery, and to leave the innocent as innocent, and to spontaneously respond without intellect. These are the natural ways of life expressed in the pregnant woman and in all other aspects of humanity and nature at all times, behind a clouds of individuation, self-orientation, fear of the unknown to the mind-identity and so on.

Birth

The expression of birth is like a fractal and microcosmic expression of the birth of Heaven energy from the mother yin of the universe (Wu); this is when Heaven begins to change. The question of why a baby comes when it does is similar to the question of why the universe went from stillness to change—at the right moment, there is birth. The child immediately becomes separate from its mother's energy field and becomes an entity in its own right, of course still connected to everything, but now in its own right as well. This is the point at which the energy is coloured by Heaven's connection—at the first breath.

At the first breath, huge changes occur. The change is from pre-Heaven energy of the mother to post-Heaven energy; hence it goes from a static state to one of change, exactly the same as the macrocosmic view. (Notice the use of word "post-Heaven" here. It is impossible for pre-Heaven to be suddenly differentiated within the mother's womb because we live in a post-Heaven universe under-pinned by pre-Heaven. However, we can say that it is representative of, or an echo of, the original universal creation process. So we say pre-Heaven can also be called pre-natal energy, and post Heaven can be called post-natal energy). At the first breath in, this is governed by the kidney energy. The first breath in ignites the baby's own Ming Men energy, which is closest to the surface between the second and third vertebrae on the back but deeper into the centre. This is perhaps where the gate for Heaven's energy is connected/released. The Ming Men or "infant gate" or "gate of Life" is the spark of life that ignites the heart fire where the Shen/"Heaven Altar" is said to reside. It is in fact the fire of the body that gives the heart its beat. Previously, this was being delivered by the mother's Ming Men and her heartbeat powered her baby's, but now the baby has its own energy to power itself, and this is Heaven's spark igniting it to life. The form is still and so is connected to Wu through relation, and the energy-Shen moves through it. Hence the yin formation of the body of the baby, through the mother's own yin, has created a vessel.

The spirit is the quality (not quantity) of energy that moves through this vessel and is connected to Wu, but is itself Change, which has its root in Stillness. Because the form is a part of a whole, the spirit that permeates it is a part of a whole, and the whole that underlies this iceberg is Wu. The difference between Wu and Shen/Heaven is that Wu is formless and changeless, whereas Heaven changes. Wu underpins Heaven. The difference between Earth and Wu is that Earth has form or shape defined by interaction with yang, where as Wu has none. Hence the Stillness of yin and the formlessness of yang make Wu; the ultimate illusion is to ignore the Wu background in Unity and see and attach to just the yin and yang.

In numerous traditions, it is said that a spirit enters the body of the baby at this time (although some say at conception, this is not as clearly understood because the vessel has not been formed yet and the foetus is actually part of its mother's body). In Taoist understanding which is echoed by all of the deeply meaningful ancient understandings, the idea of "spiritual entities" are not as clearly defined and are delineated into ghost-like souls that inhabit or enter into the system and have had many cycles of lifetimes before. This would tend to give form to something that has no form but yet is changing. The understanding is much broader. The picture is much more of a sea of energy and total integration within that sea. At the moment of birth, the energy of that time and moment colours and ignites that seeming "individual" form/person's Jing/Earth energy. That moment of change has to do with everything in the universe at that moment of change, all unified. However, birth can be considered like a dawn; this means that in order for there to be dawn, the pre-dawn structures have to be in place. We then have to consider that the moment of birth is just part of a process of change. There is no exact moment "spirit enters the body", but spirit is one with the foetus from its original conception and is all of everything at all times—there is no separation. Birth just delineates a specific point of change where we can register a level of formation that seems, only *seems*, to be a separation from one (being pregnant) to two (mother and child). This point allows us to categorise one in relation to another. The first breath holds the key to this differentiation. Hence we can suggest the colouration of the rhythm of life is all founded on that first breath. We can then relate this to the spiritual origin of the so-called individual, although as with all things, this is a total continuum. Spirit and physical are never separate; I am only expressing them differently here so we can differentiate and then unite them with clarity. Yinyang always occurs together, and body-spirit is yinyang. The spirit condenses to for the physical the physical expands to become spirit this is the way of nature.

Heaven envelops Earth, but Wu envelops all yang and yin formations. Hence change itself is a spectrum of yinyang of the changing post-Heaven universe. However, the energy originates from the nothingness of Wu. This shows us that things come into existence from the realm of Wu. Hence Jingshen or body-spirit or yinyang is a spectrum of energy and comes in and out of existence. The thing that draws it to exist or not is Wu. This can be said to be beyond spirit and body and is the eternal Atman or Buddha-Nature, or Christ-consciousness or whatever you want to call it, within. This, after a time of movement or cycles of coming in and out of yinyang form, will go back to the void it came from. This process is inevitable. The universe eventually will go back to the unity of Wu-void. However, on a smaller scale, it is the end of the cycle of Wu forming yinyang and simply not changing back into form once a life-cycle has been completed. This is called the return to Wu and will happen naturally. (Note that "return" is not a good word for what already is under the surface.). In all of this, each time the yinyang form comes into existence, it is never the same individual "soul" that does so but is part of the universal energy. We are all totally one; what happens to one happens to the whole. This is absolutely fundamental in the study of ideas like "soul", which are so preciously identified with by the attaching mind-identity that is in great fear of its own disillusion in each waking moment, and sometimes in sleep also. The clear mind sees unity in all aspects of life as the watching presence behind the rhythm of life played out by the body-spirit dance. This is known at death of the body if at no other time.

The difficulty and creating of mind-identity occurs in humans due to the over-activation of the head by the spirit. When there is the joining of Jing-body and Shen-spirit, a changing energy meets a stiller energy of form. The combination creates the possibility of clear union and acceptance (Buddha Nature/enlightenment/Natural) as we see in all of Nature, or the distortion of this as we seen in human agitated mind-identity.

In Taoism, the nine energies permeate the physical form of Earth and the myriad life-forms (All Under Heaven), so we must look at the principles relating to the Shen and the nine energies.

In the Classics of medicine the following expressions relating to the quality of the spirit and the 9 energies, or the 12 branches and 10 Stems in relation to the horoscopic perspective is not discussed. This is because these expressions belong to a broader category of expression that is known as Feng Shui. Feng Shui, as will be explained later, is really the expanded view of medicine. No longer is it about the aspects of the human body but it is more about the human within the whole of society and as an expression of a particular phase. This is why as far as constitution goes it is a very broad picture and this is very useful in our understanding of different people in relation to one another. There are 2 main expression of the person from the exterior overall view like this:- 9 energies associated with the spirit and branches and stems which relate more to the influences on the constitution of the body, or corporeal expression of the spirit. As we are considering Spirit the best way to look at these two expression is thus:-

(fig. 2.5)

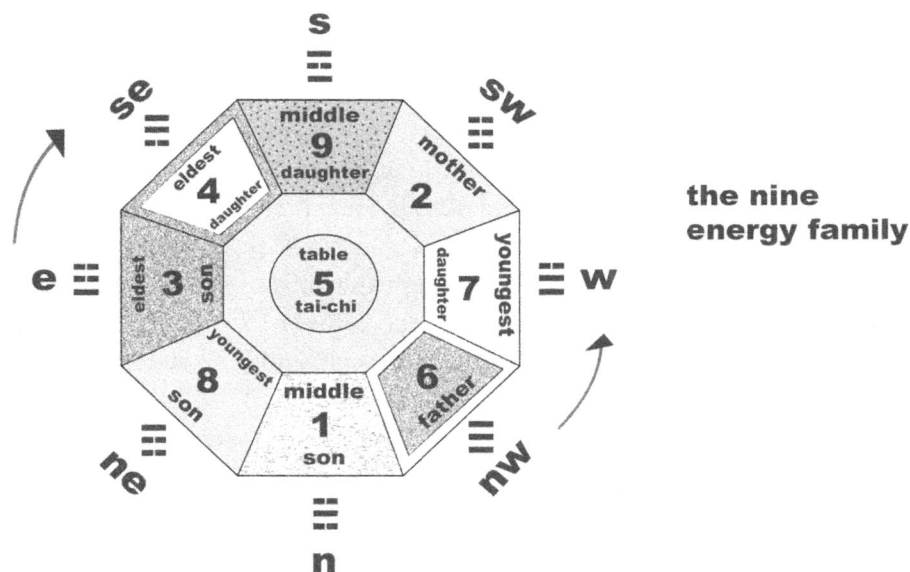

the nine
energy family

The 5 (4 directions plus centre) Jing constitutions of earth are in relation more to the 12 branches and 10 stems but still looking from the broad viewing angle, seeing the whole human as an aspect of something. Yes, it is possible to relate the branches and stems to the actual physical body but this is not to see the expression of it as it is, which is the movement of the whole of the earth with the humans in relation to it, not relating the whole of the earth to a specific human. The idea of concentric circles is always important here and to understanding things at there intended level of viewing. The Nine energies is an even bigger and broader view which underpins, but it also relates it to the yang of the human spirit and so is very important. Let's consider these things now. We will correlate the 12 branches and 10 stems to more physical and earthy associations and the 9 energies to the spirit more.

ii) The Shen Constitutional Types

When we are looking at these "types", it is important to understand what is being expressed. For all intents and purposes, this is as close to an objective expression of the non-individuated soul/ spirit-energy of a person as we can express, what we would describe as a person's true or natural state of expression. What it implies is a connection to the stillness, which reveals that the aspect of spirit being presented in the world is not the origin of spirit (Wu) but a spirit through a form in an ever-changing mix of energy. Buddha Nature is the Wu pre-Heaven within all people and the aspect that joins all this up. The nine energies are the changes of Heaven (post-Heaven) or myriad things. They are all one; this is the key to their understanding. What the following doesn't do, is explain what that feels like; it simply expresses it, as if it was in a history book, in black and white for mind-identity to gain clarity from. In all the processes we are expressing, in itself it will not help you find clarity of being yourself. It only takes you to the gate. It may show you aspects of your true Nature that are hidden by the blanket or dream of self-image, and so give you a perception, so that one can differentiate mind-identity from being. The nine energies as applied to human or any phenomena or any other principles have nothing whatsoever to do with mind-identity. Mind-identity and the phenomena that are created within it are a fragmentation of Nature. This is why the ancients used aspects of Nature that have no-mind-identifications in order to describe the true Nature of yang in the universe, which is not an aberration, i.e. mind-identity (which is a sick form of yang-Shen) attached to forms (yin), but simply an expression like the wind, water, or fire of Nature. To summarize, within all form is at the root Wu, then body-spirit, and lastly (in humans) mind-identity. When we look through Wu at the world, through the empty space we are looking through now, this is the point at which the world of forms and the Subject—or Wu—meet and greet each other. The Wu is the bigger, the world the smaller. Wu envelops the world and is at the centre of it. (See Appendix 2 for a diagrammatic representation of this.)

The ancients left this information behind for us to find our way back to Nature and to the natural state. This is therefore the intention of this work. It is not meant to become another addition to the mental confusion we live in. One is not meant to associate oneself with a "type" just as one has associated oneself with a mental image of what one ought to be. This is simply a signpost to the sense of Non-self or Self and the outer expression emanating from this like a fountain (i.e., not self, meaning mind-identified or small-self) that lies beneath the mental veil. Through that sense there is the permanence of that which has no sense and no movement, the Stillness of Wu. Other than that, the principles in this book have no deeper meaning. The only true value of this work is to differentiate, or identify, the wisdom of the clarity/ structure that the principles provide mind-identity with, making it more crystalline and less opaque so that the light of perception can break through, rendering this and all other works irrelevant, as then there would be acknowledgment of Oneness with everything and Knowingness would permeate. All the ancient classics, including the sutras, the books of Taoism, the medical classics, and all the works of meditation philosophies of the true root understanding from Christianity to Sufism, have this same notion: they take you to the door and clear or order your mind-identity; then you must let it go or burn it up.

All the energetic expressions within these first chapters express the body-spirit in pure form, unadulterated by mind-identity and in complete Oneness with Wu, the source. This of course means that they are ideal states of enlightened beings, very often a far cry from what we see in the mind-identity-attached state of the world around us. However, there is a connection that people will "know" deeply to be true about these expressions, because they touch the source of their expression. This can bring them to the source of life, Wu-Stillness. Therefore there is a depth beyond the expression of the spirit or body-spirit yinyang, and this is where the change emerges and merge to Oneness.

Each of the nine energies can be described in numerous ways. Their united picture is the circle of the energy of Heaven, which was born from Wu. The five physical constitutions could be denoted as the square of Earth, in the four directions and middle as expressed in Su Wen 12. We look at the constitutions first, as the pre-Heaven (pre-natal) formation comes before the post-natal Heaven, just as Wu is first and Heaven is second. Heaven, however, permeates all forms of Earth (yin), so Wu permeates all forms of yin-Earth and yang-Heaven, and pertains to the stillness of Earth yin.

As we get a feel for the energies in human form, I will firstly describe each one in a broad spectrum, then narrow it down to the human level, from the macrocosmic to the microcosmic. Before we start, however, it is useful to see the nine energies as a family of people. There is a mother and a father, and they have three sons and three daughters. The relationship of each of these eight members around the central table, or Earth, provides a good base from which to look at the energies in human terms. We will look more into this later, but for now:

(table 2.4)

Number Associated	Yin/Yang	Trigram of post-Heaven Ba-gua	Energetic Meaning/Translation
1	Yang	Kan / Ken/ 乾	Water (Sea)
2	Yin	Kun/ Kon/ 呻	Field (Earth)
3	Yang	Zhen/ Shin 震	Thunder (and Lightning)
4	Yin	Sun/ Son/ 巽	Wind
5	Yang (+yin)	Taiji/ Taikyoku/ 太極	Centre/Vortex (Earth-core)
6	Yang	Qian/ Kan/ 坎	Heaven (Sky)
7	Yin	Dui/ Da/ 兌	Lake
8	Yang	Gen/ Gon/ 艮	Mountain
9	Yin	Li/ Ri/ 離	Fire

(fig. 2.6)

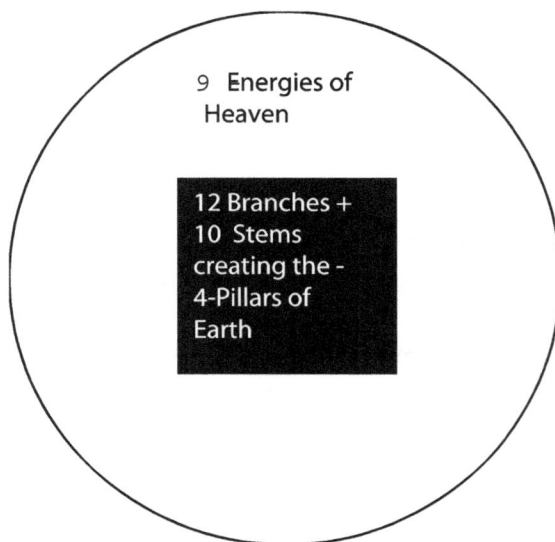

1 Kan — Sink Hole-Water/Middle Son

Trigram Line Symbology:

(Note: I will use the following terms: base line, middle line, and upper line to describe the three lines read from the base to the top/upper of each Triagraphic arrangement below.)

The trigram's most important line is the central yang line within the yin. This makes for complex type energy, as the north and yang seem in conflict because the northern concept is cold and dark without yang. The yang in the yin here is the power of this energy. It is both yin and yang and the direction of the central core yang that makes it a definite yang energy rather than a yin. The fact that it is held within the depths of yin means that the yang energy is held deep and within, like a secret, or like the light in the darkness. This kind of deep yang is very precious and rare, as usually yang is found above. To find it below means that it has made the yin potent, or infused with yang; this is a very powerful energy mix.

The water energy is a pure form and is pure yin, cold and still, but Kan is yang within yin, and this makes it a dominant energy. The yang line within it is drawn to the emptiness of the Li trigram's centre, and the coldness of the exterior yin cools the outer shell of Li's yang lines. This is like a lock and key. Note that it is Li that draws the yin of the water, and it is the water's yang (which is fire within water) which is the energy that acts to fill the centre of Li. Yang always acts, and yin is always drawn; hence this is the nature of attraction of opposites or displacement of opposite energies.

Symbolic Nature:

Water can be symbolized as a drop of water forming a stream as it passes down a mountain, eventually becoming more and more massive, till it flows into rivers and oceans. Kan really is a representation of sea or salt (salt being yang) water due to the yang within the yin expression. Water is ever enduring and the source of life. It is also the coolest of the phases and represents the winter and the north. It is fluid and gentle and yielding in Nature, yet can be drawn to be the most destructive power of all the elements, being the universal solvent. It is channelled by Earth to go deep underground where it can stay enduring, deep and dark, silent and secret. In the heavenly energies of nine, the colour associated with one is white. With the 9-star the 1-star holds the crown chakra region spiritually. This is associated with the mental realm; the water energy is bright and able to store massive information. However, in the five phases of humans, one is coloured dark blue/black.

Water is directly opposite of fire and acts to control fire. Fire burns up water through resistance, but water dampens fire through action.

Associations:

Water finds the deepest cracks to fall into and the most difficult spaces to travel through, which is specific to this element alone. It is therefore associated with aloneness, a hard life, and difficult paths travelled. Water has the ability to dissolve, to wear away, and to be enduring, which implies that water types are good at long-term struggles and keeping going no matter the hardship. Water can wear away at things with great force, things that others would find too difficult to achieve; it can have the current to shift big obstacles. Though the energy is cold, it also wants to merge with things, to dissolve things into itself. Hence there is a connective/conductive aspect of water, especially seawater, which makes it more proactive but in a colder way than stiller water, like lake energy.

It is the source of life and can be dark and dangerous; it can harbour deep treasure or deep and dark feelings. It aims to be still and to distil itself rather than to move and lose its potential energy and sexual/fundamental recourses. It prefers to wait rather than to act, but in action, it is definite and absolute about its will.

Winter as a season comes from below, percolating upwards. This is the opposite of summer, which is from the sun, percolating downwards; hence winter energy is of water's energy, which is moving into spring.

In Human Life:

These types are often quiet and gloomy rather than lively and positive, unless there is a dark time—then the fire within them can be truly warm in contrast to the exterior. They often have distant relationships with their parents and find life quite difficult during their early years, until they understand the depths of their being and gather some momentum in later life. They are fiercely independent loners but are very intelligent and talented in thought. They relate to strangers better than to people who are blood relations.

This means that they may develop a large number of acquaintances, but rarely will people know the true soul of these individuals, as they are secretive by nature. They easily cool a heated argument with wisdom and depth, expressed in a blunt fashion. They say little, but what they do say can mean a lot. Too much independence can alienate them from others, leaving them alone to philosophize, with no chance to share; in this way, they lose their will and are isolated. They have difficulty finishing things and can go on and on forever defining, philosophizing, and re-defining that which is the truth. This person is more static/ stubborn but if pushed can direct in their own way well. People often find them difficult because of their gloomy temperament and deep thought; it is too much for them, it can be too intense. They can be dictated by sexual need and desire, which can rule their internal lives. They can be seekers of the dark to bring this to light, or simply be in the darkness and relish in this. Their energy is of the early to mid-teens.

Their core mentally constructed belief, which can turn into pathology, is that they are alone and to connect with others is dangerous, as their depth will be discovered. This can lead to isolation and depression.

(Please note that mind records the past events, so the action of the spirit propels mind to record, focused in a particular way. This focus is determined by the nature of the spirit type. The time is now, and no other. Therefore, whether the person spends their time in the mental belief/ identity or united with reality is what makes the difference for them between sickness and health, fragmentation and union. If they are able to find their true expression, and one is able to get a glimpse of this truth, it is often deeply important and right at the root of the hearts and consciousness of people.)

It is the deep truth that water teaches, the thought at the depth of thoughts. Though these people are shrouded in darkness, the inner light they have is perhaps the purest of them all.

These types must use their ears to hear the sounds of the world keenly. Hence it is the expression from the heart verbally (fire) physical movement (earth) that they lack, and this leads them to mental clarity (i.e., the opposite of what they are, the weaker aspect). When put in fearful situations where others would crumble, this energy comes out and they find that they are stronger and more determined than their often cool temperament exposes. They need to be challenged in order to come into their true nature.

Water is associated with the middle son in the family of eight people (i.e., the eight directions of the nine energies, the centre being the table they sit at). It is nourished by the father (6) and it nourishes the youngest son (8), although in the human (not heavenly) association of the five phases, 8/2/5 are all aspects of Earth, so it tends to nourish 3/4 energy in humans. It is opposed by and so attracted to the middle daughter (9). Hence this very yin of phases is yang in its nature.

2 Kun — Earth-Field/ Mother

Trigram Line Symbology:

The lines here are all yin lines. Yin lines are broken lines, but in the southwest region, the yin is dense and solid like the earth, as well as soft and receptive. The lack of yang is the key thing to see here. There is no yang to potentate this energy, so she is reliant on this from elsewhere. She requires "sowing" so that she may bear fruits. This is the nature of the pure yin in this region of the Ba-gua of nine energies. Two is archetypical of Earth energy within the five phases, so the other 2 Earth related energies, 5 and 8, are secondary to it.

Symbolic Nature:

Earth is the mother, the second aspect of the universe created from Heaven, which in turn was created from Wu. The mother yin therefore has associations with the ultimate yin—like the concept of the Tao. From the mother comes all of life on earth, including humans; the mother is therefore the foundation. It is gentle, receptive, nourishing to all life, more yin than water, and female to its core—the ultimate femininity. In the heavenly energies of nine, 2 is associated with black. The black colour is associated with the square shape of Earth. This also relates to Tao and the Stillness of presence behind all things. It relates to below and the background of the chakra's spiritual energy (see Appendix 2). In the five phases of humans, it is association with the colour yellow/orange. It is yin but is within the yang in the five phases, so it has warmth. The late summer is when the sun sinks in the sky and the Earth is warmed in a gentler way than in mid-summer/Fire. This draws 2 towards the metal phase more than the fire. It is moving into darkness and there is associated with Earth and death (hence the black coloured star/heavenly energy) or an energy vibration denser and deeper than, for example, other natural forces like wind and light. What comes from Earth must go back to Earth, hence the last 30 days of each season associated with Earth shows that there is always death of the previous season involved.

Associations:

Earth has infinite associated features but is associated with the southwest and the late summer period in the five phases of humans. It is also associated with the ripening of the "fruits of the Earth", meaning all creatures and food being in abundance. This is primarily about the abundance and potential acceptance for life of the female. The female is dedicated and more inclined to serve and be receptive than to take the leadership position. Earth is constant and enduring, but unlike water which is fluid, it is firm and solid and of substance. There is the potential to ignite life within Earth, but it must absorb first.

In Human Life:

These features give rise to the motherly type who is quite simple, stubbornly set but quite self-sacrificial. This type is active behind the scenes but never the person in the lights and action. These people have depth, stability, and warmth (although yin, which is often associated with coolness), but are not overly heated. They are conservative but devotional by nature. They develop step-by-step in their own way, over time. This person is slowly adaptable due to their firm and solid quality. They find it very difficult to change quickly or do anything quickly. Change for them is imperceptible and with the seasons. They hate harshness or erratic characters. They are social and love all social activities, especially those involving food or nourishment of the physical body, but they prefer to eat than cook. They are not intellectual and find it difficult to understand complex mental activities. They can be very good at learning languages, as they love to communicate. They are unlikely to have a large vocabulary, as what they say will not be intellectually demonstrative/demanding. They are looked down upon for this, but they define themselves not by what they know but how they and others act towards them. Their methodology is often to absorb osmotically in order to create form, so the partner or situation that will provide them with most nourishment—so that they can create from this—is often what they go for. This gives them a needy or dependent disposition, unlike for example 1 or 3, who are more yang energy forms, which have direction and drive. To allow for female pregnancy/fertility (not masculine-yang/expressive/birthing energy; this relates more to thunder 3), the 2 energy must draw in before she can grow the physical form. These people like becoming more, being pregnant, and obtaining a bounty but are less likely to like the expression of energy in birth, which is more yang in expression and reduction/ catharsis. This is counter to the energy of 3, for example, which loves birth but hates the slow and gentle growth within. Also the 3 energy will cook but won't eat, where as the 2 likes to eat but not cook. This can draw the 2 towards socially recognised nourishment, such as money and food and into service of people who have a great deal of money or power. Their energy is of around 28 years old plus.

Their core mentally constructed belief, which can turn into pathology, is that they need to be nourished in order to nourish others. This can form a victim mind-identity/ mind-set or cause them to be overly needy, believing that they are useless, weak, or empty. These types are strongest at taste sensing and physical movement and kinaesthetic awareness, and so need to use the opposite sense of listening (water) and seeing with clarity (wood) to break through to the reality from a mental state of illusion. They can release turmoil within by talking socially, often without specific direction, but in a gentle and general way in order to find the right path for them, which is a general one anyway. They don't often focus and do one thing at a time but usually juggle several things in a stereotypical feminine way. They can only do this comfortably when being directed by a masculine force with strong direction himself , focusing her and activating her in the way he would like. "Life situations" causing the need for concern and possible confusion come from the exterior and are what this type responds to in their true nature.

The mother is opposed by the youngest son (8), which in human five phase terms is the yang form of herself, so this is not truly an energetic opposition. She is nourished by the middle daughter (9) and nourishes the youngest daughter (7).

3 Zhen – Thunder/Eldest Son

Trigram Line Symbology:

The most important line in this trigram is the yang line at the base of thunder. Thunder's energy "breaks through"; this is the key to its expression. It opens out and, like a shoot from the ground, pushes aside the cool yin on either side to come up. This is expressed clearly in this trigram as the energy rises through from base to top, the yang moving through. The other yin lines yield for the yang to emerge. This birthing of the yang is associated with this region of the Ba-gua. This can symbolize the breaking through the ground of new growth as well as the cracking of the frosts of winter and the rising heat through to the high places from the low, heating things up.

Symbolic Nature:

It is associated with the creativity and inspiration of explosive thunder—the injection of life force, inventive/creative eureka moments when the entire sky shudders. It is gone as fast as it came. It has the power to strike fear or to shock and is threatening and powerful, yet in the wake of its energy, life is brought forth, just as the surge of spring brings a new summer, and nourishment is fixed in the soil by lightning strikes from a thunderstorm, to give forth more life to the Earth from the sky. Thunder is of the Heavens; its light flashes awareness and clarity about situations and allows people to be awestruck by the beauty and majesty of Heaven. In comparison with 6 energy, which is also a creative force, thunder is the most pure form of creative energy, but it has no structural format. The 6 is structured and creates in the format of the physical, and the 3 creates in the format of the energetic. In the five phases, the colour associated with 3 is green. In the nine energies, the colour associated with 3 is turquoise. It has a spiritual association with the throat chakra and can express directly and effectively from this place—vocal and verbal and expressional energy. Symbolically, the thunder energy is a yang-masculine principle and is associated with the fertilization of life through the seed of the yang-masculine principle of heat and light.

The spring comes from the sky and sun; hence this energy is the harsh, clear light of the new dawn that cracks open the winter ice of the cold ground.

Associations:

It is associated with explosive energy, creativity, life-giving vibrancy, and aggressive passion. This energy breeds determination and focus but has little patience and persistence. Concentration and absorption can focus the energy, which is in bursts and is experimental and precocious. It is highly volatile and active, but this activity, like that of a cheetah, cannot be sustained for long without exhaustion.

In Human Life:

These people tend to be leaders and assertive (sometimes aggressive) commanders of enterprises, strategists, or highly intelligent planners. They have tremendous amounts of stored energy, which they can shoot out to hit their targets and goals. They are very goal orientated. They have little patience for people who restrict them and will use force rather than stand down if the situation rivals them and if the argument doesn't go on for too long. On initial meeting, unless they are the centre of attention, which they require, they can seem very quiet, but as soon as they are triggered, their energy will burst forth, which can be surprising for those who are unaware.

They are open and honest, being yang, but their open-forthrightness/ directness can be too much for other energetic phases who feel revealed from hiding by their nature; this can alienate them. However, they are life givers, creators, and the spring energy, so they are associated with birthing, children, life expression, cooking, music, dance, party atmospheres, humour, and such. Although the 2 energy is the forming of the body and the creator of form, the 3 energy is the expression of life and energy. Whereas the Earth and metal energy tend to be about moving downward, cooler and darker feeling, the opposite is true of 3; it is rising, buoyant, and positive. Their world revolves around themselves, but they are less likely to move into materialism and accumulation that the Earth and metal types who cannot see the Sky/Heaven energy. They see very little as important and relevant unless they are involved; it is what they can do or go out towards and give their expression to that is important, rather than a yin attitude of what can be received. They can be unruly especially when authority figures attempt to control them. They are inherently rebellious and independent, yet they can be good team members if they lead the team in action rather than trying to overtake the team.

They often bite off more than they can chew or outpace themselves; they are typical of the hare, while the tortoise can win through endurance. The fact that they are energy of the sky means that earthly conduct is very difficult for them; they want to fly before they can walk in earthly terms, but they actually can fly if others help them. If charged with compassion, they can be a tremendous instigator for Righteousness and bring people together to connect to each other, and by the same token when in pathology can be the opposite and drive people towards hatred and vengeance. Their energy is of a late 20-year-old.

Their core mentally constructed belief, which can turn into pathology, is that they have the right expression and right action; if they are not heard or listened to or understood, they feel hurt and disorientated. These types must use the eyes and sight/vision and expression very constantly . They need to be kinaesthetically and physically connected/ active (metal/earth), and use smell and taste sense (metal/earth), to break through to awareness from mental illusion.

They are opposed by the youngest daughter, which controls their expression and calms their activity. They nourish the eldest daughter (4) and are nourished by the youngest son (8), who in human five-phase energy (not heavenly 9), is part of Earth, so is not a true nourishment of this energy, whereas water is.

4 Sun – Wind/ Eldest Daughter
Trigram Line Symbology:

The sun trigram's most important aspect is the yin line at its base. The yin line here is soft and open, yielding, but the exterior 2 lines are yang, which means defence. The yang energies here are moving and expressive and expansive over a soft, cool core. The soft core in this region of the Ba-gua is associated with emptiness, similar to that of Li fire. There is a lack of substance at the root, which means there is defence but it is empty. If we look at Zhen, which is the opposite of this energy, we can see what this is about; the base is governed by a required yang aspect, hence is empty till it is moved, similar to 2 but not a dense material. The yang aspects therefore are not a hot energy expression but a cool one, in response to the stiller base. Hence this is wind; when wind moves, it cools, and this is the nature of wind. The phase is warmer due to the nature of the phase being in the southeast region, which is in opposition to the base of Qian for example, which although it is yang, is yang of the northwest direction and thus is cooler by nature. Hence cooling in the spring is not the same as the cooling of autumn, which is much cooler. Wind has no movement without being directed; therefore, it is dependent.

Symbolic Nature:

Wind brings about change; it rustles through the trees but is unseen and unheard on its own. Wind moves everything all over the place; it has little direction, regularity, or constancy. It can be as fierce as a hurricane or as gentle as a whispering breeze. When soft and kind, it can carry things aloft and send them to places unimaginable. Its universal change caresses all things and can easily pass through narrow gaps. In both the five phases and the nine energies, the colour associated with 4 is green. This is associated spiritually with the heart chakra, and there is a resonance with the expressional quality between form and spirit in the 4 energy. It tends to be between, as is the heart chakra energy. Although the energy is within the green of spring, it actually has the nature of metal and therefore the white colour; this is how it is expressed in this text, cloaked in green. Please consider this carefully, as this is not commonly understood within the expressions of the nine energies. One could express this as white within green.

Associations:

Cooling, gentle, easy and free in movement, communicating between things and spreading far and wide, the wind can join people and things together. It has variation in speed and strength and direction, making it elusive. Giving direction to the wind and a narrow focus to channel down forms confidence, but wind can change direction and interest at any moment. It can blow so hard that it blows itself out or so soft that it loses its core. Often hidden is a desire to acquire actualization, to be something, out of a given situation, but this is easily scattered by its energy. They rarely know what they want till they feel very strongly a flow of a direction or are directed to it.

The nature of wind is an energy that associates with autumn, but this is autumn within spring. It is a coolant for the heat of spring—almost, one could say, an antidote to the spring energy. The nature of the wind is cooling and calming to an overheated energy, yet it is not dense. It is cooling, but it comes from above, not from below (yin). It clears thing and separates; it cleans away and scatters. It is fresh and sour in its effect. It follows no direction but has the singular direction of taking physical structures apart rather than putting them together. In its wake is good communication between places, but it aims often to render things down to their bare essentials, simplifying physical structures (as well as mental structures). It does not aid fusing or creativity but cools and calms. When aggressed, it can be "cutting". It seems expressive, but the coolness of it is actually not opening, but closing down and defensive. It is not expressive of warmth easily. It creates form and structure, shaping and calming things. These are all aspects of autumn, so we see the wind energy as a part of autumn within spring, yin within yang, metal within wood.

In the spring, the wind cools down energetic yang to contain its structure more, so it doesn't burn up so easily, in relation to physical structure. It takes what is high and makes it low; it takes down things placed at heights and lets them come back down to earth; it cools down. This is the process of metal going into water, or autumn going into winter. Not only is structure blown down, it is also dispersed to form a non-structural entity ... water. In this sense, the wind is critical for structural formations; it loosens them and calms creative structure making (wood energy). However, being of the springtime, the wind mixes with the warmth of the season and creates balance by preventing yang from becoming too expansive. Wind in the autumn would be considered excessive autumn energy.

Normally, the spring energy percolates downwards from the sky, but this is autumn within spring, so it is a cooling energy that percolates from below. It is like the coolness rising from below but expanded enough to be ethereal.

In Human Life:

These people are generally tender-touching and affectionate, perhaps the most caressing of the phases and gentle in nature. They are good communicators, can do several things at once, and are appreciated by others for their calm but controlling power of communication and smoothing diplomacy. They tend to cool situations down through aversion tactics and can stand the intense heat of anger, attempting to blow it away and cool it down. However, they can often be fanning a fire, and they are very defensive. These people often defend quickly but imperceptibly, by creating an image of how they want to be seen, to gain acceptance, which is an expression of what they are, almost chameleon like. What they are is often difficult to pin down without someone else's input. This type is therefore often accused of not being genuine because they attempt to be accepted in order to be directed, which seems illusive and deceitful to those who feel that one must direct one's own life; the wind is often directionless until directed or focused on that which will connect it to others. The elusiveness is generally harmless, although they can find themselves in precarious situations when they find they cannot be what is asked of them, because of how they have morphed themselves. They can be stubborn in their requirement to be free of another energetic structure, and can also defend or blow hard to prevent structure containing them, although at the same time they long to be something more concrete than what they are which seems invisible, they can strive for this because it's what others seem to want of them. However, the invisibility is a very structural expression not just a morphological one like a net or field of energy rather than a material form.

It is important to see this energy in relation to its polar opposite, the 6 energy (Qian/Heaven). Whereas 4 is internally cool and outwardly spontaneous, 6 energy is opposite. Though 4 looks and acts without form, it is internally structured, and the mechanism that looks or seems at first to be unclear or highly erratic has an invisible mechanical engine beneath the surface—a mental clarity that it knows but which may not be clear to others from its expression. These people are very often teachers and attempt to direct in a soft way. They end up dispersing and loosening structuralism and letting go of things. They need some direction and therefore need the opposite to balance them, which is external structure and warmth within. They can be too cool and mechanical and yet too loose, this is their complex energy.

They are impulsive and tend to be very giving and emotional, but this too can be cloaked and diffusive if they want the other person to back away or avoid directness. They win the confidence of others, but often do not succeed on their own merit. Like the wind blowing the sails of a good ship to shore, so the eldest daughter guides her focus to shore. She changes her mind easily and often, and her direction can change too, but enduring she can be, if she paces herself. Sometimes these people can be very indecisive, and this can be confusing and disconcerting for others, making their relationships sometimes difficult. They are often seen as too easy by the more yang expressive energies. Deep-seated wants may be at this person's centre but may be very carefully hidden. It is true the wind can lie, in the sense that it may look from the exterior like it is going in one direction but will be going in another. This may cause turbulence with other people. Often they just don't know themselves and await direction from a higher or influential source to focus again, if only for a moment. They are adaptable but can be cold and mechanically blocked by defensiveness. Their energy is of an early 20-year-old.

The wind is cooling, so she is not as expressive as thunder's pure wood energy but is actually cooling to this. She can, if not directed otherwise, carry a cold breeze. The cloaking/hidden nature of this energy is very difficult for others to deal with. The fact that it can uncover, push apart, get into places, and separate physical entities from one another can make them highly popular and then highly unpopular. They often have enormous difficulty keeping relationships due to this. They can seem colder or secretive to the warmer energies like 3, 6, 9, and others, but are often liked by these people as they give new insight and calmer ways of being. They can become very rigid in their coolness, and if they become cold, they can isolate themselves in their own ideas. They are in direct opposition to the 6 energy; the 6 energy actually has the very opposite approach, but its environment is autumn, whereas with the sun, it is spring warmth.

These people often want to find forms of expression that are both in motion and yet render things down to essentials or clarity, making them expressive in terms of music of a minor key and quality, wistful music, or as writers, again usually where there will be a sad ending! Generally these people move on feeling and can have emotional swings, but there is a cool quality to the feeling, spring-sorrow so to speak. They can be engineers and attempt to design and streamline or be the idea engineers behind mechanical operations, usually travel-transport orientated. Generally, they are less visual, so it is not aescetics they are interested in, but motion or communication.

Their core mentally constructed belief, which can turn into pathology, is that they have the right way, direction, or movement. If they feel they are incorrect or unclear, they will feel lost and unable to move. Known for moving into extreme defensiveness with seeming arrogance and a "looking down upon" exclusivity, a stubbornness and an idiosyncratic evasivness. The mind-identity of these types can take them into dream-like internal worlds, which need to be defended at all costs because they have formed a "self" from them that must be obliterated from the exterior. These types have use of a visually absorbing (i.e., not expressional, observational with detachment) or listening ability. To break their mental illusion through to awareness of reality, they must use the opposite sense of visual opening/ engaging (wood) and smelling (metal) the environment as well as expressing vocally (fire). They must look and see experience and engage, not passively but becoming one with the exterior visual world. This type is associated with a superficial layer of aggression. However, internally the energy is metal, and so this resonates with that which causes grief from the exterior. In a situation of grief, this type is in its element, so tears and crying are associated with this type. There will be a mix of these expressions.

The wind is the eldest daughter; she is opposed to but essentially controls the father 6 energy.

The 6 energy draws her to him as she is yin, but she is metal and he is wood in the five phases of humans. Note that metal and wood opposites have two aspects, 7 and 3 as well as 6 and 4. This is also true of fire and water (which have only one aspect each), and in Earth 8 and 2, although within Earth are opposite aspects, Earth yang and yin respectively. Hence opposite balance is formed. The eldest daughter is nourished by the eldest son (3) and nourishes the middle daughter (9).

5 Taiji — Centre (Table)
Trigram Line Symbology:

There is no trigram for this energy, as it incorporates all the other trigrams within it. However, it has a nature which, in human terms, is similar to 8 energy but more powerful and deeper and spherical/ vortex rather than conical.

Symbolic Nature:

This is the primordial power of the yang that makes up the universe. When yin and yang find their partnership in the universe, then the greatest creative force occurs. This can be directed towards destruction (accumulation or like a black hole of gravity) or creation (expansive force of spinning outwards), but in itself is a whole lot of energy! It is like a whirlpool of yin and yang, which are together in this part of the universe, so yin and yang react in the same area; formation from this creates tremendous power, just like male and female create children. In both the five phases and the nine energies, the colour associated with five is yellow/orange. This relates to the solar plexus chakra spiritually and is the point of balance between the upper and lower portions of things and the expression of balance centrality and the golden mean. The 5 is the Tai chi therefore associated/ represents the origin, or central point of the universal change or spin.

Associations:

This is associated with all things that have the strength of being female and male in the same power. There is paradox in the fact that this aspect has a strong gentle characteristic as well as a strongly assertive/ aggressive one. It is usually a mixture of the two aspects. It has great controlling power over situations and great talent in its ability to be central and stable, yet adaptable and strong. This power must be harnessed constructively or it will end in being destructive to itself and even more so to others. Dictatorial and invulnerable energy emits from such a power, or it is so dense that it draws exterior energy inwards.

This is the energy of the late summer because it is in relation to the 2-Field energy, but it is actually the combined temperature of the end of each season, hence warmth or neutrality, neither too hot nor too cold. The 2 energy is archetypical Earth, or where the Earth energy is expressed to the surface, so to speak, in the seasonal calendar. The 5 energy is often seen as close in relation to 8, the yang Earth. However, there are obvious differences. The roundness of Earth is contrary to the conical mountain, hence the three Earth energies together—2, 8, and 5—give the plains, mountains, and high ground as well as circular roundness to the contours and nature of Earth and its gravity.

Five is a very important number in association with humans; the five yin or deep organs are the basis of the human perception of the universe and its relation to it. The five phases are key to the understanding of humans. Five, being the core number, shows not only its significance as a universal movement of energy—a vortex from which all energy springs universally—it is also associated with the central yang moving vortex of all phenomena, Earth and humans as well.

In Human Life:

This relates to that person who in a sense has both aspects of femininity and masculinity. They may move from one area to another, not able to focus their energy in one specific direction because there is so much to do and they can do whatever they want to, being the centre. While they don't have the capacity of the ultimate directivity and light of Zhen, for example, or the fathomless depth of Kan, they have a good balance of these aspects, and this means they don't lose qi from any particular area, as they are spherical by nature. They are deeply practical and have genuine ability to see things as they are unless the attempt analysis, which can lead to mental arrogance and strongly held mind-identity belief systems. They can be adaptable to others, but only on their terms and when they want to. They can get bored very easily and need to be constantly doing, as they have a lot of power, however although they will constantly look for new and different things they often do so in the very same way they did before, so they look like they are changing (and talk about it, alot!) but actually they are quite conservative. This type of person is still a human, so they can be seen as yin and yang together, but they will always swing one way or the other. They are not the pure spherical centre of everything, the whole universe, although they may think that they are. These people are naturally self-centric, and due to their abilities, they may be admired by many people. They need the right people around them and are constantly needing to be heard by other people. They will be able to achieve well if they focus on one area long enough. They have difficulty doing one thing only, however. They are best as controllers of flourishing businesses that change and adapt to circumstances, land owners or agriculturalists, which must do the same. They are controllers, stable and can be rigid but they keep moving and appear to be changing but change little. They are bold, and others are often disconcerted by their invulnerability. They don't really need anyone as much as the other powers do, as they are almost self-sufficient. They are not perfectly central, and the fact that they relate to the Earth centre in humans mean they may control through gravitational command. They need the less dense energies to balance them, like thunder and wind types, so they are not as all-powerful as they seem. Their nature is difficult to handle and requires patience. They believe they are right, whatever the situation, but can adapt if they find otherwise. They have a fatalistic approach to life, the Tai chi can withstand most outcomes and the time is now; this is understood by the central power as their sides are balanced. Their energy is of a 10-year-old boy, similar to but a bit older than the 8 mountain energy. The 5 pertains always to the yang and not the yin. Although there is yin expression, the yang aspect of the 5 simply draws from this yin and produces immense power. Because the nature of the human is towards the yang, people with this energy male or female body pertain to the yang at core.

Their core mentally constructed belief, which can turn into pathology, is that they are in control of physical phenomena and/or people—everything that has form or could create form. If they lose control, they feel they are failures. These types know the use of physical movement and kinaesthetic awareness. To break through to the reality from a mental state of illusion, they need to listen (water) to the world more and they need to see and look clearly (wood). They also need to cathartically express a lot verbally; they often socially connect and verbalize. If they do not do this, they will be held in and cause themselves great pain. This is in fact how they express their energy, so blockage of this expression can be disastrous. There is no specific direction to this verbalization; it is simply to be one with other people. The actual thoughts involved are often circular and "point-less", but the point is to connect, they do also become great analysts again on a "point-less" and circular fashion , interested in information and history of people and things, which they try to break down digestively, but seemingly for little reason. In truth their forte is not in the intellectual but in the practical and mechanical which is where their ability in construction and manual breaking down of things is fantastic, acting like a stomach, digesting and breaking down into parts and them moving on. The 5 and 8 energy are quite coarse or rougher in their expression and temperament. There is not a refinement but a down-to-earth quality also apparent in the 2 energy as simplicity of mind. They are not great thinkers or highly refined peoples but they are honest to what they feel and known for it. They appreciate this quality in others and don't like what they might call a smart-mouth and no action. They can loose sight often of why they are doing something and fail to see the wood for the trees. They like to have a broad perspective of things (from their viewpoint which is usually quite narrowly focused to material or individualised areas, and themselves) and often refer to past and the history that can be read from the Earth itself. They, like the 2, 8 , 6 and 7 tend to be material associated (earth and metal energies) they tend to have transactional type relationship where worth and provision of something is clearly paid for and earned, and they want back something for what they have given outwards. These people build Social structure, political and governmental ideologies if mind-identity gains control. Communication with the sky energies of 3 and 4 particularly is difficult.

These people are strong women or strong men. Seen as the centre, they are really the table around which all the others sit. If they change, others must adapt; they are the ruler of all the family, the controlling power. Though they may wish to take control of the water energy, they would damage themselves, and this energy is too strong, so they are best to give their huge supply to 7 energy, which would be nourished and appreciate the central power and can tame its high energy state.

6 Qian – Heaven/Father

Trigram Line Symbology:

This is the second most complex of the energies with sun. The Qian seems simple, as there are three yang lines so one would expect pure yang energy. However, this yang is in the northwest region of the Ba-gua, which means cooling and autumn in earthly expression. Although there are no yin lines, this is in a very yin region, so the yang lines here are very different than those in the southeast region; this yang has density. Also three yang lines means that the energy comes very much to the surface. Three yang lines also means lack of yin; however, as we know, the yin is provided by the season of autumn, and this means someone who has a lot of radiant energy but a cooled surface. The platform of autumn that it inhabits means it is a cooler phase that the energy it is at core. The yang of this energy is like a fire in autumn. It buffers the cold, whereas the 4 energy is a cool breeze in the spring and so buffers the heat. The 6 energy is yang within yin, and this adds direction or growth to yin forms, essentially turning energetic into physical formation. This is like the drawing of the Heavens-sky into the earth, forming bright creations that have warmth at the surface or are bright at the surface. Gold, gems, and the like are all associated here. Here, 6 is wood within metal, or the growth of the metal energy. The warmth within the coldness, spring within autumn, yang within yin, is not of course as warm as spring itself, but relative to the environment, it is warmer.

Symbolic Nature:

The father of the universe is heavenly. This is related to pure yang, so it has relation to the beginning of all of creation, the yang of Heaven. It is the principle giver of life from above or around us. In both the five phases and the nine energies, the 6 energy is associated with the colour white. The white quality is resonant with the 4 energy, and 4 and 6 energies associate with the heart chakra region as an axis of opposite qualities. The 6 energy balances 4, and 4 balances 6. However, this again has issues, as the 6 energy is actually green within the white, or yang of spring within autumn. This energy is warm and buoyant, not cooling, so be careful how you view this energy.

Associations:

Heaven is physically creative and charismatic, associated with clarity, constancy, perfection, brightness, awesome power, Omni-presence, Omni-potential, pride, completeness, independence, activity, and dignity. There is nothing bigger than it.

This energy is the spring within the autumn period, so it originates in the sky-Heaven but is inside the Earth like a seed of a plant pushing upwards. This energy is the sun/Heaven within the Earth (i.e., within the macrocosm is within the microcosm, and this is represented in 6).

In Human Life:

They have a dignified, noble attitude, are highly organised, and are courteous. They are calculating and often can feel uncomfortable in a social situation because they feel distant from it, rather than involved in it. However, they often put on a very fine show and can be charismatic and expressive and like to be admired by others. They can be intense and direct and forget other people's wants and needs. Most people will feel they are honest and emotionally expressive, but when you get to know them for longer, you may find that they often play their cards close to the chest in order to have a more magnanimous control. They are powerful in their understanding and ability to sum up situations, and they define and clearly concentrate away from disorder and clutter in their lives and their thoughts. They can persevere in difficult times and are solid and honourable during times of the passing of life to death, and seeing these processes through is their strength; they are strong in times of grief and will feel deeply. They are masters of structure and business strategy and have the logical coolness to create a kingdom. Creating metallic, physical structures within the landscape, or property management, is a key area for them.

Whereas sun energy will seem warm to begin with and then more enclosing and shrouded internally, this energy is opposite, and such people look as if they have their hearts on their sleeves and can come across as cool to begin, gradually warming but strong/solid. The sun types are hiding a venerable interior, which is usually cool and soft, whereas the Qian types have a harder interior, which can be quite severe but is warm. The yang within the autumn (6) and yin within the spring (4) balance each other's temperature out in each season; hence coolness in spring creates warmth, not heat, and heating in the autumn creates warmth, not cold. This shows how these two energies help to create balance. However, it can be expressed that within the five phases, 6 would belong to wood rather than metal, and 4 would belong to metal rather than wood. The 6 in the spring would add to the 3 of the spring and create double yang in the spring, moving the energy and the physical structures to creation. This would be too strong, hence 6 is a creative yang buffer for the autumn.

These people are demonstrative and may need to be in the power position more than any other energy. They do best here as judicious Kings and lord-like characters. They can be too wealth-oriented, leading them to trouble. Opposite of the 3 and 4 energies, the Heaven 6 energy likes the value of things to be defined in physical terms and is highly concerned with this. They can be generous but only when they themselves have their needs met and usually only to a courteous, worthy (in their eyes) peer who has a slightly softer touch. They are proud and dignified but only within the rules that they themselves have created—not necessarily those that are universally approved. They are often kings of their own kingdoms, which they attempt to expand ultimately. Related to the metal energy in the five phases of humans, Heaven is the yang within the Earth, the Heaven within the Earth that shines like bright gold. However, it only radiates its glory in the presence of the light of the sky, the fire of the sky; hence it is Earth-tied yang anchored by yin of Earth. Although 6 is yang, it occupies the northwest corner and as such is cooled/connected to the earth. Hence they are dominant/yang on/within the Earth/yin base, which confines its kingdom. Their energy is of a 32-plus-year-old man. Typically this energy is associated with an archetypical older, masculine energy.

One needs to look at this energy in terms of its opposite-the 4 energy. The 6 energy is the opposite of 4; internally it is spontaneous and creative, and externally it is structural. This can create the opposite of 4 energy, which tends to be dispersive or overly loose energetically; 6 tends to imprison itself, the exterior becoming ridged and too tight for the internal expression to come out. Exteriorly, the energy looks to be structurally elegant and poised, but internally there could be a storm brewing, and so there is hidden spontaneity, which can be frightening because it comes from no-where, or seems to. This spontaneity is the light of joyous expression and passion as well as assertive energy; it must come forth or the 6 energy will suffer greatly. If it doesn't get support during early childhood in establishing its internal yang expression characteristics, the 6 energy may struggle later on with a very powerful, victimized egoic nature that crushes and collapses them internally. This may propel them to look for activities that open out the body. But without inner energy coming forth due to just allowance of emotional unravelling and dissolving, they can become very austere and cool, holding resentment and aggression within. The 4 energy does the opposite, which means that even if they are not brought up very well, they have the wear with all to take direction when its given or noticed, as they have a yin base which yields to situations more easily than 6. The 4 requires the direction and passion of the 6, just as the 6 requires the loosening and calming effect of the 4. The 4 requires the inspiration and vision of the 6 to drive it or warm/ conduct its metal.

Their core mentally constructed belief, which can turn into pathology, is that they are all-powerful. If they cannot become this in a grandiose way, they feel like a failure, especially in the physical realm. These types have easy access to senses of vision-sight or smell. To break through to the reality from a mental state of illusion, they need to use the opposite senses of kinaesthetically feeling (metal/earth) and taste (earth). This will help them the most. Similarly to 4, this energy has easy emotional attachments to anger and an understanding of a situation of grief affecting them from the exterior. This type has the greatest power of aggression, as it is dense and has internally held yang, and so seems grief struck but can often be aggressed. They often need to express assertive direction instead of holding aggression or their grief, to be clear of mind-identity. The expression of grief is more important for the 4 energy, 6 is associated with grief often from the exterior, but assertion/ aggression for the 6 is derived from the interior.

The father (6) is opposed by the eldest daughter (4), who breaks up his rising forcefulness and absoluteness with wind that scatters. The youngest daughter nourishes him (7), and he nourishes the middle son (1).

7 Dui – Lake/Youngest Daughter

Trigram Line Symbology:

The key line here is the yin line at the top of this trigram. As with the Qian, the base is a solid form of yang. It is yang/bright, but the region itself, which is the west, and so cooling, consolidates it. Just as Zhen rises up through the two yin lines like rising heat, so the top cooling line here cools downwards what was the heat of late summer yang as we move into winter. It is an expression that shows cooling, and the softening principle is dominant. The Qian has the three yang lines incrementally opening to the warmer yang in the upper line. However, the Dui trigram adds the soft but dense energy of the yin above. This cooling of yang and density of yin makes for a cool and dense and soft energy like that of a lake. This is a pure form of metal energy.

Symbolic Nature:

It is like the phenomenon of a lake or clear pool that holds water that everything else can come to for nourishment. This aspect shows the beauty and serenity of the lake and the richness it gives from still and reflective water. In the nine heavenly energies, 7 is associated with the colour red. The 7 red quality is associated with the root chakra, which has much to do with basal instinctive qualities and death, being the lowest point in the body. There is often a hidden sexual quality to this energy. The redness balances the blue/turquoise of the 3 energy in the upper body associated with the throat/base of skull. The balance between these opposite poles of sacral region and cranial region is well known. Pure wood balances pure metal here in the spiritual qualities (see Appendix 2). However, in the five phases of humans, the colour is white/silver. This is the most yin of all the energies.

The lake energy is pure autumn energy and is the cooling of the lake waters and condensing of the refinement of essences into a reservoir of storage, letting go of unnecessary aspects and forming structural integrity. The energy of coolness of autumn percolates up from the Earth to the sky, making it cold.

Associations:

The 7 energy relates to the autumn. Though this is a season of downward movement and the ending of things, the 7 energy is associated with joyful expression of Nature. This is the elegance and pose and beauty of the autumn before the cruelty of the winter energy comes in. The energy is of the youngest child, and so she is the apple of her father's eye, meaning she brings joy and is pleasing to the eye, so to speak. Her surface is bright and shiny like the surface of the lake. However, it hides often deeply felt feelings, which may not be seen by onlookers who are only involved with the surface. She can hide herself in this way. It is a very female yin approach. Within these people is the deep yin that is yielding and cooling, and still she has natural pose and the ability to be still within. She is also young, and this means that she shimmers, changing form. Nervous, like ripples on a lake, the youngest daughter, though beautiful and caring, can be scared easily. One would not expect such tendencies to be associated with autumn, which one would expect to be more sedate and solemn. However, this is true of this type also, but the surface of the energy is actually bright white and shining or shimmering, which is what gives this joyous concept. Metal is always based on what light is shining onto its surface. Without the light/fire, the surface is dense, dark, and cold. Note the difference of a lake in the sun and a lake at night. This energy has yin- feminine coolness, calmness, and depth, yet at the same time, her surface shimmers and gives off light. In the playground at school, we can see that the little-girl energy can be very expressive and joyous, but seems too to be very cold. This is the seeming duality of this energy.

In Human Life:

Their personalities cause people to come to them for advice and caring. They are deeply cool and calm, yet joyful and bright superficially. They are self-interested, analytical, and critical. They are nervous and can be excellent friends in times of need. They are generally flexible, and change does not shock them. They are often drawn to be gentle, beautiful, and serene. They are quick-witted and clear and grounded, so they are easily able to understand what others are thinking. Interested in design and style, they are not as inventive as they are good at refining, or ordering and re-forming an originally created structure to give it poise and elegance. The relation of this energy to structure and design is profound. Objects and shape stimulate them more than anything. It is the physicality that they understand and admire in life and in death, hence sculpture and non-moving art and designs are their love and passion, in their creation or even their allocation and placement. They are excellent at interior design and understanding form, (like architects) and are best at creating form, which is simple to the eye and quite cooling in expression.

They are not independent and find it difficult to set a firm footing in life and need the support of others to do this. The opposite is the 3 energy in the heavenly expression, but in the five phases of humans in relation to the nine energies, she has no oppositional male counterpart, as the fire is female. The reason is that the youngest-daughter energies (as with the youngest son) have no opposition; they are too young energetically to be in a relationship of opposition. We will look at this more later.

Their core mentally constructed belief, which can turn into pathology, is that they need to be responsible adults, wise and reflective. If they to feel dismissed or childish, they feel ashamed and sad. These types have physical/kinaesthetic awareness. To break through to the reality from a mental state of illusion, they need to focus on the visual sense (wood) and expression from the heart (fire) and creatively express their energy. This will balance the passivity and open Oneness. Of all of the energetic types, these have the deepest ability to be still within themselves and connect to Unity if they want to. The mind-identity based emotion associated with metal is grief from things or people lost exteriorly to them. This type responds well in tears; if they do not respond in this way, there is often a mental blockage involved. The association with grief and the exterior will be explored more in Sections about aaetiology and pathology.

The youngest daughter is perhaps seven years old. She forms a relationship with the 8 mountain energy or 5 central energy as they support and nourish her. She gives nourishment to the father (6).

8 Gen — Mountain/Youngest Son

Trigram Line Symbology:

The clearly important line in this trigram is the upper top line. The top line is hard and solid, and the inner lines are soft. Due to the nature of being northeast, it has density and yin—as opposed to the yang region of the Ba-gua where the yin would be more Void-soft (warmed up) rather than more dense-soft. This region clearly indicates softness interiorly but hardness exteriorly. It is similar to 4, but 4 has more defensiveness although the energy is less dense than mountain. Also, the Void-ness of wind is, in the denser-yin, more of an earthly substance. The yang caps the yin like a shell or facade to the yin underneath it. The 2-field is archetypical Earth energy whereas mountain and 5 central Earth energy are secondary aspects of Earth.

Symbolic Nature:

The mountain is strong and firm, majestic. The accumulated focus of the Earth, the mountain stays in one place, immovable and penetrating. Yang and focused, heavy and physically imposing, the mountain can't be shifted unless it wants to. In the nine heavenly energies, the 8 is associated with the colour white. The white of 8 resonates with the blackness level of 2 energy, and they form the original yin/yang quality of the origin of all things below the root sacral energy in the spiritual qualities (see Appendix 2). The

8 expresses primal masculine energy or the birth of the masculine from the Origin. In the five phases, mountain energy is yellow/orange.

This energy is associated with the end of the winter period and the warmth within the Earth coming to the surface. It is also the warmth of the interior and the structural defence and hardness of the exterior. This relates and is similar to the 5 energy and is associated with the late summer period within the five phases, as it is associated with Earth. However, in the nine energies, these strong and stable structures define the northeast region. The percolating upwards of the warmth of the Earth interiorly is similar to the 6 energy, but the 6 is more refined, being of spring within the Earth, which is an life-like formation, not rock.

Associations:

It is associated with high mindedness/ aloofness; it can see the overview of a situation. It is cool at the top with deep roots to the earth. It is stubborn, strong, and lonely. People come to the mountain and it is attached to the Mother Earth (2), so he is prominent on the horizon and energy accumulates in the surrounding region, like an acupuncture needle of the Earth. People cultivating on its borders find the soil rich and fertile. It is very slow to anger and very slow to move, but accumulated tension will force it to blow its top like a volcano, which can wreck havoc, and so it can maintain power.

In Human Life:

Obstinate on the outside but gentle on the inside, their imposing nature leads to childish actions, which can be quite self-centred and seemingly aggressive or arrogant. Their minds are like rock, hard and penetrating. They focus on a point and so push people out of the way to reach that target, like a landslide. The seemingly "greedy" side of their nature is about accumulation. They accumulate to become greater and stronger and reach higher up, but this can lead to problems for others, who can see them goal focused and self-centred. When they are softened, they are calmer and less like this. When in mind-identified state jealousy and use of force become very prevalent as does self-focusedness.

They take change as it comes, being able to weather any storm. They build up big businesses slowly and become successful but need a lot of support and encouragement. Social interaction is what they love most, and they always want to be the head above the rest. They are kings of one-upmanship, but also of solidarity, loyalty, and honesty with a brave and courageous front. As with all Earth energies, they are often involved in traditions, historical interests, and conservative approaches to most endeavours, analytical but with more of a direction then the 5 energy but this direction is usually to uphold and express their own ideas and individuality.

Their core mentally constructed belief, which can turn into pathology, is that they are the social hub and tough. If they feel they are soft or not liked, they will feel isolated. These types know taste and like flavour, they also understand physical movement and have kinaesthetic awareness. To break through to the reality from a mental state of illusion, again as with 5 and 2, listening to the sounds of the world (water) is key for them and also by looking clearly and with focus (wood). Similar to the 5 energy, this type requires expression of thoughts and often needs to be listened to and heard above everyone else. This then allows them to sort through possible ways to understand confusing exterior situations and connect to the exterior. Without this, they are stuck and blocked mentally, and internal damage occurs like a volcano holding in its smoke.

The youngest son would like to be in an opposition relationship with a water energy (in relation to the five phase of humans), but they are yang/masculine in relation to his power. In the heavenly nine energies, he opposes the Mother Earth/2; he is almost the other aspect of her being, Earth moving towards the sky rather than making up the fields and plains. However, they are not energetically opposite in human terms and are actually a brother/sister relation, being both of Earth. Hence 8 must draw his attention to the 7 energy. The 8 energy has ample supply to tonify 7, and he supports her. Also, he is the same age as her; he is about eight years energetically, and so their energy combines well.

9 Li — Fire/Middle Daughter

Trigram Line Symbology:

The key line to this trigram is the open middle line. This, like the centre of a flame, is not yin-density or yin-coolness, but in the southern region, the yin quality is total emptiness or void, relating back to the Void that original yang was born from (Wu). The openness at the centre of the flame or fire is the yin quality of it. It radiates all around but has no centre from which to direct its energy. Opposite to water, the yang of water fills this void, and the exterior yin cools the yang of fire's exterior. This indicates heat and light, which are yang. However, the yin core means that it associates with yin. This type is dependent.

Symbolic Nature:

Fire brings brightness, heat, passion, and colour. It is bright in every respect and flamboyant to behold. It is radiant. It burns and crackles, reminding us of its power to emolliate and exhaust itself in a second of passion, to return to emptiness. In the nine energies, 9 is associated with the colour purple. The purple energy is associated with the crown chakra and is balanced by the white quality of the 1 water energy at the head. It is associated with bright, clear, and fast thinking, and its expression is beyond words, usually in the expression of mind-to-mind direct communication. In the five phases of humans, 9 is coloured red.

Associations:

Brightness, fame, and success are all related to fire. The exterior and the superficial aspect of things are important. As fire touches the surfaces, its heat penetrates to the core whatever is exterior and on the surface. Warming coldness and impatient, contemptuous and drawing the eye, it never burns in the same place but moves on as fast as its flames reach. It can move itself only through reproducing its own energy; a flame is still at the centre, so it is a yin entity. The flame itself is the softest of all phenomena, made of pure spirit.

The summer is of the sun, and so the sun percolates downwards to warm the whole environment;

the energy is the expression of the sun.

In Human Life:

These people can lead but do best in a secondary leadership position, as they are yin. They are a good "standard bearers" in an organisation involved in marketing or the expression of a campaign. As fire draws in fuel from the surroundings, so the information fire draws in and then emolliates in its expression is not its own. Fire often seems to take an idea or another person's creation and then expand and market it, so to speak, for the masses, rather than creating the original idea. They are therefore dependent on the "fuel" of other people's creativity and can attempt to take over a creative process in order to expand the self-image of themselves. This shows the nature of fire as an expansive and both yin-dependent, yet yang-actively perusing expansion. They have the potential to move to a climactic burning rage or to charm with burning radiance. They have little depth to their words, but their words are expressive and direct and usually fun and bring things to a party atmosphere given the chance. They can be loyal but will move on to the next partner unless they are satisfied. They can forget easily and can lose friends because they are like this and can be considered fickle. They can lie but do not do it without giving themselves away, and they may have devoted followers but can change their mind about a movement as easily as a light going out. Unless fuelled by an encouraging project that will propel them to central attention, they will move on easily. They must be the best, brightest, sharpest, and paid most attention to or they are lost, as they lose their superiority, spark, hilarious wit, and infectious laughter. Capable of generating attention in volumes, they are the most passionate phase of all. They are flexible and adaptable too, as long as conditions are favourable to them. The energy is of an early teenager. Although they attempt to teach because of their brilliant mind and big charismatic expression, they are often not so good at this because their thoughts are too fast to express and frustration sets in if the student can't keep up. Generally, they are best as expressive artists and look on as a jewel in a crown rather than a functional energy. Those of 3 and 4 energy do better at teaching because the expression is decipherable. The 9's appearance is different to its core. It seems bright and expansive but internally there is Stillness. This is true of the opposite of fire, 1 which appears cold and yin in this way, calm but actually internally there is great wilful current.

Their core mentally constructed belief, which can turn into pathology, is that they are proud, grandiose and splendid, and must constantly keep this up—being the light of existence. If they do not, they feel worthless. This type knows expressive means of dance or creative expression vocally, but to become aware of reality and break the mental illusion, they must be still and allow themselves to feel their body as a still form kinaesthetically (metal), thus allowing mind-identity to relax, also they must listen to the Stillness (water). As a natural expression, laughter is a feature of this type of person's energy; if they are not laughing a lot of the time, there can be a problem and this should naturally come from within. These types are naturally happy, so a 9 energy that is not expressing this naturally can often be internally damaging themselves and overheating. They can have very activated minds and can be very clear mentally, but usually, due to the nature of the world we live in, can become overtly mental and lose touch greatly with what is occurring around them, because of the ease with which their upper body and heart can gather energy. This can make for high emotionally-mental states of anxiety or tension.

The middle daughter is opposed by the middle son (1), the classical fire versus water reactionary relationship. Water cools fire, and fire moves up and heats water, vaporizing it. The flux of fire and water is the perfect reflection of poles of the internal body, and of yin and yang itself. The emptiness of Li is opposed by the fullness of the fire within Kan (this is important because Fire can seem quite vascular and empty which tends to feel cooler at times and water hotter), and the cooling exteriorly cools the fire exteriorly. She draws water, and he comes towards her. She is nourished by wind, as blowing winds of air fuels/encourages a firestorm, and she gives to the mother, burning the structures on the Earth, to make more Earth for the mother to accept and form.

v) Exploring Yang-Male and Yin-Female in the nine energies

Yang energies are 1, 3, 5, 6, and 8. Yin energies are 2, 4, 7, and 9.

It is very important to clarify the connection of male to yang and female to yin. It is not accurate to see "stereotypical" masculine and feminine in the idea of male and female. Here we are exploring yinyang male and female, not cultural observations or what it means to be a man or a woman. These differ in various cultures and in different parts of the world. When we explore male and female energies, we are looking at them from the perspective of yinyang, which is a far broader and more complete view. Hence masculinity in this sense means light giving, moving and energetically expansive, creating principle. The cooler, calming, receptive and darker, softer and denser/accumulative is the female energetic. These may not ideally suit the image one has of a male or female within society or a particular setting, but it provides us with a broader understanding to explore natural phenomena without judgment.

If we look at the above, the coolest darkest form is Kan, but it is male because of its hardness and momentum, whereas the softest and most energetic is female Li, which is female because of its receptiveness, softness, and dependency. This shows how there is mix within the energies. It is difficult to arrange a yin to yang spectrum of the nine energies, but this gives an arbitrary scale that can generally be applied:

Most Yang—Male

3

6

5

8

1

9

4

2

7

Most Yin—Female

Remember, this is not about stereotypes but archetypes, or principles of yinyang. Also, as we will look at further, the 9 energies distribute evenly within male and female Jing-body forms. This is vitally important. Hence this offers four overall possibilities; yin body-yin spirit, yang body-yang spirit, yin body-yang spirit and yang body-yin spirit. These four represent the whole of natural expression and we will come back to this frequently.

vi) The Age Associations of the nine energies

Each of the nine energies has age associations to a time of life. These are cycles of seven in women and eight in men. (This is from Su Wen chapter 1.) The ages and actual figures for the nine energy numbers cross over at the 8 and 7 energies of the youngest siblings. What this explains is that different ages of life have different energies associated with them, so when identifying the prime energy of a person's birth, one also needs to be aware that as time goes on the expression of this energy may change, due to age, as it mixes with the energy of that stage of growth:

(table 2.5)

9-energy F/M	7/8 (5)	9/1	4/3	2/6	(2/6)
Age in years	7/8	14/16	21/24	28/32	(35/40)
Five-phase stage of life	Wood stage	Fire stage	Earth stage	Metal stage	(Water stage)

Anything older than the full stage of growth at 28/32 is considered metal turning the decline towards death. The water therefore is the seed of the next cycle of the spiral of growth. Interestingly we see that the cycle begins and ends with 2 Kan and 6 Qian which are the pre-heaven expressions of Origin.

vii) The Nine Energies within the Five Phases

The above explore the nature of the Spirit energy within life and humans. The nine energies can relate to anything, as they are Heaven's energy and so permeate all of Earth. However, to make them more clear to human energy, they have been placed within the spectrum of the principle of five:

(fig. 2.7)

the 9 energies

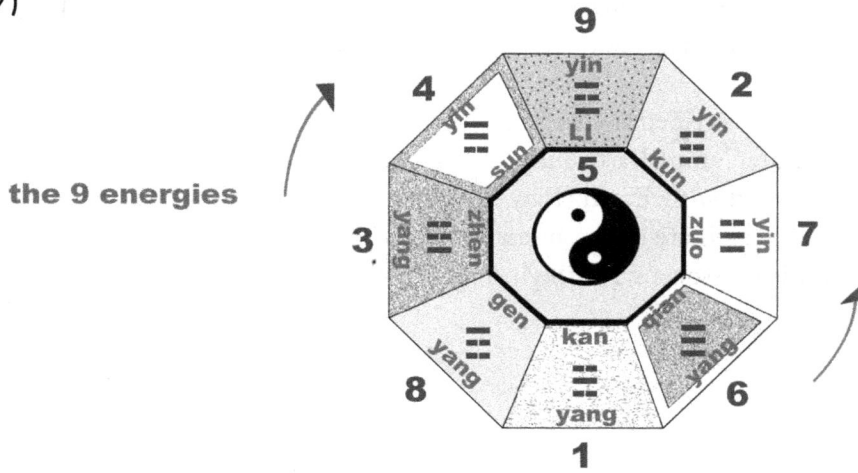

In this way, we can understand how humans relate to each other with the base energies of their character. Each human will have one of these energies at the core of his/her character, as a yinyang entirety, though what is meant by this is that one takes precedence over the rest, as they are all within all formations. This is understood and indicated by cycles of nine in coordination with the cycles of Earth and the lunisolar calendar. Cycles of nine years in relation to humans provide the one energy per year each for the yang body of male and for the yin body of the female (see appendix for details of history of how the nine energies coordinate energy to year and connections to the classics in Ling Shu chapters 77 and 79). Within the energy of Heaven, there is more yin and more yang, hence there are two cycles of change, one showing the moment of the yin within Heaven and the other showing the yang within Heaven. The yin cycle relates to women and to physical structures, so it is used in Feng Shui for the analysis of buildings/sites. For men and for non-structural aspects of things, we use the yang expression.

(fig. 2.8/ fig. 2.9)

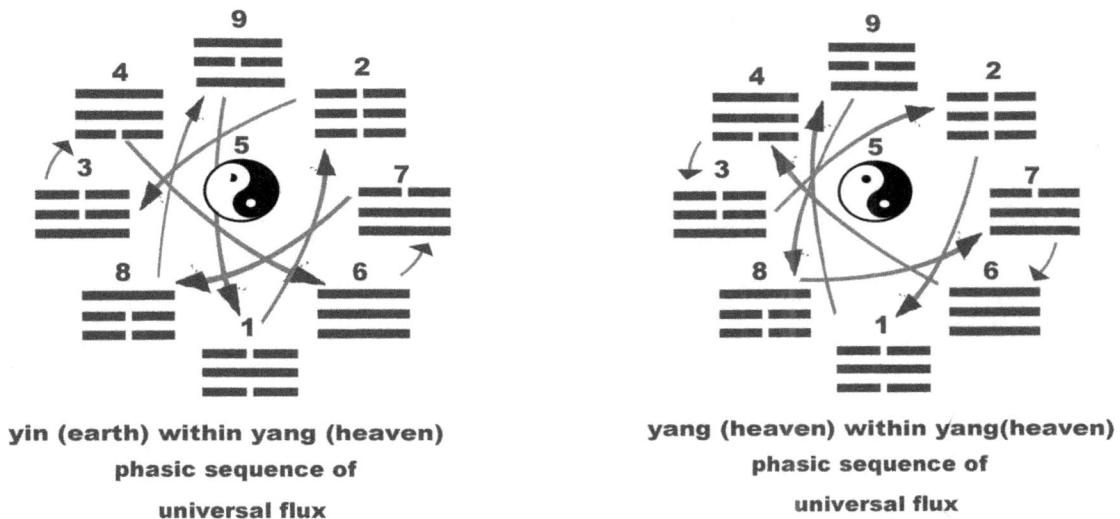

yin (earth) within yang (heaven)
phasic sequence of
universal flux

yang (heaven) within yang(heaven)
phasic sequence of
universal flux

The flux of this sequence is the flux of the universal energy and relates to all things. The cycles of nine help us to identify our core spiritual energy and then allow us to identify our basic identity within the nine energies.

The cycles of 12 have been described since ancient times, for use in agriculture and understanding of planning with the seasons. Later, animal attributions were used for the different parts of the cycles. Cycles of 12 belong and correspond with Earth. They are given a five-phase association to allow them connection with the five phases, which are of the human cycle. These are as follows (cycles of 12 follow the lunar based calendar, and this is associated with yin/night/moon/Earth):

(fig. 2.10)

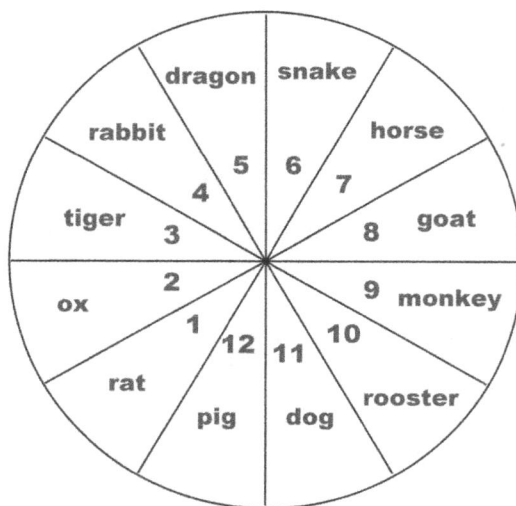

The 10 celestial stems combine with this to give the stems and branch chart incorporated below. Here again are the stems with traditional names. The stems and branches are both seen through the five phases of humans:

(fig. 2.11)

To combine the understanding of the Spirit energy, we must see the nine energies of Heaven as the broad picture and the stems and branches as the Earth within the Heaven; hence it has influence, but the 9 governs change. However, the 12 animals of the 12 branches and the 10 celestial stems (which act as a pre-Heaven reference point for the 12 branches), also influence humans but more on the form-physicality or Earth association. This can go on ad infinitum. For now we will be very basic in order to get an overall view, which is usually more appropriate anyway. If you remember the concentric cycles at the beginning, the unity is first Heaven, then Earth, then man, in order of concentric Nature. Hence we can see this as the basis for the spirit body. Primarily it is the 9 energy giving the spirit base and is associated mostly with the ethereal soul that is associated with wood energy. The second is the Earth 12 cycle (in relation to the 10 cycle) connected with the corporeal soul, mainly associated with the metal energy and the Earth. Note that all the energies are described as animals with bodies. This is very different from 9 energy talking of primordial element, and the animals are more earthbound entities. This makes up the full picture.

In order to complete this, we will go through each of the 12 animal types in order to describe their energy. The 10 celestial stems are placed in association with the five phases of humans to make use of them energetically in relation to the 12 branches. The 10 stems relate to pre-Heaven, and the 12 branches are earthly. Notice that the pre-Heaven is also yin in relative Nature, and the Earth is also yin. Hence the He-to chart was drawn in a pattern that relates to Earth, whereas the Lou-Shu chart is in a pattern relating to 9 and Heaven of the post-Heaven type. The Earth energy is central to the 12 cycle, so it follows that four signs are Earth and there is one sign in each season for Earth.

The earth only revolves in one direction and so the 12 branches and 10 stems have a single direction of movement, which is related to both male and female forms. We could say that it is more yin than Heaven so the 12 branches are more yin and earth orientated. The Nine energies are yang in relation

131

to earth, they are also non-directional (circular) as yang is non-form (earth is directional; 12 is a square format and so has innate direction), hence there are yin accumulative and yang expansive expressions of the 9 but the yang expression of the 9 is the archetype. The yin expressions of the 9 are yin within the yang.

viii) 12 Branches of Earth/Animal Types

1. Rat/Fire in Water

They can be quite miserly and introspective. They have an internal world and can delve into places that most would not. They are quick witted and passionate. They have deep emotion and cool exteriors. They tend to collect and hoard. They are very intelligent but can be nervous, ambitious, and secretive and can advance their own cause through secret methods. These energies relate to the number 1 in the nine energies.

2. Ox/Earth (of Water)

They are solid, dependable, protective, and patient. They have powerful tempers, but usually one never sees this. They never forget. They are not particularly imaginative, romantic, or exciting but are very honest and strong. These energies are close to the 8 energy in the nine energies.

3. Tiger/Wood

Dynamic, impulsive association with early spring. Expressive. Being goal-oriented, they have an association with the 3/Thunder energy of the nine energies. Easily angered, they can ignite their imagination and are dream-like creatures. They need excitement and arousal in their life to be happy; they are easily bored.

4. Rabbit/Metal in Wood

They are diplomatic and social but evasive and moderate. They can often give people answers they want to hear. Fast and flowing in movement, they can be strong-willed internally but are constantly strategizing, communicating. They are calm in their way of being but are highly strategic internally, making for a good judge of a situation. They are associated with the 4/wind energy of the nine energies above. They hate conflict and run at the sign of it.

5. Dragon/Earth (of Wood)

These energies are strong, authoritative leaders but also loners. They are hard workers and generally generous. They can be explosive when upset and have the ability to be judgmental and dominating. Relating mildly to the energy of 5, these are powerful energies.

6. Snake/Fire (Empty Fire)

Passionate and charming but toxic if crossed, these creatures have an ability to be connoisseurs of good taste. They can be dangerous and fiery if one crosses them and are ruthless if they don't get what they want. They can be mysterious and coerce people to do their will. Snakes have a relation to the yin of fire, so they actually have some connection with 1/water of the nine energies, or the root of fire.

7. Horse/Fire

Ever active, they are constant movers and workers. They have brilliantly fast minds but no time to do everything, and they miss their footing because they go too far too fast. They are ambitious and confident in their abilities and explosive in temper. These energies relate to the 9 energy in the nine energies.

8. Goat/Earth (of Fire)

Emotional, peace loving, accommodating, shy, and vulnerable. They can be stubborn to move, and they appreciate the finer things in life. They hate making decisions and avoid conflict at all costs. These energies are very close to the energy of 2 in the nine energies.

9. Monkey/Wood in Metal

Energy that pushes itself forward through sheer energetic fanatic action, they can motive others easily and can use trickery to manipulate a situation to their favour. They are not usually interested in other opinions. They are quite young in their attitude and joyous of spirit but can be childish and demanding. These energies are similar to the 6 energy of the nine energies.

10. Rooster/Metal

Lord-like and admired, they are dignified, bright, and judicious. They know what is right and wish to act in accordance, and in doing so follow rhythms and law and order to the "t". They are quite rigid and stern and can be brave and confident. They have a metallic and determined approach, attempting perfection in everything. There is some relation to the 7 energy of the nine energies here.

11. Dog /Earth (of Metal)

They are dependable and have a sense of justice and loyalty. They are intelligent and can be critical, but they are honest and always willing to help. They enjoy relaxing and are not interested in accumulating wealth for themselves but can be very hard workers. They are very social and socially responsible.

12. Pig/Water

Everybody's friend, they are geniuses and honest. They spend much time seeking out pleasure and social interaction. They are very good organisers and like to rally others. They are prone to overindulge but hold no grudges and think well of people.

ix) Unifying 9, 12, 10, and 5

(table 2.6)

	Year 1910–1969	9 Energies		Year 1970–2029	9 Energies		Associated Element	Stem	Branch	Associated Animal
		M	F		M	F				
1	Feb. 4 1910	9	6	Feb. 4 1970	3	3	Yang Metal	7. 庚	11. 戌	Dog
2	Feb. 4 1911	8	7	Feb. 4 1971	2	4	Yin Metal	8. 辛	12. 亥	Pig
3	Feb. 4 1912	7	8	Feb. 4 1972	1	5	Yang Water	9. 壬	1. 子	Rat
4	Feb. 4 1913	6	9	Feb. 4 1973	9	6	Yin Water	10. 癸	2. 丑	Ox
5	Feb. 4 1914	5	1	Feb. 4 1974	8	7	Yang Wood	1. 甲	3. 寅	Tiger
6	Feb. 4 1915	4	2	Feb. 4 1975	7	8	Yin Wood	2. 乙	4. 卯	Rabbit
7	Feb. 4 1916	3	3	Feb. 4 1976	6	9	Yang Fire	3. 丙	5. 辰	Dragon
8	Feb. 4 1917	2	4	Feb. 4 1977	5	1	Yin Fire	4. 丁	6. 巳	Snake
9	Feb. 4 1918	1	5	Feb. 4 1978	4	2	Yang Earth	5. 戊	7. 午	Horse
10	Feb. 4 1919	9	6	Feb. 4 1979	3	3	Yin Earth	6. 己	8. 未	Goat
11	Feb. 4 1920	8	7	Feb. 4 1980	2	4	Yang Metal	7. 庚	9. 申	Monkey
12	Feb. 4 1921	7	8	Feb. 4 1981	1	5	Yin Metal	8. 辛	10. 酉	Rooster
13	Feb. 4 1922	6	9	Feb. 4 1982	9	6	Yang Water	9. 壬	11. 戌	Dog
14	Feb. 4 1923	5	1	Feb. 4 1983	8	7	Yin Water	10. 癸	12. 亥	Pig
15	Feb. 4 1924	4	2	Feb. 4 1984	7	8	Yang Wood	1. 甲	1. 子	Rat
16	Feb. 4 1925	3	3	Feb. 4 1985	6	9	Yin Wood	2. 乙	2. 丑	Ox
17	Feb. 4 1926	2	4	Feb. 4 1986	5	1	Yang Fire	3. 丙	3. 寅	Tiger
18	Feb. 4 1927	1	5	Feb. 4 1987	4	2	Yin Fire	4. 丁	4. 卯	Rabbit

19	Feb. 4 1928	9	6	Feb. 4 1988	3	3	Yang Earth	5. 戊	5. 辰	Dragon
20	Feb. 4 1929	8	7	Feb. 4 1989	2	4	Yin Earth	6. 己	6. 巳	Snake
21	Feb. 4 1930	7	8	Feb. 4 1990	1	5	Yang Metal	7. 庚	7. 午	Horse
22	Feb. 4 1931	6	9	Feb. 4 1991	9	6	Yin Metal	8. 辛	8. 未	Goat
23	Feb. 4 1932	5	1	Feb. 4 1992	8	7	Yang Water	9. 壬	9. 申	Monkey
24	Feb. 4 1933	4	2	Feb. 4 1993	7	8	Yin Water	10. 癸	10. 酉	Rooster
25	Feb. 4 1934	3	3	Feb. 4 1994	6	9	Yang Wood	1. 甲	11. 戌	Dog
26	Feb. 4 1935	2	4	Feb. 4 1995	5	1	Yin Wood	2. 乙	12. 亥	Pig
27	Feb. 4 1936	1	5	Feb. 4 1996	4	2	Yang Fire	3. 丙	1. 子	Rat
28	Feb. 4 1937	9	6	Feb. 4 1997	3	3	Yin Fire	4. 丁	2. 丑	Ox
29	Feb. 4 1938	8	7	Feb. 4 1998	2	4	Yang Earth	5. 戊	3. 寅	Tiger
30	Feb. 4 1939	7	8	Feb. 4 1999	1	5	Yin Earth	6. 己	4. 卯	Rabbit
31	Feb. 4 1940	6	9	Feb. 4 2000	9	6	Yang Metal	7. 庚	5. 辰	Dragon
32	Feb. 4 1941	5	1	Feb. 4 2001	8	7	Yin Metal	8. 辛	6. 巳	Snake
33	Feb. 4 1942	4	2	Feb. 4 2002	7	8	Yang Water	9. 壬	7. 午	Horse
34	Feb. 4 1943	3	3	Feb. 4 2003	6	9	Yin Water	10. 癸	8. 未	Goat
35	Feb. 4 1944	2	4	Feb. 4 2004	5	1	Yang Wood	1. 甲	9. 申	Monkey
36	Feb. 4 1945	1	5	Feb. 4 2005	4	2	Yin Wood	2. 乙	10. 酉	Rooster
37	Feb. 4 1946	9	6	Feb. 4 2006	3	3	Yang Fire	3. 丙	11. 戌	Dog
38	Feb. 4 1947	8	7	Feb. 4 2007	2	4	Yin Fire	4. 丁	12. 亥	Pig

39	Feb. 4 1948	7	8	Feb. 4 2008	1	5	Yang Earth	5. 戊	1. 子	Rat
40	Feb. 4 1949	6	9	Feb. 4 2009	9	6	Yin Earth	6. 己	2. 丑	Ox
41	Feb. 4 1950	5	1	Feb. 4 2010	8	7	Yang Metal	7. 庚	3. 寅	Tiger
42	Feb. 4 1951	4	2	Feb. 4 2011	7	8	Yin Metal	8. 辛	4. 卯	Rabbit
43	Feb. 4 1952	3	3	Feb. 4 2012	6	9	Yang Water	9. 壬	5. 辰	Dragon
44	Feb. 4 1953	2	4	Feb. 4 2013	5	1	Yin Water	10. 癸	6. 巳	Snake
45	Feb. 4 1954	1	5	Feb. 4 2014	4	2	Yang Wood	1. 甲	7. 午	Horse
46	Feb. 4 1955	9	6	Feb. 4 2015	3	3	Yin Wood	2. 乙	8. 未	Goat
47	Feb. 4 1956	8	7	Feb. 4 2016	2	4	Yang Fire	3. 丙	9. 申	Monkey
48	Feb. 4 1957	7	8	Feb. 4 2017	1	5	Yin Fire	4. 丁	10. 酉	Rooster
49	Feb. 4 1958	6	9	Feb. 4 2018	9	6	Yang Earth	5. 戊	11. 戌	Dog
50	Feb. 4 1959	5	1	Feb. 4 2019	8	7	Yin Earth	6. 己	12. 亥	Pig
51	Feb. 4 1960	4	2	Feb. 4 2020	7	8	Yang Metal	7. 庚	1. 子	Rat
52	Feb. 4 1961	3	3	Feb. 4 2021	6	9	Yin Metal	8. 辛	2. 丑	Ox
53	Feb. 4 1962	2	4	Feb. 4 2022	5	1	Yang Water	9. 壬	3. 寅	Tiger
54	Feb. 4 1963	1	5	Feb. 4 2023	4	2	Yin Water	10. 癸	4. 卯	Rabbit
55	Feb. 4 1964	9	6	Feb. 4 2024	3	3	Yang Wood	1. 甲	5. 辰	Dragon
56	Feb. 4 1965	8	7	Feb. 4 2025	2	4	Yin Wood	2. 乙	6. 巳	Snake
57	Feb. 4 1966	7	8	Feb. 4 2026	1	5	Yang Fire	3. 丙	7. 午	Horse
58	Feb. 4 1967	6	9	Feb. 4 2027	9	6	Yin Fire	4. 丁	8. 未	Goat

59	Feb. 4 1968	5	1	Feb. 4 2028	8	7	Yang Earth	5. 戊	9. 申	Monkey
60	Feb. 4 1969	4	2	Feb. 4 2029	7	8	Yin Earth	6. 己	10. 酉	Rooster

Read across the table to see the date correlation with the nine energies and with the 10 stems and 12 branches above. Please note that the above dates respond to the solar Chinese calendar in which, for the purposes of astrological understanding, the beginning of the year corresponds not to the Chinese New Year's day (lunar calendar) but one of the 24 seasonal markers, which are 24 splits in the calendar denoting the equinoxes and the phases in between the solar phases. Hence the solar expression (solar being yang and Heaven) commands the astrological charts. Therefore, the "coming of spring" is a different point than the Chinese New Year. The "coming of spring" date will always be approximately February 4. This is what is shown above. The Chinese calendar is a combination of lunar and solar expressions 12, 10 and 9 attaching to this. The nine energies are more associated with solar and yang, and the 12 branch-stem cycle is associated more with the yin and so is lunar. This is why the Chinese New Year is the beginning of the first lunar cycle and is celebrated. This relation of the 12 to yin and the 9 to yang is very important to note.

Although this may look like a very prescriptive issue, it is not, unless one uses it out of context. The context is that each of the 5, 9, 10, or 12 in all their aspects, is represented within every so-called "individual" microcosmically, and also in the broader macrocosm. Each aspect of the whole (or "person") will however be of different quality to make up the whole expression. Remember that these are tools of exploring only the tip of the iceberg. We are exploring the world of phenomena that is yinyang, but when doing this we must always be rooted firmly in the Wu Oneness that can also see beyond it. Otherwise it ends up being a telephone directory of images, which become illusions in themselves. What the ancients explore is, of course, more profound than this. Many cultures use these understandings to choose leaders and form understanding of events prior to their physical manifestation. Hence, within context, this can be a useful tool. The nine energies allow one to expand the idea of "self" as an individual to instead be an aspect of Nature and so re-join with all of Nature. The 12 animals do this in the form of animals, again showing the direct relation to natural phenomena, not the individual narrowness of the egoic identity one has in mind-identity. Even in basic form, the attempt is always to draw one to the togetherness of Nature, at one with yinyang flow and underpinned by Wu, the Un-manifested.

The origin of these sequences was an intuitive/felt understanding that people of certain age groups have a common energetic nature. This is clear in most school classrooms. The experiential therefore is always before the theoretical. The only way one can identify the nine energies is through change-related connection. Five-phase and seasonal change seem much easier to understand because it manifests so physically or visually; however, this cycle is the same, except that it is again energetic and so seems to have little relation the physical form, but permeates it. If we understand that all physical phenomena are yin-Earth (physical or condensed/cooled energy) within the yang-Heaven (energetic or expanded/heated energy), we can see that it is all energy, and it is all one (or, all Wu). The principle of nine permeates all forms and so is highly significant in relation to humans and understanding their expression. The other significant system of understanding is that of the stems and branches. This too can be used to determine characteristics in terms of 12 animal characters and the five-phase (10 stem) relation these have. However, 12 is a cycle of the Earth and therefore is bound to earthly phenomena. One could perhaps use the argument that while the principle of nine governs the understanding and energetic change of the spirit, the system of stems and branches derives information more closely associated with one physical structure and changes on the Earth. The two parts of the ethereal spirit energy are the corporeal and the ethereal soul. One might add that the stems and branches deliver connection of the corporeal soul and the physical body, whereas the nine energies relate more to the ethereal soul and the heavenly energy. Using the combined cycles of 12 related to 10 in 60-year and 60-month cycles, the cycle of nine is applied to this. This means that the calendrical use of the cycle of nine is tied to earthly movements. Therefore, in this combined format, one is able to describe the full spectrum of more material to more ethereal energy relation. We can say that that Heaven (9) is applied to Earth (12) in this process because the cycle of 12 of Earth governs calendrical mathematics. Also, Earth energy in general is "form", so it differentiates people's natures or easily, or different parts of the whole, and describes them as bodies of form which can easily be interpreted by the mind-identity as "separate individual".

I have used the stems and branches picture to offer just the basic branch and stem for the year, but it will allow us to understand the influences on an "individual" at the end of Section A of this book. Whereas the Heavens breed the Earth, so cycles of nine create cycles of 12, and cycles of 12 create humans, which are cycles of five. Humans are between Heaven and Earth, influenced by both. Hence the cycle of yang within Earth is the cycle of 5 energy. The cycle of yin within Earth is the cycle of 12. However, the cycle of nine has 2 parts: 1-9 is the cycle of all yin in the universe, and 9-1 is all yang aspects. Because of these differences of perspective (Heaven envelops Earth), Earth has one cycle and Heaven has two. Note that humans are the yang of the Earth. Note too then that yang relates to life and yin relates to death. Within each human are yin and yang aspects, but this doesn't mean there is life and death within the body. It means that relative to aspects of the microcosm, there are cooler and hotter aspects. If we take a perspective from outside the body and look at the human versus a rock, then the human is more yang than the rock. Hence, a dead body is more yin than a live body, and thus the yin pertains (and goes back to) the Earth, whereas yang/spirit pertains to the sky and Heavens.

Even within the nine energies themselves, there are two aspects that are more yin and more yang, connecting with the universal principle of two. The energies associated with Earth and metal and water are more heavenly and earthbound. They have a tendency to be more towards the earthly phenomena (although water moves up towards the ethereal). These types are more comfortable within their physical form. The energies of wood and fire are much more ethereal (although fire moves towards the physical), they can often disregard the physical in order to be in the energetic. This can cause physical problems or what some may feel as losing touch with "reality". However, we have a clear mental and social stigma that what is "real" is what is physical, and this social focus is derived from the more earthbound energies that create society, because that is their nature. This is warped by mind-identity, which separates out. Physical forms therefore look separate from one another more easily than energetic expressions that tend to merge with the exterior (e.g., a stone as opposed to fire). We keep teaching ourselves to do this, rather than perceiving unity.

The origins of the calendrical system expressed here can be looked at in the Appendix section. What is important to understand is that the lunisolar calendar is the basis of the Chinese understanding of universal change or what might be called "space-time" (in a modern fragmented ideology), involving the combined expression of moon and sun rotations. The sun relates to yang, and the nine energies also relate to yang and ethereal energy. The 12 branches relate to earthly energy and to the yin and to the moon. Hence, the combination of 12 and nine within the context of the 10 stems (of the pre-Heaven/ He-to) made a complete system that is represented in the original classics of Chinese Medicine and is reflected in other Han Dynasty sources (see Appendix 1).

The origins of the calendrical systems and that of the nine energies go back to the Shang Dynasty and prehistoric past and are the root of shamanism and later on, Taoism. These are the precious connections we have to the origin of the feeling of change and being, which these map out for us so effectively if seen as one and neither fragmented nor complicated with many other methods. Contextualisation with the nature of Wu is imperative for understanding to be complete. All methods of understanding of the person's natal-spiritual base energy come from these basic calendrical principles of 12/10 and 9 and 5 and, as such, this forms the simple base from which one can extrapolate. Moving away from the basics often means we lose our way, so one must be careful. In modern day terminology, the nine-energy system is called part of Purple-White Star /Nine Flying Star, Feng Shui, and the 12 branches and 10 stems is called the Four Pillars method. These different schools developed from using and understanding the original principles of 9, 10, 12 and 5 and how these can be related to all aspects of internal and external life. However, it is important always to remember the root. Therefore, in the table above, the yang spirit or ethereal aspects of the energy are the nine cycles, and the more materials is the 12/10 cycle, all in relation and viewed through the five phases of humans.

The combined use and understanding of these phenomena (principle of three: Heaven/Earth/ Man) in relation to the human is explained time and again in the Huang Di Su Wen and Ling Shu; the Nei Jing. This book, along with all the aspects of the internal body, continually intersperses the understanding with constant broad views of our existence. This is to make the reader's mind-identity flexible, not rigid, in order to realise that there is an application of the completed understanding in all directions and all forms, and it can be applied to anything. The interior and exterior mirror each other macrocosmically and microcosmically.

The attempt to draw things towards unity has been lost since the times of the Han Dynasty, with people drawing away from this ancient base understanding. This is why so many schools and styles developed outside this origin. People were relying on the system, rather than the system guiding and teaching them about their own intuition for them to eventually let go and be one with the universe themselves. The energies here, though forming complex patterning, describe the spectrum of a person's physical and spiritual base. The study of the Han dynasty text is nothing to do with "ancient equals best", ideology; rather it is seen as a brief period of flowering enlightenment in the nature of humans before the "dark age" again fell. This is why we choose the term "Classical" rather than "ancient" it most cases.

Always remember that this work teaches us how to feel from the "big picture" perspective. Once we can do this and be in the Now-ness of Now, we do not need the stabilisers; the aim is always to transcend all techniques. The expressions in this work are derived from the base of the Su Wen and Ling Shu for guidance and understanding of where the external energetics of Feng Shui join with internal medicine to prove and understand its total unity. This is so important in the undertaking of study, which is the echo of this work, I hope.

I will not delve deeper into the stems and branches associations, I will not be delving too much deeper into the nine energies. I simply wish to put all of this in context for you, so that you may study further yourself. There are a great many works out there with this information in them already. The point is to clearly see the difference between the Shen and the physical and how they meet in Oneness, as well as the unity of all understanding as seen from a broad perspective. One could say the stems and branches lie between the two, whereas the cycle of nine is Heaven and the cycle of five is man. Hence we have the three aspects—Heaven, Earth, and man. All are useful, all are important for different reasons, and all unite as one.

The nine energies, as explored above, can be associated with the month, the day, the moment, and the second. However, for our purposes here, we only need a basic understanding to look at the energetic and the physical and note their interactions. When we use the 9 energies we need to realise that it is really an association with the human but it is a broader view it is connected with everything but it is also not at the level of vibration of the human. The 5 phases vibrate that the human level so the 9 energies are associated with the human energetic system but they are broader and larger than this and this must be deeply understood. What is at the human level is of a different order to the Heavenly expanse of the 9 energy of the 12 branches and 10 stems. While we can associate more easily the 12 and 10 with the body, the spirit is more easily seen as Heavenly and without form, we must be careful to understand correspondence from the actual level we are currently, as human, interacting with. Our level of order is the 5-phases and thus most of the is book is dedicated to the 5-phases energetic expression. Please always keep in mind concentric circles and the hierarchy of fractal order. Microcosm is in macrocosm and visa versa but the concentric nature is also vitally important to recognise otherwise we can easily become highly theoretical and start combining expanded ideas with internal ideas that leads away from our senses and towards a theoretical model of the universe.

The general format for an individual reading of a picture of the spirit, using the nine energies or 12 branches, is always in much more detail than presented here. What is presented in this work is the bare minimum to understand the general picture and the core basis of the spiritual structure. This would never be deemed useful enough to create a full picture. As we are using only the yearly changes, this is a very general picture; the monthly, daily, and even hourly readings greatly advance the specific-ness per "individual".

2.3 The Five-Phase Interactions in Relation to the Energy of Humans, Heaven, and Earth

The five phases are the human phasic calendar, meaning our inner calendar of inner seasons, and as such, all phenomena are seen in relation to this spectrum. The five-phase energy originates or is rooted in the depth of the human body, the five Zang organs. We need to first understand the three relationships of balance in association with this system, because we will use this not only to explore the physical body but also the spirit energies of the nine, in relation to the principle of five and also that of 12 in relation to five in the earthly phenomena in the stems and branches methodology.

An important point before we go on is that within the human body, the nine energies are in correspondence with the five Zang organs; this is why in the above section on associations with the human body we associate the spirit mainly with the organ energy. If we consider 6/Qian to be an extension of 3/Zhen, and 4/Sun to be an extension of 7/Dui, then we can understand how the spirit energy dwells in each of the Zang, of which there are only five (except for the lung, which is a meridian only as the organ isn't an aspect of the yinyang process of the essences of the body—see Section B of this work). This is opposite of the 12 branches and 10 stems, where it has more to do with Earth energy and form. Hence the meridians (connected to the Earth) are associated more than the organs (holding the spirit). This is why in this section and in the previous one, the bracketing is applied to the "associations with the human body" part of the tables.

This is a key to abbreviations I will use from here on in, referring to the energetic system of the body:

GB – Gallbladder
Liv – Liver
SI – Small Intestine
TB – Triple Burner
Pc – Pericardium
Hrt – Heart
Sp – Spleen
St – Stomach
Lu – Lung
LI – Large Intestine
Kid – Kidney
Bl – Bladder

L – Left
R – Right

i) The Ancient Characters of the Five Phases

Before we go on, let's look at the pictograms representing the five phases in order to give us a better understanding of the quality of energy we are talking about when looking into the five phases. The five-phase characters are some of the oldest of all characters (around 3000 BC and before) and are simpler than many other characters.

(fig. 2.12)

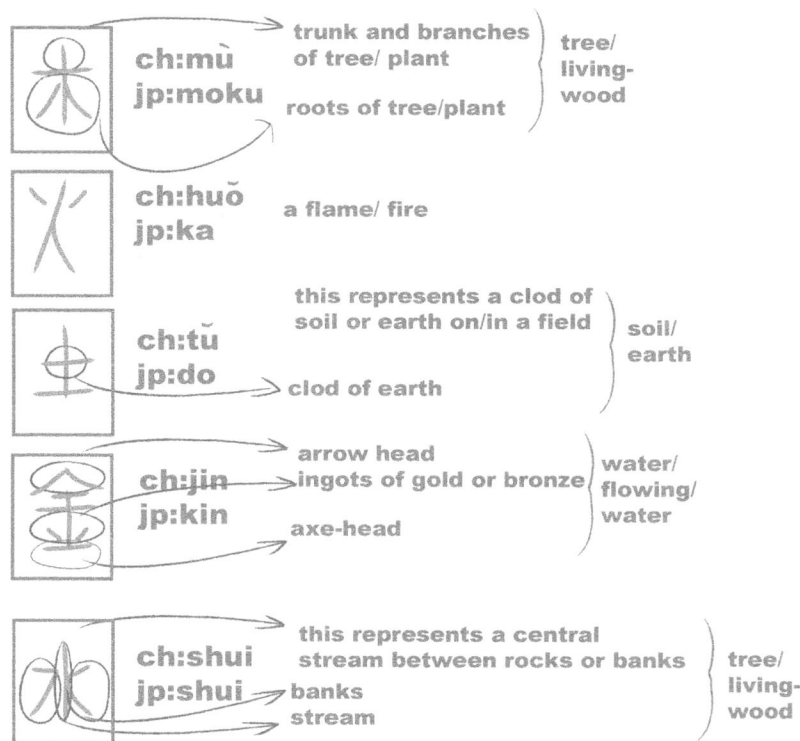

ch:mù
jp:moku — trunk and branches of tree/ plant / roots of tree/plant } tree/ living-wood

ch:huǒ
jp:ka — a flame/ fire

ch:tǔ
jp:do — this represents a clod of soil or earth on/in a field / clod of earth } soil/ earth

ch:jin
jp:kin — arrow head / ingots of gold or bronze / axe-head } water/ flowing/ water

ch:shui
jp:shui — this represents a central stream between rocks or banks / banks / stream } tree/ living-wood

Note that what we describe commonly as "wood", a dead material, is actually a living tree or plant; this perhaps is a better translation. Note too that each character is a part of a phase. Wood-tree is a growing outwards from the centre, flames burning upwards, the soil of Earth piling up not just lying flat. The metal tools are about movement not just structure, and water flows between the confines of its earthy banks. This gives us the constant idea of movement and change, never of static interaction.

ii) The Three Relationships of Five-Phase Balance

Mother-Child Interaction/ Sheng cycle

The first relationship of the three is called mother-child. The mother phase tonifies the child phase. It is a one-way relationship, and energy is simply feed to the next in line. This is sometimes translated as the creative cycle, because it is literally the changing process of energy. If we see this as a water cycle, for example, water can be turned into steam in the spring and summer, then cooled back to ice in the autumn and winter. It shows how properties of things change and merge with each other and how all things connect through forming one another.

(fig. 2.13)

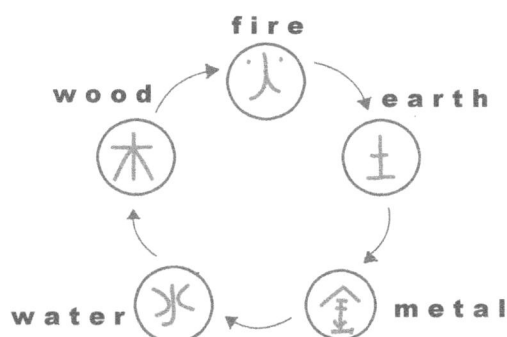

It is important to note, from the outset, that the five phases are not physical forms but energetic. Hence, wood is the energy of opening and relates to the spring. It expands to its greatest point just as trees grow upwards to the sky. When wood reaches its limit, or expansion can go no further, it bursts into flames, and the expansion forms fire, which is hot, irradiative, radiating, and upward moving. When fire cools down, its essences (ash) form part of the Earth, which is like summer reaching its peak, and the heat and moisture coming downwards a little. Then condensation starts to come back down to Earth, and this too forms the Earth. When the Earth energy cools and condenses and becomes harder within itself, metal is formed, or minerals which are dense and heavy. This is the metal energy, and it relates to the drawing inwards of autumn. When metal is purified and refined to the highest level through further cooling, it finds its way to the deep earth and still water is here, which it mineralizes. This is water energy. When water energy flows and is drawn by the energy of the spring, it evaporates and opens into wood, which is drawn up by the trees.

As you can see, the physical forms of the ideas of the phases are simply expressions of its energy. Be very careful not to get hooked on the form or word and thereby not see the energetic expression behind it.

Mother-child relationships, either Jing body- or Shen body-related issues, hold to the same principles—the mother gives energy to the child. The mother needs to be very strong for the child to accept and draw food from her. It is a one-way relationship, so the mother's energy cannot expect anything back from the child in exchange; it is a purely one-way transfer, and this allows the child to prosper. It is a giving and expressing outwards relationship in terms of the mother's role. The child needs to be accepting and able to absorb. In human relationships, this is most appropriate in an actual mother-child relationship. However, if put in the context of a partnership, one can immediately see how this can cause difficulties. The mother energy will become very tired and despondent, perhaps resentful, and the child could become too heavily dependent.

Brother-Sister Interaction

A brother-sister or friendship relationship is within the same phase but of different aspects. For example physically this can be expressed as male-female twins. They clearly create balance within the birth/ phase. Another description is a couple that are very similar in nature but are of different sexes. The relation of this is understanding, mental acceptance, and reliance. There is easy communication and regularity, and there is peacefulness with this relationship.

Again, either a Jing or Shen expression will have these issues. A brother-sister relationship is very easy as the energies do not polarize and are in constant agreement. The saying "Birds of a feather flock together" is one where we can see that it's one unit, like a school of fish or other groups of similar energy. The difficulty with this relationship is that there is little change. Nothing changes in the understanding between the brother and sister; they need the dynamism of opposite energy to open out themselves, so that it isn't an enclosed world of similar ways of being, thinking, and behaving. This is a friendship in human relationships based on similarity and understanding. However, without opposition, this can become stagnant and boring (i.e., without moment) if this is all there is.

It is also wise not to overlook the importance of the brother-sister relationship. This relationship is the archetype of friendship, fellowship, or kinship. It is about unity in its basal expression, whereas mother-child and husband-wife is about seeming differences. In fact, brother-sister could be considered the background to all true relationships. It is the foundation for any relationship based in Oneness, and that similarity is known above and beyond the differences that are seemingly present. There is great relevance in this as part of an understanding of Oneness. In all harmonious relationships, friendship is derived as the foundation, as this has endurance. It is what the ancient people did in the form of "match-making"—the art of finding partners that were matched, not opposite. More importance needs to be derived from seeing this as the base in all situations of conflict. When differences of two opposing forces are understood only to be superficial, and friendship or Oneness is known to be the foundation, all manner of difficulties fade into the background, and life carries on without resistance to opposition but acknowledging it for what it is.

Husband-Wife Interaction/ Ke cycle

This involves a dominant force taking control/ directing a submissive force, but keeping its distance so as to create harmonious control, rather than being overly forceful and destroying the situation. This relationship is difficult and complex. The opposites attract each other and respond to each other, as they need each other to create a more universal balance effect outside of their own phasic arena. However, this is about acceptance by the submissive energy/ yin to the yang force and drive. The yang is tempered by the submissive nature of his partner, the expanse of her energy to absorb his impact. The active energy is yang and is the one that attempts to control; the submissive is yin and is controlled. The balance in forces occurs when the energy of the control doesn't over dominate and the energy of the controlled force is not too large to be controlled. The yin aspect protects by being firm and solid, the yang by pushing or drawing towards its partner. Notice therefore that each element has a yin and yang aspect, dependant on the quality of the energies involved. Water is drawn by fire, and wood draws metal, just as these energies move to overcome or void the difference. For example, the dryness of fire makes water come to it easily, the softness of Earth makes wood energy easy to make root, and the suppleness of wood makes it easy to cut with metal (which is actually an act of cooling that is drawn by wood). In all phenomena, this balancing is often complex and delicate, as it requires very little to shift it to a place where there is massive imbalance and collapse. Usually the partner in the dominant position can damage the connection most easily, because with natural position of dominance comes the situation of the possibility of over-control or abuse if there is a pathological state of mind-identity present.

(fig. 2.14/ fig. 2.15)

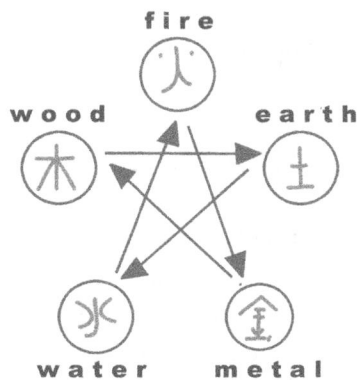

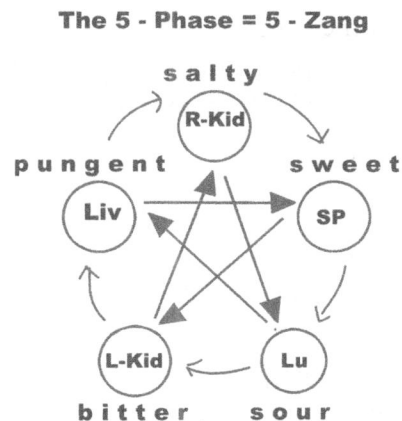

As with the generating cycle, the energetics of this are as follows: The vigorous expansion of spring controls the warm-mild and rounded condensation of Earth. The warmth of the Earth controls the coldness of ice water. Ice water is drawn to cool down the fire of summer. The fire of summer controls the accumulation and cooling of the autumn. The cool accumulation of the autumn is drawn by the spring's expansiveness to calm it. Please note that the yang expressions of wood, fire, and Earth act; they have action, yang. On the other hand, the cooling energies of autumn and winter are drawn by the opposite season, or aspect of the body, to create balance; because they are yin, they do not act, but are acted upon. (Just as energy is first and form is second, yang within the post-Heaven universe creates and spreads yin within it).

These relationships are in body and spirit and are the most complex, for they involve opposing forces coming to meet each other. This is an attraction relationship; one loves the passionate response of opposition (attraction). But the control and power struggle that ensues between people (and inside the organs of the body) when there is an imbalance of power, which inevitably there will be at some time, can cause destruction to the greatest degree. At the same time, balance can be the greatest pleasure. It is a complex balance of energies and timing that allows these relationships to last for a long time; usually in ke-cycle they can be very acute.

The five energies can be explored: the winter and autumn represent yin moving into the yang of the spring; the summer and late-summer period of wood, fire and earth. The metal and water energies are the yin, hence they follow the yang. This is true even in oppositional relationships. We see above that water controls fire, but it actually does so by the draw of fire; fire draws water towards it. The same is true of metal to wood; the wood draws metal towards it to control it. This is because the energies of the yin do not act in expansion, but they act to accumulate and cool; they are still, but the yang is moving. The yin displaces the yang, but the yang actually directs itself towards the yin; this is the nature of yin and yang.

Thus we can understand that the fire energy in relation to water acts as an attractive/radiant yang energy that pulls water to it. This change would be considered water's yang. Wood, acting as attractive/radiant yin energy, draws metal to it; this would be considered metal yang. Hence the action of the yin energies is different from the yang energies. Also the wood-yang will be very much more yang in its effect and energy than the yang of metal—which is a form of yin (in relation to the whole spectrum) that is being drawn by its opposite. In this way, we can see how the whole process of change occurs because there is yang in relation to yin. If there is no yang, there is no change, no life, and no yin; there it goes back to the enveloping Wu.

Within the body's organs (5 Zang, 5 Fu, plus 1 fu is exchangeable with S. I/ associated with the heart network; TB) and within the nine energies, simple principles are clear for mother-child and brother-sister relationships. Basically, an energy that is yang or yin that is the next phase of the five-phase cycle is the child energy, and the giver/previous phase is the mother. The yang energy would be more expressive or masculine in its giving nature; the yin energies would be more motherly, feeding the deeper aspects of the child energy. In brother-sister relationships, there is little to say, as there is simply the same energy and little happens except reflection of the phase within itself. What follows are the five phases within the context of the human body, the nine energies, and the 12 animal associations.

iii) The Five Phases Applied to the 12 Branches of Earth

Humans, as entities, have little to do with direction, as direction has to do with that which is outside the body and relates to seasonal change in the Earth. Hence, the above diagram of Earth in the centre and the four other phases around it should be seen in the sequence of change related to the 12 energies of Earth. There will be more about this later. It is more important to see that the five-phase sequence relates to the body and its internal change more than to the exterior and its change, as the human relates more to the sequence of five and the Earth changes (weather or seasons) of the 12 earthly branches:

(fig. 2.16)

The 5 - Phase = 5 - Zang

f i r e

snake/horse

e a r t h

goat

w o o d

m e t a l

dragon

**monkey
rooster**

**rabbit
tiger**

dog

ox   rat/pig

w a t e r

this is the 12 cycle of animals placed
into the 5 phase cycle

The seasons themselves are created by the earthly changes in relation to Heaven (nine energies: see Ling Shu 79). Hence the weather/seasons are related to 9 and 12 together, but the arrangement of the Earth as central means that this diagrammatic formation is associated with the earthly 12 branches—the Earth in every season or the last 30 days of each season. Please differentiate the five phases of humans, the 12 branches of Earth, and the nine energies of Heaven in all cases. The picture above can look confusing, but this is simply a conversion of 12 branches to five phases, but still with 12-branch bias.

Please note that the association of the five phases to the 12 animal energies expresses the notion of the 30 days at the end of each season as associated with the Earth phase. As we can understand the 12 energies associated with Earth, the four earthly seasons rotate around Earth, so to speak, or the Earth is affected by these seasons (i.e., Earth is a background for the seasons). In each of the seasons of the 12 animals, there are two aspects for each season and one each for Earth energy, which is late in that season (four Earth energies per year, four often associated with the four sides of a square of the Earth), the pivotal point between the swing into the next season. (This pattern relates to the five constitutions of the Jing-Earth body, which we looked at previously; Su Wen 12):

(fig. 2.17)

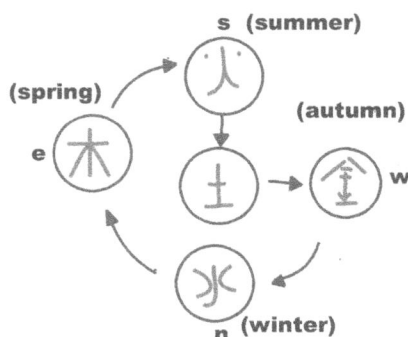

earthly charge in relation to
principle of 12 (seasonal)

This shows why the above map is associated with the 12 energies, whereas the pentographic five phases, where the Earth is in the right corner, are associated with the human (yang-energetic), not Earth (yin-form), and is not tied to geographic directions. Notice however that the archetypical association for the five-phase energy of Earth is in the upper right region, as this relates to the most earthy aspect of Earth (derived from the Kun trigram of the post-Heaven sequence).

In opposite relations of the Jing-Earth body, one can express the below interactions. These are a general implication of chapter 12 of the Su Wen and chapters 50, 59, 60, 64, 65, 72, 73, and 79 of the Ling Shu, corresponding geographical location to physical structure. Hence the following expressions are of the five-phase interactions of the Jing body. Notice that these implications are very broadly viewed and general (Feng Shui rather than Internal view), and I am looking at this from a global perspective, where the directions of north, south, east, and west will flip if viewed from the northern or southern hemisphere. Because China is in the northern hemisphere, everything is based on the south being yang. From the southern hemisphere, the directions themselves would be opposite, but everything else would remain the same.

A Note on the Phase of Earth

Earth is both a central phase and occupies the upper-right position. There are a number of reasons for this. The central position is associated with geographic direction, and so the cycles of 12 and the movement of Earth. This is also associated with the geography of the physical body, which is more associated with Earth than the more ethereal energy of the body (which is more associated to spirit and yang). Therefore, the Earth phase as central also associates with an Earth phase in each season and so the four directions. The upper right/ south- west region is associated with the energy of Kun/ Field-Earth, or 2 in the nine energies. This description has to do with Heaven and spirit and so with the post-Heaven Ba-gua (9-energies) but, the direction itself is actually a layering of direction associated with pre-Heaven Ba-gua (8 directions = 2 x 4 directions), which is more yin/ structural, and about form. Post-Heaven has no direction; it is energetic and ethereal and the mixing of energy to transmute into another. Hence, the upper right/ SW is 2 which is Kun, or pure yin, and this is the archetype for Earth energy, rather than "Earth direction" which is a pre-Heaven description not and energetic/ Post-heaven description. The lower left/ NE corner of the post-Heaven Ba-gua and the centre relate to Earth energetically. However, these are secondary because they have elements of yang within them, whereas the upper right is yin. Eight and 5 therefore are extra to the expression of Earth, whereas 2 is archetypical, just as 7 is archetypical of metal and 6 is extra (and complex), or 3 is archetypical of wood and 4 is extra (and complex). What this means is that on a physical, earthly energy plain (relating to 8 or 4 directions), the still pivot of movement which we call "earth" occupies the centre (Note that kun as the energy of earth, as we know in the post-heaven universe is not created in pre-heaven, hence although the North or bottom trigram is Kun in the pre-heaven it doesn't mean the same thing in the post heaven expression), but on an energetic plain (9 energies), it associates with the upper right of the post-Heaven Ba-gua, or with the five phases the upper-right point of the pentagram. These things will become clearer as we look further into the nature of the Earth energy and the difference in the spectrum of energy between the more physical (Jing) and more ethereal (Shen). The point that this brings up is the origin or transmutation of the circle of Heaven, or the nine energies, into the triangle of human or the five phases. Heaven (and Earth) are prior to human, therefore the nine energies transmute to five rather than the other way around. Both are energetic and non-directional. Of course we must note that the "cross" shape of the four directions (associated with 12 and earthly movement) is connected to the square shape; this is transforming into yang (the cycle and triangle of Heaven and human) just as the yang is transforming into the yin of Earth. It is a phasic process in itself, and as always, it's not either or, but both aspects which give the full picture:

> Earth = yin = 12, 8 and 4 = ■ = "cross" formation of phases, directional-geographic-form-based, has central pole but no movement.

> Heaven = yang = 9 and 5 = O and △= "pentograph-circular" formation of phases, non-directional, energetic, no centre but moves.

Please note that the Earth energies of 2/ Kun ,5/ Taiji, and 8/ Gen cuts diagonally across the 9-energy ba-gua, showing the split between yin and yang energies and also therefore showing that Earth energy has both yin and yang aspects within it. This split shows the end of summer and beginning of winter. When we move to the Earth association with 12 cycles, we can see that there is a cross formed from the Earth energy showing the end of each season; this is simply and extension of 4 (12 being a square/ even number expression also). Hence in relation to 9 transmuted to 5, the following diagram expresses this:

(fig. 2.18)

**The formation of
5 phases (human) from
9 energies (heaven)**

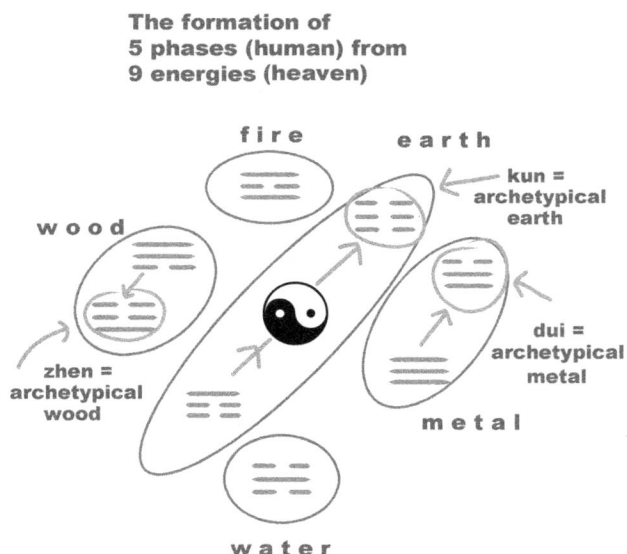

Earth yang is not true Earth, just as metal yang is not true metal and wood yin is not true wood. There is a mix. However the archetypes of each element are the basis of the five, and when expressed as the five phases, they hold within the other aspect, just as yinyang holds within it the entire spectrum of the "myriad thing" as Lao Tzu terms all forms.

The 5 Jing-body Relationship Interactions

The following are ideological expressions of the different interactions of the physical constitutions of peoples of the world. This is simply an example of interactions of all of humanity as One, there are no absolutes in this, and it is highly general to the point of abstraction:

The Jing Body in Mother-Child Tonification:

- Wood males tonify fire males
- Fire males tonify Earth males
- Earth males tonify Metal males
- Metal males tonify Water males
- Water males tonify Wood males

And …

- Wood females tonify Fire females
- Fire females tonify Earth females
- Earth females tonify Metal females
- Metal females tonify Water females
- Water females tonify Wood females

Notice that yang tonifies the yang, and yin tonifies the yin. The yin can also manage to tonify the yang and vice versa, but this is more complex, as the energy has to change from yin to yang to do so. Generally, male children belong to the males, and female children to the females; this is seen in most traditional cultures. It is most important for a boy child to have a father's energy and form into this, just as the mother and daughter relationship is most important. Of course this does not relate to the Oneness of either part or their spirit energy, which can alter the balance and is the driving force of the expression through the body; hence, this is all very general.

The Jing Body in a Brother-Sister Relationship:

- Wood male, Wood female
- Fire male, Fire female
- Earth male, Earth female
- Metal male, Metal female
  Water male, Water female

Brother-brother and sister-sister are also possible but does not complement within the phase. Each phase has five phases or yinyang partners, hence within each phase are complementary opposites.

The Jing Body in Opposition:

- Wood type males direct Earth type females
- Fire type males direct Metal type females
- Earth type males direct Water type females
- Metal type males direct Wood type females
- Water type males direct Fire type females

These are huge generalizations, but body types follow this balancing pattern because they are so different that they combine to create very powerful offspring (although in very ancient times this would not occur because these tribes would never meet, hence the balance of polar opposites usually occurs within brother-sister based communities). The energy follows the five-phase logic, with the male always directing in this case, as the male body is set up to do so. Note however that the metal and water males, being yin, will be drawn by their partner, as they are generally stiller and quieter.

If we used the 12 animals in relation to the five phases of humans, the relation to the body can be derived from this, always the 5-phase organ expression of humans is basal, also please realize we apply the notion of the macrocosm onto the microcosm here and this therefore is correspondence, but we must realise that the notion of the earth's movement is much broader and larger than the human energetic expression, just as a bacteria has a lifetime of a few hours, a human several decades and the earth millions of years, there is a concentric circle effect here and the fractal and one level may express the microcosm in the macrocosm but the scales are different (also please note that this is not to do with timing of a specific meridian's correspondence this is simply association and resonance, it is not part of the calendrical system, this will be discussed later).

(table 2.7)

12 – Branches	Application of the 5-phases to the 12 branches	Application of the 5-phase organs with the 12 branches	Associations to the meridians
1 Rat	Water yang	R-Kidney – Yang	R-kidney (organ) TB meridian, BL meridian (organ)
2 Ox	Earth yin	Spleen – Yin	Spleen (organ) – meridian
3 Tiger	Wood yang	Liver – Yang	Liver (organ)
4 Rabbit	Wood yin	Liver – Yin	GB (organ) –
5 Dragon	Earth yang	Spleen – Yang	ST meridian (organ)
6 Snake	Fire yin	Hrt – Yin	HRT (organ) PC meridian
7 Horse	Fire yang	Heart Network – Yang	SI meridian (organ)
8 Goat	Earth yin	Spleen – Yin	Spleen (organ) – meridian
9 Monkey	Metal yang	Lung – Yang	Lung meridian, LI meridian (organ)
10 Rooster	Metal yin	Lung – Yin	(Lung organ)
11 Dog	Earth yang	Spleen – Yang	ST meridian (organ)
12 Pig	Water yin	L-Kidney – Yin	L-kidney (organ) and meridians

(fig. 2.19)

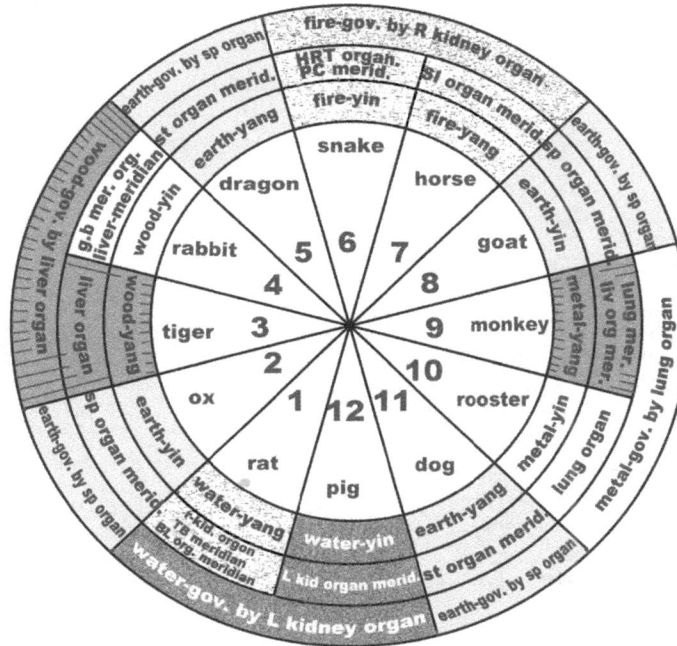

As we can see from above, the metal yang and wood yin have obvious opposition in this exploration of the five phases. This is particularly expressed in the classics of medicine, especially in chapters 33 and 64 of the Nan Jing, which warn us NOT to consider metal and wood as PURE energies, but as mixes of qualities of each other. We will continue to explore this throughout this whole text, as it is a vital axis for understanding the nature of metal and wood. Water yang and fire yin are opposite, as one is full while the other is empty. This, too, is important and shows again the dynamic interaction of yinyang in their various expressions; here yin is associated with emptiness, whereas in a wood-metal relationship yin is associated more with cooling. The relation of branches to individual meridians is different from what is being expressed here, which is the relation of five phases to the 12 branches. We will look at this more in later chapters.

(fig. 2.20)

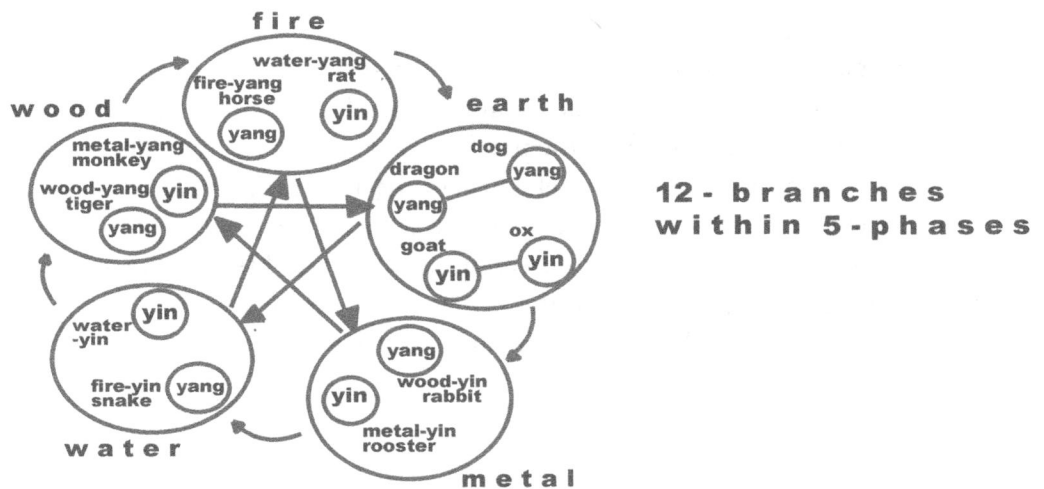

12-branches within 5-phases

(fig. 2.21)

12- branch interactions within 5 phases

Fire yin is empty; hence it cannot make a move to dominate water yang. However, put into the five-phase picture, we can see that fire yin is yang to water energy, and water yang is yin to fire energy; hence there is balance derived. Note that if the five phases were to be used instead of the 10 stems, the yin and yang aspects of each phase above would become one, as expressed by the colour in each phase above. So in the five phases of the 10 stems above, we see that we are not looking at individual aspects, but in fact the broader perspective of the five phases.

iv) The Five Phases Applied to the 10 Celestial Stems

As we saw in the second part of this Section of the book, the 10 celestial stems are originally of pre-Heaven, and the five phases are applied to it in order to render them useful in understanding the origin when associating with the 12 branches in the calendrical system. The 10 stems act almost as a reference point for the earthly change from the pre-Heaven perspective, although stars, which must be in the post-Heaven universe, are used to give reference to the 10 influences. It is very important to see that the 10 celestial stems are NOT the five phases split. It is the integration of viewing the 10 stems through the five phases of humans. When the five phases of humans (post-Heaven) are applied to the 10 stems, we have the following format, expressed in the names of 10 pre-Heaven energies. These are used in correlation with the 12 branches of Earth to give a mix of pre-Heaven association and the 12 branches of Earth.

(table 2.8)

10 celestial stems of the pre-Heaven He-To*	Application of the 5 phases of humans to the 10 Stems	Names of the stems	Application of the five-phase organs to the 10 stems	Associations in the body
3 – "Wood" first created	1 – Wood yang	Jia	Liver – Yang	Liver (organ),
8 – "Wood" second created	2 – Wood yin	Yi/ I		
2 – "Fire" first created	3 – Fire yang	Bing/ Ping	Heart Network – Yang	SI meridian (organ)
7 – "Fire" second created	4 – Fire yin	Ding	Hrt – Yin	HRT (organ) PC meridian
5 – "Earth" first created	5 – Earth yang	Wu	Spleen – Yang	ST meridian (organ)
10 – "Earth" second created	6 – Earth yin	Ji	Spleen – Yin	Spleen (organ) – meridian
4 – "Metal" first created	7 – Metal yang	Geng/ keng	Lung – Yang	Lung meridian, LI meridian (organ)
9 – "Metal" second created	8 – Metal yin	Xin/ Hsin	Lung – Yin	(Lung organ)
1 – "Water" first created	9 – Water yang	Ren	R-Kidney – Yang	R-kidney (organ) TB meridian, BL meridian (organ)
6 – "Water" second created	10 – Water yin	Gui		

Note: the phases have not been created yet, as they are not moving. Hence the north, south, east, or west or central region dictates the position of the explanation of the phase relation in forming the five-phase viewing angle.

Notice the yang aspect is the first to be created from Wu. The above expresses the 10 celestial stems, but please note that the 10 stems below have been applied to the five phases of humans. Hence the 10 stems are not of humans but of the celestial aspect, one could say the almost-pre-Heaven aspect within humans. The 10 stems, 12 branches of Earth, and nine energies of post-Heaven can all be translated into a five-phase picture, but this may cause discrepancies as the five phases of humans may not fit the model.

Above we can see that wood and metal energies are mixed, as this relates to wood yin and wood yang. Water yang and fire yin are opposite, as one is full and the other is empty. If we describe the situation in terms of five, then all the above aspects relating to Liv, R-Kid, SP, Lu, and L-kid would be associated together, which is what we will see in section B. In 10 stems, associations are mixed due to the nature of mixing the expression into 10 subparts. Similar expressions are found when we apply the five phases to the nine energies and to the 12 branches (see diagrams below). Also note again that the 10 stems, in context of the five phases of humans, can be looked at as; per phase, as paired yinyang aspects per phase, or as aspects of the 10-stem interaction without 5 phases. All are correct but are looking at narrower-stems and broader-phases perspectives. The five phases are always based on the Zang organs, being the basis for humans. Humans are a mix of earth and heaven but are foundationally yang and more "heaven" associated. The 10 stems is a yin number and as such has relation to earth and so the body.

(fig. 2.22/ fig. 2.23)

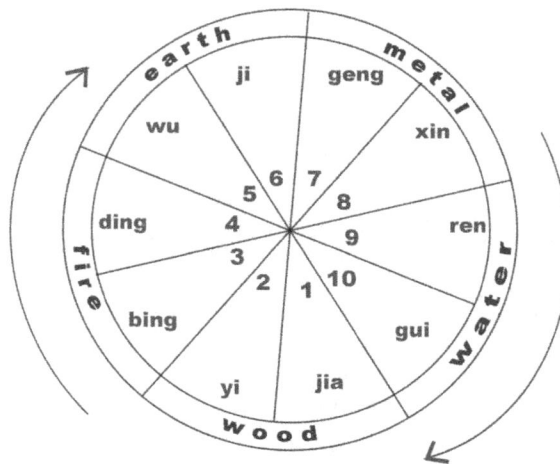

the 5 phases and 10 stems
nan jing 33

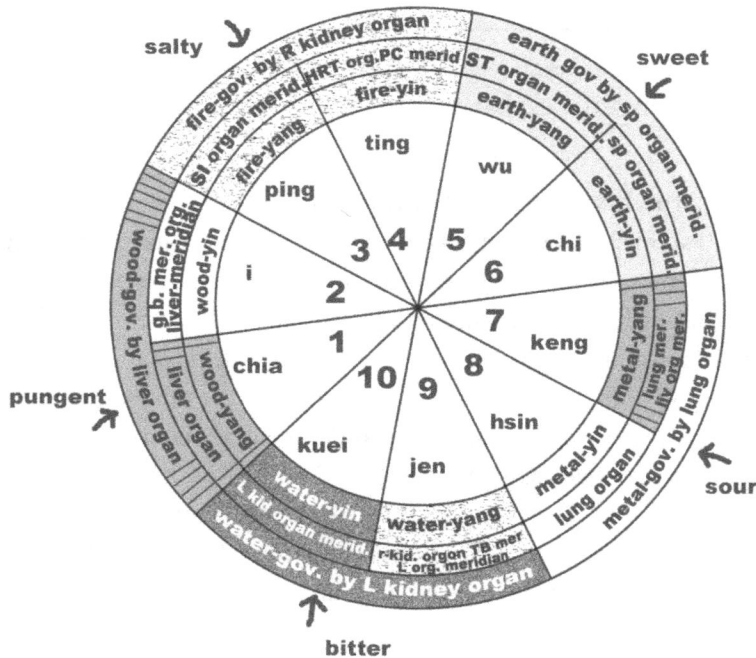

(fig. 2.24 / fig. 2.25)

10 - stems within 5 - phases

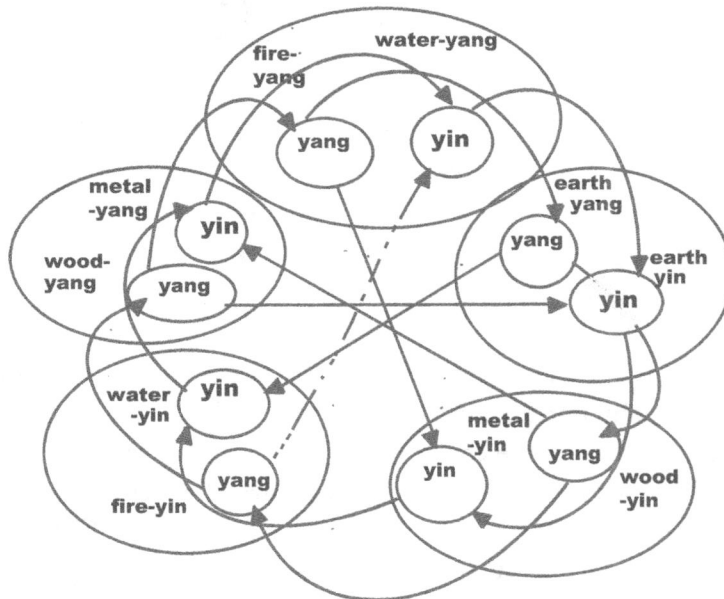

10 - stem interactions within 5 phases

v) The Five Phases Applied to the Nine Energies of Heaven

From here, one can look at the nine energies through the eyes of the five phases of humans. (table 2.9)

5 phases	Combined aspects of the 9
Wood	3 and 4
Fire	9
Earth	2, 5, and 8
Metal	7 and 6
Water	1

Represented overlaying the 9 sequence, we can do the following:

(fig. 2.26)

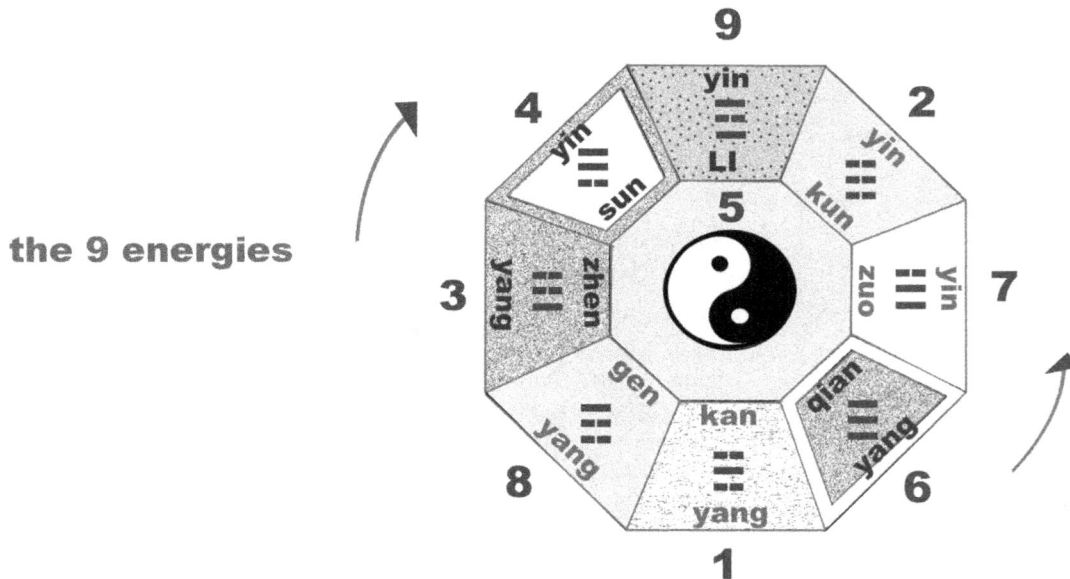

the 9 energies

Wood and metal have two parts to them in the nine aspects; these balance each other. Fire and water have one aspect each, balancing again. Earth relates to the centre, and 2, 5, and 8, forming an axis through the other energies. For humans, Earth is the centre of the universe, and so 5 have the role of being similar to the centre of the Earth. However, in a universal perspective, the 5 energy represents the centre of the universe or the original combining of yin and yang. As explained above Earth can either represent the centre or part of the five-phase cycles. The five-phase arrangement of humans is:

(fig. 2.27)

If the 5-phase expression is applied to the 9 energies, please note I will not relate the 9 energies to the human body as I did with the branches and stems because the branches and stems are yin number and associate with the physical . The 9 energy is different, it relates to that which has no form and so I will relate it to this in relation to the human:

(table 2.10)

9 energies	5-phases applied to 9 energies expression	Spirit relation
1	Water-yang	Zhi
2	Earth-yin	Yi
3	Wood-yang	Hun
4	Wood-yin	Hun-(P'o)
5	Centre Earth-yang	Yi
6	Metal-yang	P'o-(Hun)
7	Metal-yin	P'o
8	Earth-yang	Yi
9	Fire-yin	Shen

Once again, as with the 10 stems, the phases mix with the added yin and yang expressions. This is vitally important to understand in relation to the natures of the energies involved; it will be discussed in the constitution section of this work. The wood-yin is metal within a wood environment; the metal-yang is wood within a metal environment. Hence the energies of these, as well as the 1 water, are a great mix of energies, and fire is an empty yin energy, which balances this. Please note their complexity. Below we have the interactions of the five phases of the Zang of humans, then the interactions as a result of the nine aspects of the phases. The phases work in fives, and the nine energies work in nines, hence there are complexities. One is a narrower perspective, and the other is a broader perspective; both are correct.

(fig. 2.28/ fig 2.29)

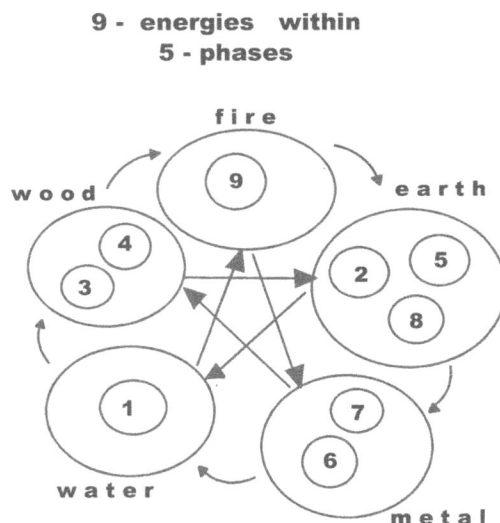

9 - energies within
5 - phases

From the above diagrams:

(table 2.11)

Relationship type in the 5 phases	Variation of combination:-	
	5-phase Combinations	9-energy through 5-phases combinations
Mother-Child (one way creative growth)	1 to 3/4, 3/4 to 9, 9 to 2/5/8, 2/5/8 to 7/4, 7/4 to 1	1 to 3, 9 to 2, 2 to 7
Brother-Sister (reflective friendship)	1 and 1, 3 and 3/4, 6 and 6/7, 9 and 9, 2 and 2/5/8, 5 and 2/5/8, 8 and 5/2/8, 4 and 4/3, 7 and 7/6	same
Husband-Wife (partnership) – (harmonic / balanced type)	1 and 9, 3/4 and 2/5/8, 9 and 4/3, 2/5/8 and 1, 4/3 and 7/6.	1 and 9, 3 and 6 and 4.

From the perspective of 9 energies there are far more energetic nuances and complexities about what constitutes a mother-child relationship, brother-sister and husband wife because there are so many extra parts that the simple 5 phases do not have. Hence we must consider both the 5-phase dynamic and the 9 energies just as with all the other explanations above.

Difficult or problematic relationships are to do with mind-identity attempting to force something which is not meant to last for long periods. In mother-child when the child of mother is too forcefully taking or forcefully giving, in all of the above situations of mother-child this is possible. (Collaborates with Nan Jing chapters 33 and 64 by implication, is associated with the notions above.)

9 Energy Mother-Child Relationship complications

These are relatively straightforward, as we are applying the 9 energies through that of the five phases, but there are some issues. The yang energy tonifies the yang, and the yin energy tonifies the yin in sequence. Th is means that sometimes it is not possible for the yin mother to tonify a yin child, or for the yang child, it is not possible to have the yang mother (father) energy. This is due to the complexity of using the nine energies within the five-phase principles. However this relates often to the natural energy-personality of these energetic forms in some way. This presents as 3 not being able to find a yin mother; 7 cannot form a yin child energy. This shows 3's resilience and 7's youth.

In brother-sister type situations problems associate with stagnation and lack of movement in the relationship, in all the above situations this is possible .

9 Energy Brother-Sister Relationship (Stagnation-based Relationship)

The above shows that wood and metal presents with both brother and sister energies. Within this, 3 is more yang than 4, and 6 is more yang than 7. Hence these form pairings within the five phases. The 1 and 9 have no brother and sister energy, so this is difficult for these types. The 2/5/8 are brother-sister without problems; in fact, there are three earth energies symbolising how earth energy forms society's human social structure. It is very important to note that the brother-sister or brother-brother, sister-sister pairing is very often idealized within films and drama as the soul-mate type relationship where the partners are friends and lovers at the same time. Overall, we can see the brilliance of this within such an isolated society creating the nuclear idea of family. However, even in ancient times, this was considered and still is to be (along with mother-child) the ideal relationship for longevity. If partners understand each other, then it is much more likely for stability to reign, but of course the issue with this is that the spark and passion is less there. Very often, these relationships tend to be for those who want children and to settle into a life, usually now over the age of 30 or 35. Prior to this, generally people will move naturally to the passion and difficulty of early relationships involving opposites. The importance of the brother-sister relationship cannot be overstated. It is vitally important to recognize one's own energy in relation to a patient and also to attempt to "match-make" patient to practitioner. Those practitioners who have the same spirit essence understand each other, and the resonance makes for very effective treatment. We will look at this more in later sections.

However, the relationship within the element is imbalanced overall or in the larger perspective, and this causes stagnation, lack of movement or expression of the same movements within the relationship. This leads to severe difficulties of the long-term chronic nature; long and flat movement can lead to many other difficulties. This relationship, especially sister-sister or brother-brother, will constantly cause issues of stagnancy.

In husband wife situations of control occur in 4 forms seen from the more complex perspective of the 9 energies through the 5-phases which is based on the idea that all relationships outside of the balanced opposites of 3 and 2, 1 and 9 and 6 and 4 will have yinyang imbalance:

A. Abusive control

The husband-wife relations above can generally be called balanced/ harmonic opposition. They are when yin and yang are balanced, so one yields to the other BUT -when the yield force is too yielding or the dominating force is too dominating and yang, this causes an abuse of the yin principle. So for example 3 can over dominate 2, 6 can over dominate 4, 1 can over dominate 9.

B. Passive control relationship

The relationship here is the spirit 2 which is a yin spirit attempting to dominate a water energy 1 and is a yang spirit. The female energy attempts to dominate; this goes against the yin yang principle. If it is forced it is created by the mind-identity. However the pathology of it is that the yin cannot control the yang, as she has no focusing, penetrative power to do so. Th is creates an attempt to passively control him, which causes damage to both, but mainly to her.

C. Passive stagnation relationship

This is 9 attempting to control 7 or 7 trying to control 4. It is basically two yin energies, with one attempting to take control. It is a very passive aggressive power struggle and often ends with neither side being able to feel satisfied or being able to move each other.

D. Attack opposition relationship

Th is is perhaps the most destructive and acute of all. It would be 5 or 8 controlling 1, 3 controlling 5/8, and 3 controlling 6. Basically there are two yang powers, one attacking the other and the other firing back. Th is is an aggressive relationship and often includes one of the two energies compromising themselves and futile force being applied in order to find bridging of unity between each partner. These are often over before they have begun.

In all of the above there is no right and wrong. All these forces are at work in nature. However for smooth flow and going with the flow the attack opposition relationship is often something that causes much more pain and struggle. It is often a mentally constructed need to abuse or be abused within a relationship....i.e. it is not something that relates to unity but that which attempts to go against it; as this type of combination requires force, it will create sufferance within the relation to itself and to surrounding parties. The Tao always speaks of going with the flow rather than being a static brittle force against the flow.

The way to see the various relationships above is as a three-legged stool. One's life needs to be balanced between time spent (sat in) three areas:

1. just being and doing on ones own and with friends of a similar disposition (brother sister relationship),
2. with family or situations of giving/ offering and receiving (mother child relationship), and
3. In relation to ones opposite/ partnership.

To again balance between these three portions is to allow for harmony in all aspects of the world. There is a larger understanding here that makes up the circle of life. Some people will have tendency to be more on their own than with other people, more "cliquey" than others by nature. Some will be more involved in family and giving or recieving, others in opposition and relationships, however the balance of all of life together as one whole will always form a circle. The quality of the energy determines the balance. Generally however if the 3 legged stool of relationship is too skewed then legs start to break. If too much pressure is applied to 2 portions rather than 3 or 1 portion rather than 3 this causes issues in whichever the area one is focused. For the most part the mind-identity devised nuclear family focuses too strongly on the partnership and as a result problems emerge because one person requires to be a total world in relation to the other. While stress may be applied to one area rather than another based on natural expression, pathology is when 2 or 1 leg is being applied to take the whole weight, this is very much the basis of why marriages and relationship become a source of major pathological sufferance and cause much dis-ease in today's society, as well as major lack of understanding of the nature of sexuality (Please see the book "Sex at Dawn" in the Annotated bibliography for the true nature of the human sexual animal).

Energetically, relationships balance through authentic (contextualized-mind) approach to different people. If we notice we can spend 5 minutes with a person and with others several hours, there is usually mind-identity at work but there is also more deeply an energetic interaction behind the scene, which is actually the foundation of the mind-identity anyway. Body-spirit was before mind-identity so mind-identity is always fundamentally enveloped by body-spirit and as a result even though it can attempt control, body-spirit and different constructional energetics play a large part in relating to one another. Social norms and ideology derived form mind-identity often hold these in check, but fundamentally if we allowed ourselves to be directed towards people we resonate with naturally then we would find natural balance of the 3 pointed balance of relationships happening without "doing" anything. Some energy doesn't mix so well with others (like oil and water) for long periods of time and works best for longer periods, all is necessarily in the mix but all needs to respond to each other as it is, and so, even in difficult situations where there is contention, one can see through to the naturalness of this and the realisation that it will not last but resolve and move to calmness, as confrontation of high energy can only last for a short time before they burn each other out. Other situations have more longevity. There is acceptance for all, and no-one is in control of it even if the individualist's dream is that they are in control. If we allow for no-control, then we allow for the possibility of connection and order spontaneously arriving through natural balance being seen, this will occur anyway but to see it and accept it is to go with life's flow, which is easier. The 3 legged stool is idealistic but it is an expression of nature finding balance.

The above seems very formulaic because of the expression of not really being able to fit the nine energies into the five phases so there is perfect balance. However, this perfect balance can be simply viewed by looking at the energies as whole units. For example, 4 and 7 as one- metal energy, are in husband-wife relationship with 3 and 6 as one –wood energy, or water as one of the five phases tonifies all of wood, not just the 3 energy. The point is that when we are being specific about cycles of nine, 10 and 12, we need to be clear as to what they are as parts within the five phases and the combined energies associated with each phase that make it up in relation to these cycles. In this way, we can use them as one whole, not as fragmented aspect. Deeper than this, all of these unify with the deep stillness of Wu, which is the root of and permeates all the myriad expressions here. If one is in touch with this, then the intuition of being knows all that is expressed, and to a depth that is not limited by the abstractions of mind-identity. Please beware of believing anything mental, as being anything but signposts to truth. The signpost is not truth itself; it, too, is as smoke. In fact, used clinically, the above expressions are never incorrect if seen from the perspective of unity.

Notice an important point here that we will see later too: in compatibility issues, yin within yin women, which means women spirits within female form, are attracted to yang within yang males. This means that those who are mixed also create balance. Hence yin within yang males is attractive to yang within yin females. This is important, as when we look at the unity of the body and spirit below, the male and female form is only one expression. The spirit within is the driving force, and it is this internal aspect that needs to balance as well as the male-female balance that is more physically obvious. Yang within a female will be more yin than yang within a male, and the same within yin. The body acts as a buffer for the spirit energy, one might say, so this means that with yin-within-male type and yang-within-female types, they often play to aspects of themselves that are more yang or more yin. Total unity is achieved only with acceptance of it all together and all now.

All the systems of understanding of the ancient wisdom expressed here will only act as a guide, as anything that becomes specific and acute is narrow and doesn't see the context. All relationships and interactions have no right and wrong attached, whether to food, to people, to the environment, to anything. All relationships of energy have an effect. If we can understand the effect through feeling, then all systems are simply stabilisers of mind-identity, of which we can let go in the moment of "knowingness". These expressions, such as the principles of 2, 3, 4, 5, 10, 12, and 64, are simple methods to train ourselves to let them all go. This said, we can see them in context and know the power of what they are. Nothing is a replacement for your own sense of being, as these things only train mind-identity to be flexible and see broader. If they create narrowness, then they are not being used effectively.

2.4 Accepting the Unity of the Jing and Shen Bodies

One of the central themes of this text is to bring about the acceptance and the conjoining process of notions of the body and the spirit. We have seen the constitutions of the bodies in the five categories, and we have seen the nine personalities of the Shen and associated them with the principles and interactions of five, as applied to human life. We have also noted the interactions of the cycles of 12 in the interaction of bodyspirit expression.

It is vital to understand that all nine spirits reside within each person, as do all five phases. When we are considering constitution, we are looking at the strongest and therefore weakest areas of a system, so that we can understand its general patterns and interactions within the full spectrum. We are always looking at concentric circles of understanding the human in a particular; more yin relative to more yang, or more yang relative to more yin expression.

i) The Complexities of the Four Combinations

Now we can look at the complexities of the four combinations that follow:
- Yang spirit within yang (male) body

- Yin spirit within yin (female) body

- Yang spirit within yin (female) body

- Yin spirit within yang (male) body

If we consider that, of the physical constitutions, there are five males and five females, 10 times 2 for yin and yang spirit combinations makes 20 possible combinations of spirit and body.
In fact, there are four yin energies and five yang energies. Hence the total combination is 10 x 9, which is 90 possible combinations:

(table 2.12)

5 Phases	9 Energies								
	9	8	7	6	5	4	3	2	1
Wood	M/F	M/F	M/F	M/F	M/F	M/F	M/F	M/F	M/F
Fire	M/F	M/F	M/F	M/F	M/F	M/F	M/F	M/F	M/F
Earth	M/F	M/F	M/F	M/F	M/F	M/F	M/F	M/F	M/F
Metal	M/F	M/F	M/F	M/F	M/F	M/F	M/F	M/F	M/F
Water	M/F	M/F	M/F	M/F	M/F	M/F	M/F	M/F	M/F

The numbers represent the nine Shen spirits and the five phases of male/female represent the 10 physical constitutions.

An important point to consider is that the spirits within a physical framework are in fact one unit of the body-spirit, having a particular subjective environment. Or, we could say that the human formation is a collection of energy—a pillar of energy we call body-spirit. It is a unified expression like yinyang and is never divided. Even in death, the energy simply changes form. Humans can only have this view of the universe from the perspective of looking out from the inside. Although there are many other levels of communication/unity, from the sub-atomic to the galactic, the human level is where communication/unity of the human world is mainly associated. This perspective, unless one is aware of what is beyond its physical borders and is able to feel it, can cause mind-identity to latch on to this and form the concept of separation from the whole. The birth of the "individual" through a mental perspective is due to the process of being yang/spirit within the yin/body. Mind-identity cannot see beyond this, so it believes that there is a confinement of the spirit within the structure. This encourages the process of attempting to expand the self, oneself, in order to escape the confinement or become bigger, like a tree pushing up through the ground. This anxious expansion—and over-focusing and overemphasising the individual and its expansion—creates massive mental egos in people, which puts greater and greater demands on the seeming "exterior" environment, as they see it as a fight for survival to simply "be" in their separate shell. However, this is not the case and is never the case in the animal kingdom, where what looks like a fight for survival is in fact not a fight at all, but a simple expression of Nature as it is without human ideas of separation. Personification or humanization of the natural world is applying mind-identity-based structures onto something, which carries on perfectly well without it. Another example of this is the Chameleon or the hover-fly that "looks like a wasp", these creatures are seen as being "fake" or lying to the audience of nature…. far from it, nature never lies, it is always totally honest and innocently clear and without cause or reason. There is no intention around these phenomena, not like the re-meditated ideas of the human mind-identity. The natural state of these animals is to adapt as expressions of nature, electromagnetically morphing and changing to express themselves in the environment. They move with all else creating all the different forms we see, some looking like each other because thy simply have formed this way because of the infusion of energetics that IS all of nature at any given time. Evolution is a process but it isn't, as Darwin knew based on survival ("Survival of the fittest" was not Darwin's ideology but actually the Economist/ philosopher Herbert Spencer! Relayed in his Principles of Biology, 1864, Internet Reference 4), but in fact on "natural selection" or better natural transmutation, the Oneness of nature moving to create different forms through the changing of all of it together, this is evolution, evolution it is not a personal issue!

As a general rule, the yang spirit energies can more easily draw these mind-identified conclusions unless they're very much in touch with their senses. The stronger the yang and the heavier the yin bodies, the greater the pressure. This is interesting, as generally yang is said to expand, but expanding from within a form is like a pressure cooker. The pressure of being yang within a yin body is the difficulty to overcome, and the challenge is to see not only the unity of body-spirit and Earth and Heaven, but also of each of us together as One. Yin spirit energy has the ability to be stiller and connect more easily to sense of Wu, not trying to drive outwards while allowing the exterior to come in. Hence the 1/3/5/8/6 energies tend to have more mentally constructed egoism and are hence ego-centric, while the 2/4/7/9 types are more connected and represent yin energy within both male and female body structure but especially yin body. Again, male and female structure will further augment this, the female being the most connected and therefore often most intuitive/ sensitive or broad.

This relates to growth also. During youth, the energy is often hotter and moving outward. One has little chance of hearing the exterior change and listening to its energy moving towards us. It is for this reason that children are naturally energetic, but they are also receptive too, as their physical construction is not that powerful. Hence they get the best of both worlds, and it is often this that the exercise of Tai chi and Qi gong attempt to draw the body towards, being soft and flexible and sensitive, yet having ample energy. It is, in a sense, re-forming youth within the body, or gathering energy rather than dissipating it. As the body becomes older and tougher there is more power but also more resistance, arrogance can develop, or "self" image. Towards the end of life there is toughness but no power, this often means a continuation of what occurred or was learned during youth but sometimes with greater acceptance rather than fight/ struggle. Gradually letting go occurs towards death.

The next section of the book is about these 90 types, or the joining of the nine energies and the five phases and finding out what they are like. Each one has a particular combination of strengths and weaknesses and possible constitutional variations of difficulty, which could create dis-ease states. This will be discussed. Note that the cycles of 12 (stems and branches) are incorporated into this exploration and will be discussed also, after this section. If we placed them into this format of explanation, there would be 90 patterns x 12 animals, making 1,080 patterns, which I could not develop fully within the parameters of this book. The principle is that Earth is within Heaven, so in the next section, there will be nine categories for male five bodies (9–1) and female five bodies (1–9). We need to be aware of the power of the Heaven energy permeating all things and mastering all energy within the moving universe, which is why the categorisation follows this order. To order it as the five energies or 12 branches would give a confused picture.

As the five phases deal with the body, and the nine energies with the more ethereal-yang aspects of the spirit, the more yin-corporeal aspects of the spirit delivered by the stems and branches affect and influence both body and spirit and are between both, so it is important to note its influence (see end of this section). As a very general interest point, note too that for the energies of 1, 2, 5, 6, 7, and 8, perhaps the 12 animals have more importance than for 3, 4, and 9. This is because the previous energies have more of a relation to the energy of the earth. The 12 animals have closer relation to the Earthy qualities of the spirit. Therefore the lives of these people can be more clearly understood as associated with the sequence of 12 animal than of nine energies, whereas the 3, 4, and 9 spirits relate more to the ethereal, so the nine energies dictate most aspects. (6 is complex, as it is yang but earthbound, so tends not to be spirit associated; 4 is yin but Heaven bound, so can be associated with spirit but is too cooling; 6 and 4 can be somewhat interchangeable.) Notice that the nine energies, when inside the physical structure, will associate themselves with the region (governed by an organ) that is most like their own energy. These are only general associations; they are not specific locations of the spirit, which permeates throughout the system and is the colour of the whole energy. But one could say that it is where the energies house themselves and have association with the ethereal aspects of the spirit within each yin organ. Note too that some organs have more of an earthly or a more spirit/heavenly association, as described in Chinese Medicine. Again, we need to remember that we have all of these aspects in us, just as we have five phases. We also have the nine aspects of the spirit in us, but the constitution is the degree of one above the others, which gives the person a particular "spin" in life.

The yin-corporeal aspects of the Shen/spirit are more associated with the 12 branches. However, the nine energies associated with the 1, 2, (4), 5, 6, 7, and 8, which are more yin, do also have connection with the yin-corporeal aspects of the energy. One needs to see this as a spectrum of energy that blends, as in the Heaven energy of the 9 and Earth energy of the 12 join and blend, so the associations are in some way arbitrary as specific units, and only clear as overall pictures of unified interaction of 9 and 12, and all of the other sequences discussed in part 2 of this section of the book.

In this part of the book on each type, issues arising from the cohesion of body and spirit, or seeming opposition of body and spirit, will be looked at briefly. However, it is important to note that the spiritual energy has the stronghold of power. Even if the body is one shape and the energy is opposite expression, the spiritual energy can drive and mal-form the body to suit its mind-identified wilfulness/abuse. This is a form of lack of acceptance. In order to be clear and well, we need to accept both body and energy and find the mid-place where, moved by the spiritual energy and beneath this Wu, this way the body can express itself without severe damage or consequence. In real terms, this is simply "be yourself", but to be to totally authentic, you can't choose parts you don't like, and also there is no "you" to be anything, so essentially this is a phrase pointing towards letting go or acceptance of the present state of things. Anything else is all the work of mind-identity, which is the pathology of fragmentation. In an understanding of Oneness, life and death is no longer a dualism but a continuum. This is neither good or bad, right or wrong; it is just a way that allows the whole body-spirit to be one with the process of universal flow, rather than fighting it with mind-identity.

ii) Other Important Aspects When Considering Constitution

Age

The body goes through different parts of the cycle of life, just as with the seasons and the five-phase chart that showed the life process of the body, beginning with wood and ending with water. When looking at the following expression, one must note that the age will determine some change of expression in the body. The spirit energy, however, is quite constant, although it too is slowly evolving throughout the process of being within the structure of the body.

Proportions/Size of the Jing body

The quantity of energy that forms whatever structure of body can vary in size. However, as a general rule, the wood and fire are more expanded, and the earth, metal, and water are more accumulated. Even here there is variation; water types can be very large people, and fire very small. It just depends on variations in populations and cross-cultural breeding. Remember, there is no rule; this is all a spectrographic view. Males are larger than females—again, a general observation, but further exploration of this follows.

iii) The Ruling Strength of the Nine Energies over Earth (Heaven Envelops Earth; Mind-identity Develops)

The energies of the 9 drive the body. This is because the post-Heaven energy envelops the earth. Those who have more of a yin-corporeal earth/metal/water base to their spirit energy will create more interesting earthly phenomena in life, but they are still governed by the Heaven energy.

Hence the energies of 3, 4, and 9 represent the yang.

The energies 1, 2, 5, 6, 7, and 8 are more of the yin. As a general principle, the energies of the yang do better with body types that are in mother-child correspondence with them. The spirit energies of the yin are best supported by the mother constitution of their bodies. This is due to the fact that fire and wood tend to get too overheated. Hence the need to cool off, or express themselves, and be dispersed. The yin energies tend to be cooler anyway; therefore, they need often to be supported and strengthened, rather than drawn from or dispersed. However, one also finds that 1 energy, as it is moving into the yang, needs to be drawn out or dispersed, as this helps its energy out to spring. The Earth energies of 5 and 8 need to be dispersed also, as they are yang Earth, which is close to fire. Hence those that do best with a mother role are 1, 3, 4, 9, 5, and 8, and those best in a child role are 2, 6, and 7. The husband-wife spirit-body connections are considered to be difficult, while the brother-sister connections are considered easier:

> 1st — Child or mother
> 2nd — Brother-sister
> 3rd — Husband-wife

This is in order of comfort of connection, body and spirit. The husband-wife connection is therefore the hardest combination to accept. In all cases, the fragmented mind-identity and pathology set in only when the body-spirit is not a uniformly accepted being, and instead the spirit attempts to rule the body (mind-identity). This creates sufferance.

iv) Natural Balance and the Interaction of Body-Spirit Combinations as a Whole

Essentially there are four categories of male-female body-spirit combinations:
(table 2.13)

	Yin spirit	Yang spirit
Female body	2, 4, 7 (Yin spirit in Yin-Zang organs) – Earth/Metal-Wu	1, 3, 5, 6, 8, (9) (Yang spirit in yin – Fu organs and heart) - water
Male body	2, 4, 7, (1) (Yin spirit in yang – extra-ordinary fu) - fire	3, 5, 6, 8, (9) (Yang spirit in yang – the meridians and surfaces) -Wood

The above is the basic way of looking at body-spirit combinations, just using yinyang rather than the 5 phases as a base. Here we can represent the whole of humanity as a singular organism. By looking into the way the different expressions of the body react; we can see how if we look at the broad perspective of the nature of the body-spirit of humans in combination, they represent different aspects of the one organism that is the human. The yang within yang male dominates the surfaces and the defence and protection. His concern is to protect the yin within yin at all costs, and he balances her inner quality. The Fu/ yang organs of the body are internal, yet they are moving exteriorly so they belong to the female quality that has this nature and a water expression. At the same level is the extraordinarily fu, such as the gallbladder which is cooling but still on the same energetic level (cooling) than the Zang. This organ is moving inwards, whereas the rest of the fu move outwards.

There is a balancing going on between the water and the fire qualities in the above table. Please note that here we look at fire being the masculine because the body is yang but the spirit is yin. Therefore we consider them to be yin, whereas the opposite is true of the female yang within yin. The body or shell is yin, but in inner quality is yang; these two are hidden. The key is to see that the spirit rules or uses the body vessel, so it is key beyond the body's physicality. If we are just talking theoretically or physically only, then the male represents the water and the female the fire, or the male the Heaven and female the Earth. This is a stereotypical ideology, but here we are looking at a more actual/ non-theoretical picture. The deepest layer inwards is the yin within yin. These people are without surface covering and need the protection of others, and that is what they find. They are fundamentally right to be still and quiet and internal, and their resources are used by others, most notably by the yang with yang expression. These are the organs of life. The yin within yin is the closest to Wu and so the underlying quality of these people is at the root of life. Not to listen to the root is to forget the whole tree. In fact, most of society is dominated by the surfaces, the yang within yang. The female "emancipation" comes from the Fu organs, the yang in the yin. They try to push outwards at the over-domination of the yang within yang type's aggression and pulling of the yin within yin types. The yin within yang males are the "new man" types, which essentially are those who will tolerate the yang of the stronger female, whereas the yang within yang types will not, instead attempting to force them into submission, something that doesn't work. If all of society could see their part in the whole, this would create unity of understanding.

An important point to note is the different nature of the 2 forms of what in present day terms is called "femininity". The Yin within yin expression is true femininity it has no real voice but in a sense doesnt not necessarily want one. They are utterly yielding and feminine and in the mind-identified masculine world do get heavily abused and sacrificed in various forms but still continue to yield. The other form of feminine is actually not feminine but female body with masculine spirit. This feminine is rivalled and aggressed by the male domination because they, rather than protecting the other female energy, attempt to vie for power with the yang within yang expression. This is very often what happens when yang within yin women attempt to dominate situations and say that they are protectors of what they see as being yin and "defenders of the faith" so to speak. They are involved in just as much self-promotion and individualism as the yang within yang expression themselves. It is important that in the context of this book the word feminine is used to denote the true yin, which is the female with no voice and no direction and is the source of the yin. The females who are yang internally are a form of yang energy; the males who are yin internally are a form of yin energy. This is about looking beyond the surface.

v) The four body-spirit categorizations explored

The nine energies are vital for our understanding of each other in relationships. If we simply look at the yang spirit qualities—1, 3, 5, 6, and 8—all these qualities will take the lead and direct relationships. They will be the prime mover of the relationship. That means these people will direct through intention where the relationship takes place, how it takes place, the circumstances, and the situation. They will also formulate the structure and the boundaries of the relationship. It's all about what these people want and their direction to get it. Hence these people are the leaders. They have no choice but to lead. In fact, it is a heavenly ordained leadership; it is their purpose to lead.

For the yin 2, 4, 7, and 9—for health, these people need to take the position of being led. These are the resources of life and the core, and so for these people, there is little choice in the pattern of being led. They are led and are moved by any form of yang energy, so it is very often that these energies, if the yang energy is leading them towards danger, will follow. The issues for the yin are discernment and placement. These two things are vital for the yin to be correctly suited to their environment. Discernment means to yield to the yang who is moving from a sense of truth within, not from a mind-identified state. This is coupled by placing oneself into a situation where the yin will be noticed and energy will be directed towards her. Most yin energies, when there is no yang presence for a long while, attempt to create the exterior yang to form protection. They need to let go, as often this anxious tendency can draw them into manipulation and an attempt to control masculine energy to get what they require or to protect themselves from an abusive one. Energy needs to be focused into being exactly what they are but finding situations where what they are is clearly wanted and their resource is needed. Both in the male and the female with yin spirit, the tendency is to "fall in with" the wrong crowd or "be taken in by" people who are not supportive. This can become a habit. The process of connection to the core of the yin is to feel and discern the influence in their life that they want and need, making access to them easy for this kind of energy and more difficult for the destructive mind-identified quality, which will be abusive generally. His is the art of placement, something that is expressed perfectly in Feng Shui. For yin energies, the study of Feng Shui can help to understand the quality of making a house of a yin energy or a resource suitable and right for the inhabitant merely though the art of placement and balance.

Hence it follows balance that yin in yin energy needs to be balanced by yang in yang energy. And yang in yin balances yin in yang. This is a fundamental rule. When there are two yin energies, there is no movement (stagnation relationship). One of the two energies must attempt to perform yang, and does so very badly and often with a lot of stress and tension, as the movement will be one of force. This will happen very gradually and will not result in balanced comfort. If there are two yang energies, one must take the alpha position and the other must yield to it (compromise relationship). This means one gets to express its energetic power and the other has to submit to this. This end therefore is one person ruling the other, which often results in stagnation of energy and heat and tension for the partner in submission. This situation is eradicated if the partners are opposite energetically. So for example, 3 and 5 are wood and earth yang, or 9 and 7 are fire and metal yin; this causes even more problems as the opposition relationship means little communication, as well as there being an imbalance of the natural energetic of yinyang. These are vital basics for understanding how relationships need to work out. As the spirit rules the body this occurs in all relationship no matter weather the physical body is met energetically exteriorly by opposition of similarity (for a discussion of heterosexual/ homo sexual understanding please see Appendix 2, and the book "Sex at Dawn" – Annotated Bibliography).

The relationship is always governed by the spirit at the core which drives and expresses from the vessel of the body and which either goes along with the spirit or opposes it, as shown in the first section of this book. In relation to each other, the relation of yang spirits will cause a relationship of compromise; one will always have to yield to the other and therefore feel overcome or constrained by the other, never able to express the full expansion of themselves. Hence the relationship always becomes about compromise, and very often people in this relationship will always talk about compromise being the key to "success" in relationships! In order to balance this, one person has to be more yin than the other, and this is very often the female partner who will yield. If she does so, she will be filled with resentment for this for long periods of time, which eventually will cause dis-ease or "break up" the relationship explosively at some time in the future. Whichever partner yields will feel squashed by the other, and so it renders one of them in a "victimised position". This is a movement of the female to a yin position or a male to a yin position, but in either case, it's a victimisation movement, not a true sense of expression. The other kind of relationship, which is problematic, is the relation of yin spirit to yin spirit. Here, no movement goes on, and the relationship is one of stagnation. The two cannot move without another exterior force of yang being applied, so it tends to get inert and stuck, and the life is taken out of it. In this situation, one of the two partners needs to activate the relationship and move it; this tends to be the yang male partner of the two. However, it is not within his nature to do this, so he invests a lot of energy, trying to be masculine. This causes much stress and tension and eventual exhaustion of the energy. This is a movement to "domination" which doesn't work long term. It is important that we are not looking at specific qualities (i.e. specific qualities of the 5 phase energetics of body or 9 expressions of the spirit) other than basic yinyang of the spirit here, because the balance can be different forms of opposition relationships, or a mother-child and brother-sister relationship. A yang male and yang female relationship can kind of be balanced a bit by being the controlling partner, for example a 3 male (wood) and a 5 female (earth). But in fact, both are yang, and as a result the fundamental yinyang balance has to be taken into account as primary. Movements of relationship into "compromise" —victimisation or dominance-stagnation—are keys to understanding health in relationships at this very basic level before we get into qualities.

It is also important to note that couples are not the only type of relationship. Why people are relating is a hugely complex subject, but at the root, many people may be keeping a relationship the way it is, and many internal and exterior situations may make a particular relationship work. However, what equates balances is when the life within the "individual" is being expressed fully. This then has an impact on the people being connected to a broader picture, and the health of a society and the balance of it is always conducted originally from within. Hence the relationship of harmonic opposition is yin within yin and yang within yang, or yin within yang and yang within yin balancing each other. These are an ideal direction, which means health is emanating from the "individual" and therefore he or she finds himself unifying with all other seeming "individuals" into the body of humanity. In this body, a person cannot be an individual cell but is part of a whole, and so as this, in health, will respond to the other people around it appropriately based on instinct. Everything spoken here is not about trying to balance out patterns of illness on the exterior world or balance out problems of the interior, but is an instinctual movement of the healthy person who senses himself or herself and therefore moves towards people in ways that allow them to be fully what they are to the greatest extent. This also therefore has no political or social morays such as monogamy, marriage, the nuclear family ideology or any of this type of socially identified thinking. The fundamental of yinyang is that is it not something that can work in bits and pieces it is a total exposure of what IS, AS IT IS, that means an opening to everything, without exception, Oneness is unconditional.

(table 2.14)

Energy Type	Opposition Relationship	Mother-child	Friendship
1	9	7	
2	3	6	5 / 8
3	2	9	4
4	6	1	3
5		7 / 9	2
6	4	2	7
7		1 / 8 / 5	6
8		7 / 9	2
9	1	8 / 5	

As you can see from the above, this does not include all forms of relation, just the ones that are totally balanced yinyang as well as in the five phases. We can see from above that certain energies will always have issues seemingly unresolved in their lives. Finding true friendship or balanced opposition relationship or even provision for some energies from a mother-like energy will be very difficult, and so this shapes the energetic quality of the energies. Earth is very friendship and social orientated. Water tends to be a loner apart from the opposition relationship, and lake is all about the family relationship. The others tend to be mixes of the different qualities. These are idealistic, but they are vitally important and clear directions for harmony within relationship.

(fig. 2.30/ fig 2.31)

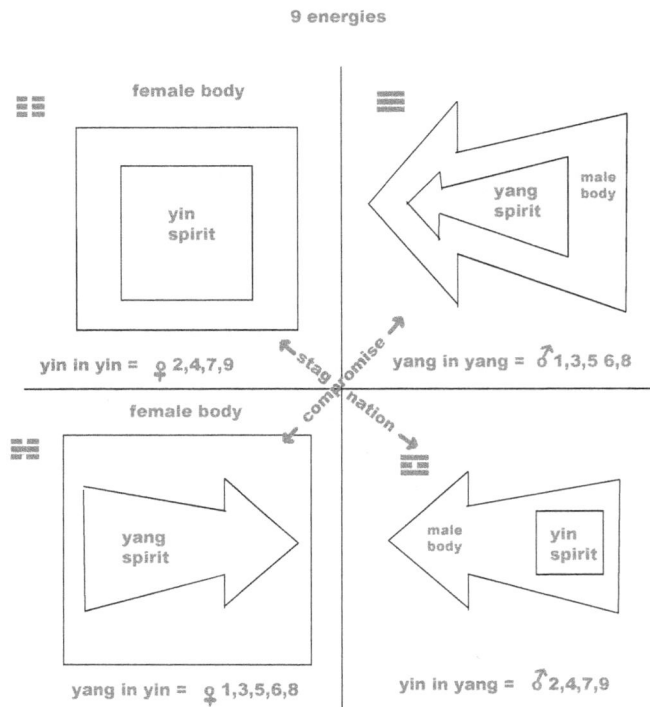

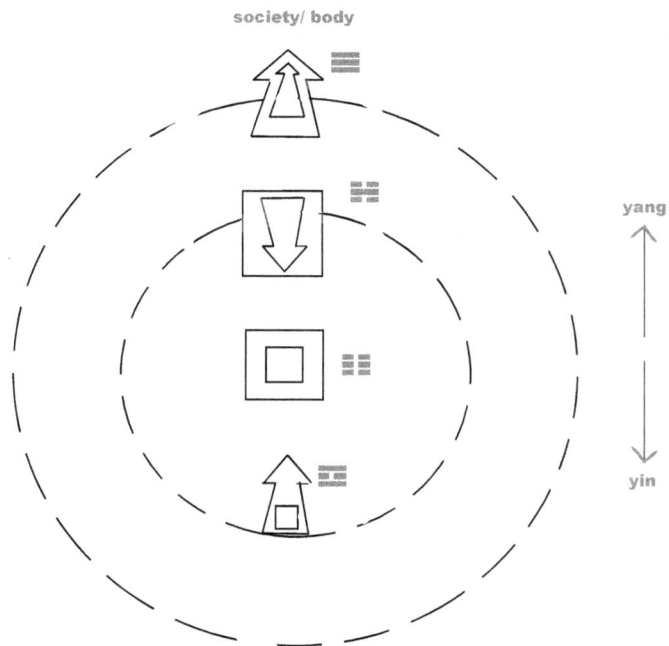

9 energies

female body

yin
spirit

yang
spirit

male
body

yin in yin = ♀ 2,4,7,9

←stagnation→
←compromise→

yang in yang = ♂ 1,3,5 6,8

female body

yang
spirit

male
body

yin
spirit

yang in yin = ♀ 1,3,5,6,8

yin in yang = ♂ 2,4,7,9

society/ body

yang

yin

The diagram above expresses the whole of society as well as the whole body in relation to these four expressions of yinyang. At the centre, there is the yin. This is the vital-organs of the body, the deepest place of yin. These people are like the organs of society, the resources to be used but not to be abused. Circling this on the outer ring is the yang within yang, primarily focused to be the king and the direction of the whole movement, also the upper and the outer, the protective surface for the yin within yin who will feel completely unprotected and vulnerable without them. The yang within yang can abuse the yin within yin, which is like an autoimmune effect of attacking the self and destroying the insides of the body/society (the mind-identified state). The outer cycle is the meridians of the body, and the muscles are the head and upper body. The in-between are represented here by the yin within yang and yang within yin. The male yin types are the gallbladder and extraordinary fu energy cooling but also within the yang relative to the core. The female yang is the fu organs, still internal yet moving to the exterior. This represents the communication systems between exterior and interior. The male yin goes from exterior to interior; the female yang goes from interior to exterior. This builds an internal balance, and these people are complex expressions, whereas the interior and exterior of yin within yin and yang within yang are simple extensions. Simplicity balances simplicity, and complexity balances complexity here. So this is how people are and also how the organs of the body are and the whole of society.

Looking beyond just the physical structure, we can see how those who seem to be groups of people who are obscure and almost ostracised are usually ostracised by the yang within yang types, who judge based on simple projections. For those they box as "homosexual" or "coloured" or of a particular "class" or "creed"—these are usually complex expressions of mixes of yin and yang which are just not understood by the rigidity of mind-identity but wholly accepted by the body/society and the yin as a whole. This is very important, as the whole purpose of all these expressions is to see the equality of each expression and the value of it as "fingers of the same hand", of which we are. It only seems we are individual. The seemingness is the illusion, and the illusion is most strongly held by the yang within yang. The yang types have the response-ability to connect to the yin and see their place as the expression of direction for the yin, and the yin as resource for them. To see oneself in the scheme of things is inevitably to join with Oneness and so end individuality, superstition, suppression, judgment, and all forms of separation.

The yang-spirit male and the yin-spirit female are simple expressions. They might see things very clearly as roles of male and female. The yin-female naturally will accept leadership only from a male because to her all males must be of the masculine nature. They see men as fulfilling a certain energetic role. Yang-male is the same, seeing women in stereotypical roles and often within this society falling into stereotypically "dominating" positions. Interestingly these males will expect all women to behave to some degree submissively. It comes to a surprise and woe for both the yin-female and yang-male when they come across a yin-male or yang-female. This "grey-area" is very difficult for the pure polar energies of yin-female/ yang-male to engage with. If a yang-female tries to take control of a situation, often the yin-female will resist because she isn't male, often the yin-female will consider another female a challenge to her male attention if she is more yang-female and this can cause a kind of jealousy pattern, which the yang-female is not aware of, as she is independent of the male. Another example of judgement is when a yin-male follows a yang-male or a yang-female, he is seen as weak or stupid. The yin female is dependant on yang-male energy and is polarized by it, waiting for it's movement, or hypnotised by it's direction, the yang-female isn't and is independent of it moving into thinking that yin-females are "weak". The yin-males are dependent of yang energy also but not the strength of the yang-male which will be often resisted to and considered "forceful" (which it often can be if their demands are not met) a certain extent, yet they will follow the command of a yang-female more easily. This judgmental pattern of the yin-female/ yang-male often comes about because something "doesn't fit" with social norms based on polar opposites.

Generally society is greatly missing to see the yin, what is considered female-rights is often an attempt to de-feminize women to be more "like men". Today the yang-female has been acknowledged and also to some extent the yin-male but there is never a time when all 4 expressions have been accepted as one whole. When one understands that the simple type expression of yang-male and yin-female find it very difficult to actually understand the combination of qualities of the yin-male and yang-female, then one is able to realise that often these types are talking at cross purposes. It's an important point when observing and interacting. Generally the yin-females will group together and consider all females as of the same ilk, the yang-males will do the same, they will not expect men to behave any differently to how they behave. For the yang-female yin-male this is very different, there is a complex and "equality" of interaction but is in fact still about leadership and yielding: the male leads with his body the female with her spirit hence there is balance. This is very different from the type of expression that can be associated with the yang-male- yin-female type interaction, which is much simpler of the male leading with body-spirit and the female just follows this. Neither is right or wrong, all threes 4 make up the big-picture of our interactions.

As a general point, it is usually the case that the yin needs to realise that they are the source and to stay internal to see themselves and be internal expressions. They will often be told by the external world that they are wrong and be tainted by that, especially by the masculine expressions that fly in the face of what is naturally a masculine role for yang within yang males, through total individualism and judgement. This judgment will also come from the yang within females. This, in both cases, is abuse of the yin. The yin qualities need to be seen and respected, and these people need to look within for the answer about what they feel—never to the exterior. The yang need to look outwards. The answer is not held in their beings, but they need to connect to the yin for both male and female body types holding yang expression. The answer for them is without and is into the re-connection to the source, which they find through seeking and investigation of the yin. If they do not respect the yin and abuse it, they will be harming themselves, like cutting of one's nose to spite one's face.

The yin is the source. People you know who have a yin spirit expression need to be acknowledged. These are the hidden aspects of society. When the yin is listened to, all things come to order. When the yang then leads with acknowledgment of the yin, they are moving from Oneness. For the yin, it is always a matter of staying internal and not being affected by the chaos. This is called steadfastness in the I Ching. It means the quality of constancy and stillness. The yang are like children and will act in this way. When they connect to the yin and sense the truth that the yin realise, which is that there is always underlying Oneness, then they are freed from the idea of a separate self. This allows them to lead without abuse of the yin and with openness to direction from Heaven. The yang doesn't need to yield; the yang needs to listen, to understand and to know their place. If they are the head, they are no better than the tail. This is the key to balance. The Tao Te Ching is aimed at the yang principle to allow it to re-connect to the yin. Those of the energies 1, 3, 5, 6, and 8 need to read this as a basis of life direction. Those who are 2, 4, 7, and 9 need to realise this in themselves. The narrowness of vision and complexity within this narrowness of the yang will always be undermined by the simplicity and yield of the yin. This is expressed in combat where the masculine fighter, perfected in all the skills of the external arts finds himself at a loss when dealing with the yielding of a yin female who knows her self and body (this natural body movement formulated in Taiji Chuan). He wears himself out as she yields to every step he takes. Finally he collapses, and she gets on with what she was doing before. This is the Tao of yin, and this is the truth to really sense.

No way is better, but the way that predominates for most of human history due to its already yang moving nature is the yang, so the yin is the cure, and it always will be. It's the simplicity behind the complex foreground of life. Much of this book has dealt with the complexity of the yang expression of medicine. But the pure and simple yin is its root and is constantly expressed as the Wu background. Yang is the complex foreground ever seeking; yin is the endless background ever steadfast. In the I Ching's first hexagram, the text describes the state when all the yang lines are seen together. It says:

> *When all the lines are nine (solid) [i.e., all six lines are yang lines], it means:*
> *There appears a flight of dragons without heads.*
> *Good fortune.*
> *(Wilhelm, 1951)*

The power of this expression is insurmountable. When everything is in a yang state, there is a possibility of total abuse and annihilation of the yin, so the text states that it is only when the dragons are headless, meaning it is being conducted from Oneness not from individuality, is there fundamental "good" or balanced expression in Nature. This means acknowledgment of the yin. The second hexagram, Kun, is really the most important of the I Ching as it expresses the nature of the fundamental yin, and as such the foundation of Taoist philosophy.

Please note: in discussion of the various types below are always talking about CP not DP, which means no yang deficient conditions will be looked at, all the patterns below will be within yin deficiency of the body for male and female body types. Also please realize that the CP's are associated with specific environments that they are found in. When one is talking about a wood CP for example this relates to the Eastern region associated with Su Wen 12 and all the environment of that region, it is a growth form of a particular climate, we are therefore considering these CPs to be within their particular climatic environment.

Part 3

The Male Body-Spirit Combinations

3.1 The Male Body and the Nine Energies

As we are using the Heaven principle 9, we can see that yang is born first, and so Heaven is before Earth. Hence we look at the yang numbers first and in the format of the 9 male body types. Yin is only considered first within the context of the Earth when the mother-yin (representative of Void or Wu) forms Jing, and Shen is of Heaven but within man. Therefore man is within Earth. I refer you to the diagrams in the beginning of part 2 of this section, the introduction sections, and to the cover of this book. We are constantly looking to see things in context and therefore to see the whole and not the parts.

i) The Male Jing Body

Before we move on to the categories, we should make note of the male body structure. The main obvious difference between the sexes is the reproductive system. The reproductive systems are an expression of the form of the energy being held within the structure. The masculine body is harder on the surfaces, and the musculature relative to the equivalent female is usually larger. This may be slightly different in the earth, metal, and water types of construction because the yin of these energies means that the yang is held down. So females and males may be of similar proportion, but certainly for the yang body contractions of fire and wood, these will be male-dominant/female-submissive body structures and there will be a noticeable size and muscular difference. Generally, the energy is more on the outside. Generally too, the legs will be weaker than in the equivalent woman, size for size. The upper body will be stronger and more powerful, as the yang energy rises and expresses in the upper body.

Men live a shorter length of time due to the energy being expressed more to the exterior and used up faster. In sexual contact, the man expresses energy in semen; this draws directly from his physical constructional energy and can exhaust his bodily energy if constantly used. In fact, most practises in Chinese martial arts focus on using the physical sexual energy and cycling it through movement; if done for general health and simply to express the body energy this can allow for a strong body and enhance clarity of mind/brain, as this energy also cools and forms the bone marrow and brain. Forceful practices of sexual abstinence and the ideology of longevity as a goal are all about control and mind-identity, as is sexual addiction. If allowed, everything is regulated by nature.

The female body receives the male's energy; this is why the female body is entered and stimulated by the masculine energy, and also why the masculine energy is given to the female to create life within her (yang within her yin). This life is born from yin and yang energies combining, but the exterior energy of the man provides this energy. We can see that the male body is yang, and it moves; if it doesn't move, it becomes overheated and exhausted. Exhaustion for the male body comes from yin/cooling resources being depleted, from overworking it, or from not moving. The masculine body is therefore the "expressive body type".

Within the understanding of the masculine energy, one must consider that the body of the man needs to be in dominance of the female body. Hence the masculine spirit within masculine body types will feel comfortable within their body structure. On the other hand, a female spirit within a male leads to difficulties, due to the society around them and the mental fixation on the physical. Usually, these types of males push forward more forceful aspects of their masculine nature, which would naturally not be present; in order to front an expression that matches what mind-identity accepts as masculine behaviour in the world. The yin within the yang male expression is complex because the sexual act itself requires yang dominance from the masculine to a certain extent, and so if the spirit is softer it can seem like a loss of potency, but simply it is a loss of direct forcefulness, something which the yang female expression greatly desires, so there is always a balance found. However if the yang within yin female or the yang within yang male, which is the front of expectation in the world, expects to be feminized by a yin-male, then this will be very difficult as he will not be able to dominate her and make her feel like yin within yin female for long, this is all a product of mind-identification as soon as we lose our way from acceptance, as it is.

Outside these however, there will be a mix of yin and yang in these energy types that can create quite a balanced human being, if one accepts the unity. For men, there is often a degree of difficulty in accepting yin within themselves in relation to social norms without presenting themselves as being "weak" males somehow—males that have mixed energy rather than a deficiency of yang-ness. Yin is solidness and heaviness and creates density of feeling and consideration before action. This is lacking especially in the yang within yang type male so they are typically more volatile than women. This can cause them to be far less likely to accept and allow things to be, as opposed to going with mind-identity or the heat of the upper body and living in this way. Hence men and mind-identity go together, which is why men and mind-identity and egotism go together, and wars often develop from this process. Fragmentation comes from mind-identity.

Feminists (interestingly often yang within yin women, generally) often talk of a male-dominated society. It is very clear that this is true; the masculine fragmentary idea has overtaken the female yin, cooler, listening and understanding a larger picture, way of being. In the Tao Te Ching, the readers, who were bound to be men at the time, were taught to look towards the yin principle to understand their existence and accept it. It is very clear that if society managed to control the yang principle, much suffering would be alleviated and many fragmentary ways of behaving would be altered. Most meditation and ways of accepting are therefore more directed at men, for it is literally within the body of a woman to be naturally clear, if she is aware of it. For men, it is natural to go towards a situation of creating fragmentation. With the heat rising in their systems, filling their brains, they can easily lose touch with the Earth, which is destructive for everything, including shortening their own lives. Not that this is right or wrong, good or bad; in the end, none of this matters. Please remember this throughout. Also note that whenever we say "spirit", it means a more expanded energy, and when we say "body", we mean a more condensed energy form. All things are unified, so we are looking at expressions of the body-spirit. This means that the coarser, more structural Earth energy and the rarefied spirit are a spectrum of yinyang, and this yinyang is the layer of expression above the Origin of life—Stillness, which is Wu or Void.

The mind-identity of men tends to be very, very active and ever-narrowing. Whereas women are much more of the body, men are of the spirit. This means men are of the change principle, which is the activity of mind-identity. Mental formation occurs when the spirit and body do not accept the unity of themselves and of creation as a whole and the formless reality behind what seems like separate forms. In men, there is far more focus on the perception of change, and so the physical structure of the body seems to get in the way and be an unacceptable obstacle to expansion. This causes the greater problem of mind-identity in men than in women, who, because they are naturally related to stillness (earth-yin-body), can contact it more easily and are less influenced by change. Spirit, in the context I use it here, has to do with yang; this is not "spiritual", as this is associating with connection to Stillness. Hence one could say women have a much better spirit-sense connection, due to their being less related to what I have called Shen-spirit. Most spiritual practises therefore are arranged for men, as we expressed, due to their natural lack of this and the fact that they need to be grounded in the reality of Wu. Masculine mind-identified ideology is that bigger is better, more power is better, and interesting the crown, is seen as more power, the top of the head where the yang energy dominates, the more powerful or bigger this region the better. Intellect is seen as power. Women, considered lesser, smaller, less upper and more lower energy, in fact is more rooted which makes the mind function more clearly and have more bodyspirit unity where as the male is often top heavy, and toppling!

The male mind-identity (ego), though fragile and easy to break up, change, or influence, has little structural format, which is associated with practises that attempt to break the ego quickly, such as Rinzi Zen practises. The ego can break easily when challenged by a brighter clarity, but heavy mental use over a long time using entrenched patterns can make mind-identity almost impenetrable.

If there is a sickness, it is mind-identity, and if there is one who harbours this sickness, it is the masculine body. This is represented as the snake who gave the apple to Eve in the Garden of Eden. The base principle that corrupted reality and formed mind-identity or the "tree of knowledge" is masculine. However, it is not mind-identity that is the problem, but the lack of context that focus and identity with it creates; this is the inherent dis-ease. Interestingly, the devil is synonymous with the snake and the phallic expression of the detached/cold/poisonous expression of mind-identity. The meaning of devil in the ancient Greek diabolous (from diaballein) is to "throw across" or to divide. The divider, the separator, is simply the tool of mind which then displaced the human being.

3.2 The Quality of Fire Energy within the Male Body
i) 9/Li /Fire

The fire is a female energy within a male body, so it is termed "complex". The fire energy is complex itself. It has aspects of male energy, being the exterior heat and expression, but its core is female, as it cannot move other than through reproducing itself to another stationary place. The yin energies always have an aspect, which is of the yin in its quality. Men of this type tend to overplay the fact that this energy is expressive and explosive, and they have charm and charismatic wit and humour. They often overlook the central emptiness or yin within their centre. Though to begin with these people can seem bright and radiant this is quite external, as interiorly there is an empty quality, which is ready to absorb and use something given to it for the exterior. This is a very yin quality, and so can seem cool. This is opposed to 1-water energy, which is cool in appearance but has a warm heart. Looking at a flame, we can see that right at its centre it is empty and that a flame is actually very soft—perhaps the softest of all energies. However, the brilliance that it gives off can be alarming and dominate, and this is what the masculine body likes to feel, so it moves towards this expression. For these energy types, it is often important for them to show off in some way, perhaps by buying expressive clothes or being lavish in ways that can seem slightly inappropriate on certain occasions. They are generally colourful and flamboyant in what they wear and way of expressing; this is to hide an actually quite placid and calm inner nature that they perceive as still and boring to be exposed. There is an empty quality with an external show of heat.

ii) 9 within the Wood Body: Lung Deficiency, (Kidney Yin Deficiency), Blood Excess

The wood body is strong and has a plentiful supply of blood and wood energy. Wood supplies fire, and so the wood body allows fire energy to burn brightly and with a strong drive. The wood body is yang; it supports the spirit as a mother and can be drawn on to provide its will. It can be a very volatile combination, as the wood body has blood that can be hot, and with a spirit that is fiery, this can make for a very angry person or someone who is in a dominant role and will take no other. Likely to be demanding and expressive, this body structure is a pioneering one, and so with the brightness of fire, this is the makeup of a leader or someone who believes they should be a leader and will possibly push themselves to this position. However, due to their nature, being a yin energy, their ability to lead and direct with effect is less firm than the yang energies. They do better as a second in command.

The main difficulty will be high blood pressure, as well as fluid exhaustion problems. Also, addiction problems can occur with substances that help to relax tension, such as alcohol or marijuana, rather than psychedelics, as they are already quite manic. They can develop depression if they stop moving for any length of time, as the body has stores of heat within the liver. If this is freed up and the aggression is directed towards physical movement, they can be quite spirit and "spiritual" orientated, although the fire energy does tend to forget this in today's society and drive towards financial and earthly wealth. They need to use their bodies but will be very intelligent, especially at hunting or seeking out. The liver energy relates to the genital function in men, and so the powerful expression of the wood energy here and the want to express this through the heart from the sexual energy makes this a very sexual combination, and sex orientated in life, when the head gets involved though this type can be overly focused in thought and can lose contact with genital function causing erectile problems.

iii) 9 within the Fire Body: Kidney Yin Deficiency, heart heat

This is fire within fire. This I call a "simple" type, where the body and energy match. The energy is very fiery, and it is likely that this person will live a very short but very full existence, one filled with excitement, exploration, and imagination. They will be naturally charismatic and flamboyant, perhaps slightly more accepting of the yin nature of their energy than the previous, but only just. This person can also attempt to be too pushing of their fiery energy outwards and so cause themselves difficulties. They will likely have high blood pressure, and this heat can go into the digestive system and cause wasting and thirst patterns (secondary diabetes like pattern). They can have manic illnesses and psychological anxiety and tension. They can be very fearful when taken to the limit of their anxiety. Heart problems are the key issue, and they are most likely to die of this if they are not careful. They have a spiritual connection rather than a purely physical one. They will have very fast minds and be good at music, language, and abstract mathematical approaches. They love being at the centre and in the limelight.

iv) 9 within the Earth Body: Spleen Yin Deficiency

These people can gain weight very easily and can become obese quickly. They will have a passion for food, as the passion of the spirit is driven into a body that is heavy and Earth-like, hence the spirit will be drawn to satisfy this body. Wasting and thirst patterns (diabetes like pattern) can be a problem, and again, overheating of the system can occur easily. The fire feeds the body's desire, so this person is likely to be a pleasure seeker, especially of the sexual and of touch. They could be highly intellectual, absorbing a lot of books and literature. Their life could be about "gathering together" or "gathering" things, and it is someone who loves social contact and being in the limelight. The ambassadorial role or company boss or second in command are very good positions for this one.

v) 9 within the Metal Body: Liver Yin Deficiency

This is a complex structural type, meaning the body and spirit are opposite within the system. This has the effect of holding in the spiritual energy, and this can cause difficulties. The energy of the body structure is built for longevity, but the spirit is built for speed and expression. This can cause the person to become agitated. The skin doesn't let out heat as fast as the expression comes. This can be a barrier of expression, and skin problems may occur with emotionally related issues. This is more common in Japanese and in some Chinese people. If the spiritual energy was not in opposition, the passage of energy would be smoother. They will tend to have bursts of energy, a kind of manic-depressive tendency, like the opening and closing of a lid of a pressurised container. Hypertension problems and possibly cancers can occur. They want to be expressive but don't feel that they can be as easily as a 9 with fire body. However, following just their spiritual path can lead to exhaustion of the body at a slower rate than the fire type; due to the body not being able to respond so fast, this can be frustrating. Acceptance of the unity of body and spirit is key for this person.

vi) 9 within the Water Body: Kidney Yin Deficiency, Bladder heat

Again, this is a complex body structure. The legs are heavy and solid, and the whole body is much denser and stronger than fire would wish to be. A fire frame is a much faster build, with responsive muscles and developed musculature of the upper body. This one is much more heavy-boned and slower; it makes the spirit much more held down and heavy, which feels like a handicap. In a sense, it is like an anchor on fire. This can be very useful, unless the fire energy of the spirit resists its body shape, and this can cause problems. A light and fiery mind and spirit feels trapped by this body and overcome by it. This can cause depression, high blood pressure, urinary problems, prostate problems, urethritis/nephritis, kidney failure, and infections. Also skin problems can occur due to the want of the spiral energy to move and push the body to action and overheating it.

3.3 The Quality of Mountain Energy within the Male Body
i) Mountain/Gen/8/Earth Yang

The mountain energy is a male energy within a male, so this is a "simple" formation. The energy is related to the youngest of the family of eight, sitting around the central table, in the 9-energy description. Eight is the youngest son in a family of three boys, three girls, a mother and father. He is therefore the second youngest of all, as his youngest sister—the 7 energy—is a girl and therefore seen as younger due to the Heaven energy being born first, the yang of the universe, Earth, the yin within this, being created second. Wu is eternal to the universal change.

This energy has a phallic symbol attached, the penetration of the sky from the Earth. Earth is usually a yin energy, but here we see a Tower of Babel in the form of a mountain reaching up to touch the Heavens. In this respect, it is very related to the forming egoic mind-identity and the rising up of power. However, it is constantly attached to its mother, the Earth, and never to be moved, which is unusual for a yang energy. Here, the yang is the hard protection and aggressive stance on the horizon, the nature of being the best and the first and an ominous power to contend with. These aspects fit a stereotypical male prowess. Note too that the mountains, of all of the nine energies, are amongst the last to be formed as far as the energy of the cosmos and formation of the planet are concerned. Water, fire, etc., are more basic, whereas a mountain is quite a complex structure. He is often associated with a rough diamond type character, with boyish good looks and charm, sociability connecting to the surroundings, and lording over a small community of his own. He is the lord of a small kingdom. He can have good overview of a situation, is practical and not spiritually related. Power, strength, and social interaction are his favourites; he would be captain of the football team, for example. He can accumulate much knowledge, skill, experience, and pride after gradually getting better and better, and slowly, truly reaching a pinnacle that is hard for others to reach unless by the same approach, so they can be experts in their field. However, often there is a physical and body-orientated component. Unlike the 9 energy, which has a more spiritual connectedness, this energy is heavy and of the body and the earth, so their physical structure and form is important to them. They can be more accepting of the physical realm, whereas the wood and fire energies often want to be elsewhere, higher up at least. These people can be soldiers, policemen, people involved in enforcing the physical power of constraint. They can also be ambulance drivers, train conductors, and people in charge of construction sites and the like. They can provide good understanding of the physical body and are attached to its function, and therefore often train their physical form.

They can be chauvinistic, but they often have a bigger mouth than their true intention and heart, which although is very self-centred, is also honest and loyal and internally they are softer. A James Bond type with a look to match, rugged yet cool, they are image conscious as are all the yang expressions and associate with the surfaces more. They are forever young at heart, happy to "play" with their new toys at whatever age, and biggest is best!

ii) 8 within the Wood Body

The 8 energy is based on accumulation and slow change. In the wood type body, massive energy is directed into the liver. This is the opposite structure of the mountain energy and can cause some problems. This is a complex formation. Having a strong liver energy doesn't help these types; it acts like a pressure cooker. The liver boils up the blood till the energy can hold it in no more, and then it explodes like a volcano. This can be a lifetime of accumulation or simply a volatile temper. If they hold onto temper rather than express it, which is more likely to be physical, they will get high blood pressure, heart attacks, and strokes. Cancerous type formation is also common. The solution is to reconcile the situation. The way this type can do this is with physical exercise and movement; use of the body is very important. The wood type body is the strongest, as it holds much blood and energy. Hence the mountain energy has to be able to expand its energy out, rather than blowing its top. This will cause calmness; otherwise, this person may be severely disliked. They may be considered rude, impetuous, angry, and violent in a very physical way. This can lead to isolation and deep depression, or alcoholism; many Modern Western people can have this energy issue. They can be highly sexual and more focused on the act of sex than the spiritual association, a very penetrative combination. This can also be the male shovanitic type.

iii) 8 within the Fire Body

This can create a much more harmonious relation than the above. In this energy structure, the mountain's energy is allowed to radiate. The mountain feels strong and comfortable in this body structure, and it is most beneficial to the spirit of this energy to grow and prosper. This union can allow the fire body to be used up creatively, and the heart of this person will not be so taxed as a fire within fire type. The key is that the body's energy is harnessing the accumulative and upward structure of the mountain. There can be issues of over-consumption of food, however, and this may express itself as generosity and social kindness. However, on his terms, he will be confident and charismatic, physical and sensually interested, especially sexually, as their energy tends to be directed towards penetrative expressions. Food connoisseurs and enthusiasts, they are very interested in volume and quality, but volume is more important to them.

iv) 8 within the Earth Body

This is a simple formation type. This combination is smoothly connected. The body and the energy are

Earthbound; they are completely unified. This person will be very practical, of few words, not a particularly good analytical thinker, but able to devour massive quantities of books and information, which is broken down coarsely and stored. They love food and eating, and this can lead to obesity and greed. Accumulative dis-ease is their main problem, such as cancer and heat dis-eases, due to food or alcohol and lack of exercise. They move slowly, so unless interested in a sport, they can become too heavy and stagnated. The difficulty with the Earth types is that they are body- and sense-oriented, and they do like moving their body, but only slowly, so they often lose interest in sports that would be most helpful to them, in their youth, and so accumulation sets in around the thirties or forties.

v) 8 within the Metal Body

The ways of the body this time are benefited by the ways of the spirit, again a good balance for the body and spirit. The metal body is also slower and methodical. This leads to a good combination. Achievement comes after long periods, and this body is suited to this kind of process. In fact, whereas the fire body gives a spiritual and lighter connection, the metal body is more sullen and tougher. It can achieve much, but stagnation patterns and habits will be hard to break. This cooler temperament makes for a longer life, but this person must be careful of not holding in emotional expression, as this means accumulative illness is present. Overall, they are more sullen and more introverted, a cooler mountain.

vi) 8 within the Water Body

As we have seen, 8 energy is associated with Earth in the five phases, but as an energetic formation, it makes up Earth as associated with the cooler period between the beginning of spring and mid-winter. As a result, the water body does conflict somewhat with the Earth energy, but not entirely. The water body is associated with the Ming Men and the right kidney. This powers the digestive system, which is the residence of the spirit energy of the 8 type, just as the fire type resides mostly in the heart of the person. The water body, rather than providing nourishment via the heart fire, provides nourishment via the Ming Men fire. It is the difference between the action of the pericardium, of moving the energy down back to the Ming Men then up to the digestion, or directly through the kidney yang to the digestion (see section B). The 8 energy therefore thrives on this too, but in a much calmer and less irradiative way. This is the heating energy of deep inside, so it creates a warm base for this energy to grow. It supports the growth in a similar way that fire would, but with more power and anchoring, and less exterior heat, which was expressed as showing off and needing to be number one. This type will become number one through their own hard work; they are a very hardy opponent and a very powerful body and spirit combination. However, Earth energy is connected/originating in the southwest region and the centre, so this extenuates the isolation of the mountain, the winter mountain. Hence these types are prone to depression, isolation, and loneliness, especially if they do create a massive internal heat, which can turn into a pressure. High blood pressure is therefore a very common problem. They can be cold and miserly and withhold, which does little for them socially. Overall, this is not an easy mountain energy combination.

3.4 The Quality of Lake Energy within the Male Body
i) Lake/Dui/7/Metal Yin

This is a complex formation. It is the yin energy of the youngest daughter and perhaps one of the most difficult for the male body to interact with. This energy is very complex by nature; it is between the seasons of the end of summer and winter, and so it is in flux. The energy is very placid and soft but still made of a substance that can form metallic shapes with a soft metal quality about them. This energy is considered the most dainty, tantalizing in some ways, just as the shimmering and playfulness of the surface of a lake can be deceiving. This energy is manipulative and deceptive at worst, but can be logical and meditatively clear. They are deep and reflective people, often conscious of design and shape. In men, this energy manifests itself as a player and a charmer. They can charm and even "tickle" their way through situations, as they are playful and charismatic. They often try to push too much masculinity to the forefront in male company, but with women, they feel much more comfortable. They can be a tease and can play games with women they meet. They do not aim to harm at all, but they like to have things their own way.

They can be very good judges and can be involved in law and education. They can also be good surgeons or people who are skilled with knives and blades. They are best, however, at visual art associated with design and manufacture, creating simple and elegant forms and very "cut" styles; fashion is another forte. However generally they will wear very plain cloths, simple and elegant but not excessively showy, being yin. They are somewhat refined and like the best of everything, if they can have it. As one can see, the image is not of the general masculine dominant position, so more yang women like them because there are fewer dominance issues. However, there they gain control, by attempting to calculate and be logically cool and critical. This gives them a kind of power base that suits their bodies better, it seems.

Prostitution is often associated with the number 7 energy in ancient times, and the change into the darker world in autumn brings this kind of energy of money and the Earth and sex together. These energies are deep and dark, not bad. The metal energies can have a much cooler vibration, and metallic energy has a much more detached way of being, so this can be very good in circumstances relating to law and order of society. These people can appear to be selfish and cool, and also calculating and suspicious. Sometimes this is true, but often they are following a morality, which they feel is unclear concerning what is "right". They are very good at categorising this for themselves. They are the youngest of all nine energies.

They can be quite self-righteous, an attitude which could be found in company bosses, but they are not very good at this role and would do better in second command or doing a specifically allocated job. However, men will rise to this challenge. Metallic energy finds its sea to swim in computers and technology, and these types will create a connection to objects as well as people.

These people can turn easily to a meditative path and towards stillness and awareness in association with monastic type traditions. This is because they not only feel different, being yin within yang, but also have the great ability to be still within themselves and so can find deep peace in this way. However, in modern society, this stillness renders itself as an association with form and shape, or is pushed away as it feels too "feminine". Attachment to form can turn in on itself, and these people can become superficially expressive and shiny but, deeply depressed and conflicted internally, holding emotion within to "save face" at all costs. This leads to further sufferance.

ii) 7 within the Wood Body

This is a complex variation. The mix of the highly potent wood energy and the spirit of the soft and complex lake energy is a opposition of types. This time, the body is stronger and the spirit is softer and cooler. This means that lack of activity can cause this person problems. The calmer, stiller energy of a lake doesn't know how to control the turbulence of strong wood energy, and so a withholding of expression often occurs. There is no reason for the energy of the body to be used, so it can stagnate and cause accumulation-based dis-ease, such as cancers.

There is a desperate need for the spiritual energy to control its body, and this brings a sexual element into it. There will possibly be a withholding of the sexual energy and an attempt to control it, or self-control to some degree. This can be useful, but usually, without training, this can lead to prostate cancer and accumulation within the genitals. The wood body type needs constant sex for satisfaction and is the most sexually strong type. It needs sexual activity to allow the energy to flow. However, the lake energy itself, although it is about very physically form-orientated sex, is not about passionate expression. It is more about form and creating unity of shape. Hence, these two form complexes and erratic sexual binges, or augmented fetishism can occur if there is not an outlet. Here, the body and spirit have to find acceptance for there to be allowance of freedom if they are to find peace.

iii) 7 within the Fire Body

Again, this is a complex match. The fire body is fast, energetic, and warm, and the opposite is true of the lake energy. The lake wants to move towards stillness, the fire towards the Heavens, expanding out. Hence the fire attempts to overcome the spirit. This can mean that the qualities of grounded-ness and the subtlety of the metal can feel burned and bruised by their over-energetic system. They cannot settle, and this can cause them to become anxious. In men, this anxiety often expresses as a frantic nature or a person who is very busy and without time. They can be quite short and sharp, hurried, and constantly attempting to do things neatly in an uncluttered manner. They want to be organised and can be judgmental and critical very quickly. The fire almost propels the lake to become expressive, even though it doesn't want to be, and this is what the expression is—almost a withholding being forced out. If it is held in, this can result in a more forceful outburst when the bubble breaks, and this can be quite explosive.

iv) 7 within the Earth Body

This is a harmonious relationship. The spiritual energy uses the Earth body to form deeper connections to the physical energy, and there is an overall bodily comfort. The masculinity here will come out in a far more fatherly style, but with motherly attributes. There will be a much greater confidence with this construction, and tasteful food will be presented well and considered. Connoisseur chefs can have this combination. They will feel at ease. This type can live for a long time without any problems, perhaps only lack of moment, but they are often logical enough to do something about this. They can be prone to accumulation dis-ease.

v) 7 within the Metal Body

This is a simple combination of body and spirit. The metal body is a very good home for the lake energy. The slow and subtle formations of the metal body, which tends to be smaller and subtler than its fire counterparts, perfectly combine with the lake energy. Subtle hands and feet make for models, designers, and form experts of every style (interior designers, for example). They tend to live long and can sometimes be cool and sullen, quite cat-like in approach, and very distant. These men are not really family people. This energy can make them become distant, as they find its lack of masculine feeling disconcerting in a society that demands other expressions. These people are quiet and do their own thing.

vi) 7 within the Water Body

The water body is heavy and compact and powerful. It is perhaps too dense for the lake energy to feel in moving, but the density of it means there is a good weighting down feeling that allows the lake energy to feel comfortable. They have a long-lived expression, but the internal heat of the water body can be a little too much for the lake spirit, whose waters are cool. However, the water body endures, and the lake energy can reside without being fired to the exterior, unlike the fire energy type. This, however, does not help the exterior coolness (associated with the environment of the water body type, i.e. northern region), and the exterior can be very cool with this combination, leading people to believe this is a cold person. They are softer than imagined. This aids the masculine outlook of being a cool person.

3.5 The Quality of Heaven Energy within the Male Body
i) Heaven/Qian/6/Metal Yang

The 6 energy is made up in the triagraphic representation of three pure yang lines; it is therefore the archetype expression of Heaven/ sky energy or yang energy. However, it has a very yin position in the nine-energy sequence, being the governance of the late autumn or the northwest region. The content of it is yang-spring energy within the Metal. It is associated with metal, but its energy is consolidated yang energy or firm yang energy, not expanded—very much Heaven within the Earthly realm. It is given the name Heaven, for it has the power for great expansion, just as the seed of yang is within yin in the Tai chi symbol. This energy is truly masculine, but not necessarily in an extroverted way; it is more in a dignified, all-powerful way. They will dress for power, often wearing gold or sliver and showing this off, modestly! Usually with suits or business looking expression.

This energy has conviction that it is masculine and therefore rarely needs to prove its masculinity. These are people who are entrenched in it; it is heavy masculinity, just as a sword is. A metal sword is strong and hard, yet bright and a symbol of power and justice. Metal itself is very yin and heavy but the shaping of it and turning it into a masculine shape is all done with fire. This is the expression of this energy. Within the male, it suits well, comfortably fitting into male roles associated with fatherhood, being the elder of a tribe, or a leader or king. These energies often have strong egos and hard attitudes, and it is always their way that is right. So, males take on this energy far more easily and comfortably than women do. They have a sexual prowess, not because they are necessarily interested in an spirit connection, but because the power of sex joins them to a harem of women, which makes them feel more powerful or more successful.

Materialism is important; these people buy refined suits and refined cars and refined everything, if they can. They are tough in a fight, but their underbelly is weak. This is because they lack yin, which means that recovery from situations is a problem if they are hurt or damaged at an early age. Psychologically, the brittleness of the yang can cause them severe repercussions later in life, and they may not recover unless focused on doing so. They are very body-oriented, with smell and touch being important. They can be aware of shape and space, making them extraordinary architects and designers of space and change. They can be cool and mechanical in feeling. They are highly creative and inspirational in the physical realm, which makes them good inventors. They come up with mechanical solutions, although it is usually beneath them, in their own eyes, to create the process themselves, and so they usually employ others to do more menial tasks. They take risks, particularly in business, and look for the way to create the biggest gain, even if they risk others to fulfil their own desire.

It can be lonely at the top, and these people know that. They rarely trust someone enough to reach beyond their surface, and they can be very manipulative and decisive, attempting to take over using force if necessary. Their clear judgement is usually misplaced by their own egoism. They can be good listeners if there is something they want to hear, and they can be very magnanimous and generous in their efforts. However, there is usually a reward for them at the end. Men of this type have been very good at sports and physical performance, showing off their body form, or in games where they can be captain or have a leadership role. Socially, they are very expressive and well connected.

These energies are yang spring within the autumn, so they have associations with very assertive tendencies and attitudes that attempt to control. However, within the yin realm, aspects that are not material are immaterial to these types, hence they are often associated with masculine archetypes of insensitivity, "bolshie" attitudes, cars, money, and sex, which are all creative and yang in association. Hugely charismatic charmers and can sell ice to Eskimos, so to speak! Usually there will be some attempt to "gain" whatever they do. They are the different to 3 energy, which is about connectedness of sense and expressional energetics, not physical—this is opposite; the 6 can feel the energetic easily but understand the physical creativity and mechanistic approach more easily. 3 and 6 is all in fact the same, just a continuum of spring energy. 6 is more yin-physical, 3 is more yang-energetic.

ii) 6 within the Wood Body

The wood body is well suited to the 6 energy because the 6 is actually a spring energy although found in autumn. The wood body provides ample power and physicality to express the tendencies that this type has. This body is active and energetic. If used effectively, is has structural and muscular power, all of which is associated with the 6 energy. This body has high sexual power in the form of blood in ample amounts, which can be used to sexually express this type's energy. The 6 and the wood body are a perfect match, so there is very little problem, except that they can overdo it and so create high blood pressure problems and overheating of the system. However, if they exercise constantly, which this type tends to do, they will not have these problems.

iii) 6 within the Fire Body

The fire body suits the 6 energy also. It is more energetic and expressive of the power of creativity within, so this is like an outlet valve. They will revel in it. However, the raw physical power of the 6 energy will be difficult for the fire body to sustain, so it can drain the fire body easily, especially sexually, as this is the weakness of this body. Essentially, it like throwing logs on the fire, so it makes the fire body overwork and over-live, having a short but energetic life. The difficulties here will be from high blood pressure and heart problems.

iv) 6 within the Earth Body

This is not a harmonious relationship. The Earth body is too static and heavy for the 6 energy. There is not enough change and creative expression for the energy to expand, so this body becomes too cumbersome and static for the expression. There can be depression or aggression from this type because they cannot express themselves directly. Issues will be fatigue, lethargy, and impotence. They will feel like they are trapped in a heavy and rounded body, whereas their energy is much more about erecting creative expression in the world. This is a difficult combination.

v) 6 within a Metal Body

Though the 6-energy type is often connected with the autumn and metal, its expression is actually spring within autumn, so it counteracts this energy. This is probably the worst combination for this type. This body cannot withstand the physicality of creative expression of the 6 type; it is too fragile for that. It can easily weaken with massive sexual output, and these body structures tend to be cooler and calmer and less expansive, more yielding and yin. Hence this spirit will be highly over-controlled and feel trapped and isolated. Depression is the key aspect, but also hypertensive blood pressure, skin problems, and constipation, to name a few, as a result of the tendency of holding in.

vi) 6 within a Water Body

This combination is perhaps the best for the 6 energy. The water cools and controls the passion of the 6 energy and holds it down, giving it foundation and strength. There is a strong sexual capacity, which means this type will not drain its physical form; this is beneficial. It is slower and steadier than spring, but from this springboard, the 6 energy can find balance and less harmful expression. Problems can occur, such as urinary tract heat-based infections and backaches.

3.6 The Quality of Taiji-centre Energy within the Male Body

i) Yinyang/Taiji/5/Earth Yang

This 5 energy represents the vortex of power at the centre of the universe, and in the microcosm, the vortex of power within the Earth's centre, and in the human, the vortex of power within the digestive system. The centre in this case is something that has a core of magnetic potential that draws things towards the centre and yet provides a stable base for all things to grow. In the masculine body, this energy is about power—the control and understanding and enhancement of this power. It is a self-sufficient power, and often these men will do things in their own right, having their own business or being in management.

They are social and interact with others easily but often try to gain control over them in some way. They have a clownish or slapstick sense of humour and are often very humorous; they often play the fool, but often in order to be the centre of attention. They are a very stable force, though they change constantly within their own limited parameters. They have a great deal of pride and can be seen as pretentious because of it, but actually they really mean it. They have a very powerful sense of a separate self in mind-identity and are the most egocentric of all the energy types. They are very comfortable within the male body and are very physical. Everything they do will somehow involve their physical self or control of their physical nature. This power is again used in the sexual realm; they make sex very physical and are very non-spirit orientated about their connectivity. This can mean women especially of the yang type find them disagreeable. They see most things in mechanical or practical terms and are often too caustic to see through the strata of different people's energetics, again limiting their progress.

They often feel this limitation and are prone to becoming and feeling isolated because they do not ask for help, as they feel they should be self-sufficient and hold in their emotional expression. They have a huge appetite for knowledge, but do not have analytical ways of thinking. Their mind is not their primary strength; it is more the body that has the greatest ability. They can be skilled, however, in languages and in expression because they want to communicate, not necessarily because they are communicators and teachers, but because they like to show off their abilities and therefore need to express this in many forms. They tend to look rugged but smart and somehow are able to blend the two. They are very image conscious generally and want to be seen as being powerful and looking good especially by women.

Structurally, they are expert, natural energisers and mechanics, not inventors, but what they can create can be very efficient because they do not wish to spend any more money than is absolutely necessary. They are grasping in nature and often want to take more than they give most often; they store things and money and hold back often. Men of this type are always in total control of their family and their life. They seem well structured and well organised in their approach to things, but because they cannot let go of their own egoism, they can become aggressive dictators and can physically lash out at opponents, or at least threaten them. This can make them liked by yin women who are attracted to powerful men and see them as good providers—which they can be when inside their household, but you are owned too. This lack of freedom is like our requirement for gravity; it is necessary, but too much can alienate.

Tradition has it that one should never go towards the energy of 5 but always allow it to draw you in if it wishes, or repel you away. This is because movement towards this energy can draw strongly incoming energy through sucking the energy out; they are controlling powers and will not be controlled! This makes them the most powerful of all energies, and if inhumane in consciousness due to drawing in and defining their ego so greatly, they can be destructive in action and fragment greatly. However, of course, the opposite is true if they can see they have a role, and no more or less than this.

ii) 5 within the Wood Body

The wood body has a great deal of yang within its liver, and this conflicts in explosive measure with the 5 spirit. The out-of-control energy of the liver, which is spring like and expressive, does not give the 5 spiritual control, and it therefore acts like an irritant to a powerful energy. This simply drives the 5 energy stronger and makes for an irritated or agitated body. They can accumulate anger and become very aggressive physically, expressing in forceful ways of behaving towards others. If they do not have a physical outlet for their expression, it can become internally destructive, particularly causing depression. Serious illness often doesn't come, as they are body-conscious and behave as one with it. They do not ignore the body as much as the more ethereal energies like 3, 4, and sometimes 9 do. They can be very demanding and dictatorial in a Napoleonic way. They can also be frustrated and determined to take control. They can be a force to be reckoned with, but often one that can drive others away.

iii) 5 within the Fire Body

This works well in so far as comfort for the spirit, which feels at ease within an energy form that is feeding it. However, this can simply drive the egoism and power of the 5 energy further and further to become stronger. This is a problem, as the 5 needs to express more rather than accumulate for the benefit of its own role in the scheme of things. This makes a very dominant energy. The fire expression does make this energy stronger and express it's feeling, but its feeling will often be too demanding and constantly in motion. The masculine energy expands and grows stronger too, and often this type will have many children or be a family person—a father, but one that is in utter control of his children and can be dictatorial in this regard. He has a very healthy appetite and adores good food, which is a passion. He sees he has the right to own as large and mighty a body as possible, so heavy consumption is a key factor with 5 energy. Collections of items, usually of financial value, often go along with this type, as well as the 8 energy, although this will be more child-boyish collection. This combination can create a very long-lived body troubled only by high blood pressure and possibly wasting and thirst patterns (diabetes like pattern).

iv) 5 within the Earth Body

A very comfortable combination of energies means this type may become too heavy and motionless. They can be overweight and greedy for food, as well as content in themselves, which leads to difficult situations. Earth within Earth is so heavy that it is hard for this energy to become anything different than it is. It is very stubborn and doesn't like change, unless it is through its own judgement, and this will be slow. Diabetes is the main complaint, although this type can have high blood pressure too, if overweight.

Again, accumulation of foods and things can be their main difficulty, but they can be softer and more generous than the fire body and less aggressive than the wood body. The Earth-core moves, but it does so slowly. Double Earth energy both in body and spirit results in slow change.

v) 5 within the Metal Body

The metal body is the best and most harmonious for this energy. It draws away the excess energy of the 5 energy so it doesn't become too strong, and so allows the 5 energy to see itself rather than be introverted. The metal creates form driven by the Earth spirit, so it makes for a wiser and more pragmatic person. They can be more aware of an exterior to their energy, and they will be more refined in their desires. The power of the yang draws to create a stronger, more defined body, so they can be very defined and strong physically. They can be models and the like, showing off their body publicly and being praised for this. They can be egotistical too and colder than the fire and earth, concealing themselves emotionally sometimes. This can lead to sexual issues, but there are fewer desires in this type. There is more of a need to connect physically than spiritually, but much less of a need than the wood body has.

vi) 5 within the Water Body

This is a very powerful body for the 5 energy to be in command of, and as such it suits its purpose in some ways. The 5 is similar to the 8 energy in this body. The water body is an opposite/interior version of the fire body; hence there is a source of internal power to this body. However, the 5 energy is more expressive and less cool than the water, and this makes for a difficult balance. The water draws inwards, and the 5 can draw inwards also; this can cause a problem, as it leads to internal control and depression. It will be less damaging to the exterior, but more damaging to the interior of this type, and it may drive the 5 to deeply isolate themselves. In some ways, they will be very self-sufficient people, but usually this is because they are hard and cold and move away from others. The interior heat is key with this body, so high blood pressure can accumulate, especially due to the control that the Earth spirit has over this. It can cause the body to have internal problems, prostrate issues, and urinary tract problems.

3.7 The Quality of Wind Energy within the Male Body
i) Sun/Wind/4/Wood Yin

This is a complex pattern, as the yin nature of 4 is within the male. As we have expressed earlier, the 4 energy is somewhat complex because it is autumn with spring. This makes it a cooling energy and yet part of a warm season. The yin quality of the wind makes it open and gentle, but the coolness of it is somewhat disconcerting, as it is both gentle and cooling, which is very yin, making it strange as it is immaterial (yang) but at the same time cool (yin). However, wind is one of the softest energies, being invisible and immaterial, unlike the 2 or 7 which do have some material weighting. Hence, for men of this type, the difficulty is being stable and rooted in their being.

Masculine roles in society often seem to require a great physicality about them. Archetypal masculine roles tend to associate with the 6 energy types. The yin of the 4 energy is difficult for people to understand because it is not structural and it is not firm, yet can seem to have direction and purpose. These people are often communicators and can express musically (usually in a melancholic or minor tone) or verbally (can be critical or anarchic) or in other ways that are usually a combination of cool-spirit and mental vibrations of energy. You can hear the wind rather than see it, unlike lightning and thunder, which you see first before you hear, so these people tend to be sound-associated rather than visual. This is generally associated with the autumn-winter, whereas 3 and 6 energies tend to be much more visual and denote spring. As a result the 4 energy tends to wear awkward mixes of things, they sometimes look highly independent, not necessarily in a way that is well constructed but almost put together last minute. They are not concerned overly with the exterior unless other people are watching, then they will try to fit in as much as possible. Sometimes they can be idiosyncratic in their way of expression and this can be an endearing feature to others as they are kind of "wired" looking, and they kind of make a style out of it, but this is to express in a sense that they don't care much about the exterior, not that they are trying to say something impressive.

People with 4 energy are usually very gentle in touch; this often makes people respond to them as if they were feminine, which makes this type sometime over-compensate their masculine traits in order to balance. The problem is that they can change their minds and feelings fast, and this can be disconcerting for those who are more solid and structural. Also, people of these more solid types lose faith in the wind energy. As a general rule, these men need to be allowed to feel the impulses from within, because placed in crowds of strong emotion and conviction, they can very easily lose themselves and be swept along with the tide. They can easily be influenced by other people's emotional states and will change and adapt like a chameleon to suit different situations. They are often afraid to come into direct conflict with people and attempt to cool/evade such a situation, unless a limit is reached. People who are more structural will look at them as weak or shy or manipulative (i.e. with intent to gain), but in fact the energy is simply soft. The difficulty for a man is to be able to accept this within a society that does not accept it, except in very specific classifications like being a gay man or something of this nature, which is a mental concept (see Appendix 2 of this section).

Men of the wind type need to develop a listening ear to their own being and form themselves on this basis. They need to realise that they should not involve their energy in pursuits that do not develop their ability to be free thinkers and workers. They are often involved in communication—as writers, musicians, inventors, and expressers of various kinds. This method seems to be the way they can be themselves, but of all the energies, this one is perhaps the most complex mix and the one that can be most lost. They must constantly be made aware that they have a body, for they constantly forget this fact and can move almost as if they do not have one. This is true too of 3 and 9, but for 4 types more so, because the wind energy can be cooler in temperament and so not rise up as easily. These energies are of the sky and Heavens, so they are almost alien to the rest. They can have such strange ideas of Earth life that others shun their approach. They are often isolated because of this.

ii) 4 within the Wood Body

This body would seem correct for the 4 energy because it is so-called a wood energy. However, since 4 is autumn within spring, this body is actually the most difficult expression of the 4 energy. This body is very powerful and sexual, whereas the 4 energy is much more gentle and also ethereal. The physicality of and the need to express sexual energy is very draining to the wind energy. Often they can attempt to disassociate from this and become very mental, ignoring the body and the needs it has, and in doing so cause digressional/ detachment/ dissociation illness and mental problems. They can become violently angry if the heat held within the body is not expressed; they often need to do exercise to prevent problems. This energy is like holding in an explosion, so one must be careful when knowing this type, as their energy may be very erratic and seem mentally confused, especially in old age. Blood pressure and blood stagnation problems can ensue, as well as circulation issues.

iii) 4 within the Fire Body

The wind is a cooling energy. The heat of the fire renders the coolant useless, and so that of the fire body again overpowers and overheats this temperament. It is too energetic and expressive for the wind energy, which is more self-protective and enclosed, so again, pressures build up internally as the wind attempts to cool the fire. This struggle often makes the fire body stagnate and overheat, and the wind energy cannot feel cool and clear enough to express its clarity of feeling and caress the surfaces. Overheating affects the heart, which can cause high blood pressure and issues of this nature.

iv) 4 within the Earth Body

The Earth body actually suits the 4 energy very well. It tonifies the physicality of the metal coolness of the wood as it goes from late summer to autumn. The wind energy is fed by this energy, which provides the earthy coolness from which to express itself. These people will be creative as chefs and attempt to make simple and tasteful foods. The metal energy softly cools the Earth's; they can move towards a slower metabolism and have a tendency to store, thus putting on weight.

v) 4 within the Metal Body

This is also a very ideal physical structure of the wind energy. The Japanese body structure for example, is similar, and the cooling nature of it is very easy for the 4 energy express in. The nature of courteousness and non-aggressive action in social relations works very well with this type, which leans towards an overall quieter tendency. The wind energy is cool and relates to this temperament, so they can feel comfortable and do not need to feel as if their expression is being forced or blocked; they can be as they are. These types tend to be very long lived. They can relate to their physicality well in this body, and they tend not to be overly sexual, and so conserve their energy and are able to control over-expenditure of it. They can be very good with money, saving it very effectively, but make very bad investors as they don't know how to take risks easily (unlike 3 and 6, for example).

vi) 4 within the Water Body

This physical structure is again a simple one for this energy. This body has more physical power but is steady and has good ability to help the 4 energy express its nature in mental application. They have a solid bone structure, allowing the wind to feel anchored into the body, as opposed to feeling like it is being burned off the surface, as with the fire body. This combination is also very long lived.

3.8 The Quality of Thunder Energy within the Male Body
i) Zhen/Thunder/3/Wood Yang

This energy is considered to be yang within yang. The most masculine of all the energy forms, it has both the expansive quality of yang and is within the yang in the daily/seasonal change. Thunder means not only the quality of the sound but also the lightning and the visual expression of a storm. The energy cracks open the dawn and the spring and spreads clarity and light into dark places, so that all can be revealed in an instant, similar to the light of fire but brighter, more direct and purposeful, more terrible and awesome. This energy releases freedom of expression and opens outwards, more direct than wind but just as moving, if not more so. This energy has the impetus to electrify or charge to create change.

These energies are leaders by nature but very different from the 6 energy. This energy has the capacity to change and cause a revolution, to inspire and uplift, charismatically extending itself. They are entrepreneurs and pioneers. Thought provoking, intelligent, dramatic, they have lightning wit and mental intelligence. They look for adventure and thrill and attempt to be the pinnacle or the first, never the last. Often their biggest obstacle is the physical reality that stops them from moving as fast as they would like; this causes frustration and anger to build up, which is very damaging to them. They can be assertive or even angry people, and this is the masculine trait of yang in its element, often pushing themselves ahead of the competition, and not able to wait for anything. They must express now, no matter what timing is right for anyone else. This can place them at a great advantage in situations where impulse is important. Directing others towards a cause or change is perhaps their ultimate ability. They can ride the wave of emotional charge to destroy or to perform miracles; it is a double-edged sword. These people generally have great spiritual inclination and are much more likely to shout/scream and use "energetic charge" to overcome opponents rather than physical aggression. They have a tendency to move into manic or anxious states, but for men this will express often as anger. They can be very boisterous and noisy much of the time, needing smaller amounts of time to be still and quiet, yet deeply needing this to store charge for the next expression. Within the male body, the energy is suited perfectly.

They tend to be conscious of their looks and wear things that allow them to move and feel free but at the same time are very expressive; sometimes with a shock factor or a style that people admire, they want attention generally, even if it is person that they are with who is the main feature of what they are "wearing". The Yang within yang male expression is the instigator of the objectification of women and in fact everything in their life. People and things seem to be all associated with themselves and are objects to be moved here and there, without expression of their own. This generally is ok for yin within yin expression whose main focus is to connect and emphasis the light of the yang through there being yin, but for the yang within yin female, this is a threat and also seems to feel like it's obnoxious and aimed at them as being nothing but a body, and the spirit (the yang in their case) is not respected or looked at. This trait we will find very much in couples where there is an yang within yang and yang with yin partnership leading to great difficulties.

ii) 3 within the Wood Body

A harmonious combination, this joining allows for full expression of the liver's energy to the exterior and so allows the body to be very healthy in its feel. The wood body creates tension, and the spirit distributes this and expresses it so there is a constant flow between interior and exterior. This is very useful, as it means nothing can stay trapped for long, and expression is easily expressed from the system. This stops stagnation and so prevents anger and aggression from accumulating, so the body will be very wood in its nature, with a full expression of the spirit. A healthy system results, and perhaps exhaustion would be the main cause of dis-ease here, or illnesses associated with nerve over-charging. The do have to be physical though. There is a great tendency for all the energy to go up to the head and throat and the body below to be forgotten about, if this happens energy accumulation pattern can develop which in this person could be serious as the energy is so expansive.

iii) 3 within the Fire Body

Fire and thunder work together to create lightning, and so this again is a very expressive combination. Very well-expressed thunder energy is much better than held-in thunder, which can be internally or externally explosive. Here the thunder is drained by the fire, which helped it to move and lose tension. This is useful and creative, and this person would be accompanied by insomnia and exhaustion but little—if ever—accumulative illness. Psychological mania is possible, as are high anxiety states. This body creates a lot of heat and is fuelled by the wood spirit, so overheating can be a problem too. This will be a sexually expansive/expressive energy combination.

iv) 3 within the Earth Body

The Earth body presents an obstacle to thunder, and the thunder will force through it. The body will block the spirit's flow through the system, which can be very frustrating and annoying; as it cannot express the way it wishes to. In such circumstances, the thunder will attempt to force the body to do its will, and this can create heavy exhaustion patterns and tiredness for the structure, which cannot be easily moved to action. These people are seen to be very hard on themselves, and often they will take this out on others in aggressive feelings or shouting.

v) 3 within the Metal Body

Here the skin and outside act as barriers to the thunder energy. Trapped emotion, repression, depression, and feelings of suicidal rage or aggression can result. The person, unless allowed to express himself, will explode or implode, causing massive damage to the body physically and emotionally. Cancer can stem from this, as well as accumulation illnesses, strokes, and other nervous/high blood pressure type stress systems. Blood pressure issues will be due to the withholding of the system's energy within. Sexual repression can lead to control issues around sex, which can be damaging to this person.

vi) 3 within the Water Body

As with the 4 energy, the water body acts like a springboard for this energy and can also be a firm anchor. Allowing for a more controlled expression than the fire body, this construction allows for expansion without going too crazy. There is great power, so the body can last a great deal longer as a result, in this way. Emotionally, it will subdue the yang expression somewhat but not in a negative controlled way, more in a anchored way, which allows for more body awareness, which is strongly lacking in this type. They are much more sensitive however. Sexually, this too allows for a less damaging start position, having a strong foundation.

3.9 The Quality of Field Energy within the Male Body
i) Kun/Field/2/Earth Yin

This is another complex combination. Kun has three yin lines, a pure yin energy formation. It is however associated with the southwest, whereas 7 is associated with the west, so is perhaps more yin. However, the fact that there are no yang lines in this formation makes for a very yin energy form. These men therefore have quite a difficult time in some ways, as they are unable to feel deeply masculine. This feeling is often deeply protected. They will be very sociable and connected to people. Most people will like them and find them accommodating. They are physically-oriented people who somehow use their body in their work. Mentally, they are slow, which creates difficulty if they focus on using their mind. However, they can be logical and follow patterns, so they can be involved in banking or the storing of commodities or some process of storing a "nutritive" resource. They are providers of care and can make excellent nurses.

These people care and often are self-sacrificial in their way of being. They are often people of slow actions rather than words, although they like to talk a lot, but more about social matters than highly intellectual events. They are very mothering, and this can be the aspect that makes them attractive. For women, these men seem the perfect father, who will spend a great deal of time with children and have a lot of patience. They tend to easily be dressed by the yang females around them, as do all the yin expressions in the male, and as a result they can look good but not of her own accord, generally they are not bothered about the exterior and prefer to be less exposed as possible.

These men often have difficulty sexually. The dominance of being a man can consume them, and they can fall into patterns of constantly attempting to be the stronger or more powerful energy of the two, and attempt to take control. This is against their nature, for they are yin. However, their action is to attempt to be more like a 5 or 8 energy. They cannot do this for long and eventually end up again being dependent, that thing which they fear and often run from. The difficulty can be that the masculine expression for them is too harsh, and so in order to be like that, they overcompensate and lose their true identity. This leads them to alcohol and all manner of self-abusive behaviour. They are far less likely to lash out than to place frustration and aggression inwards and destroy themselves. This is a big problem for them.

To carry the most yin of energies, a great deal of acceptance is needed by these men, not through being feminine, but by doing what they feel. Often, the best situation for these men is to be directed by a yang energy in female form. This helps them build themselves and their confidence and become part of a team. They do less well in management and being in charge. They do very well in diplomatic situations where their ability to talk and to smooth social relations is very effective.

They can be excellent diplomats, and politicians often have this configuration. The key thing is to accept the softness within their nature and act to make this present in their life, rather than fighting a masculine worldview. They are solid and strong and have traditional opinions that often need to be expanded and broadened, and they can do this, but only over time. Instability causes them to feel uneasy, and this can be again a slight on the masculine performance pressure. Sexually, they find it hard to connect spiritually, unless with a partner that is directing them. If they are the ruling force, inevitably it will just become a physical act, more to do with power than connection. They can be very good at growing things (making them excellent gardeners) and in the role of nurturing a project or a plan (not necessarily from themselves) and bringing it to full bloom and fruition. They can do the same for people's health and are excellent in the nursing role, rather than doctors who involve themselves with diagnosis, which is more of the role of the wood energies, or surgeons that tend to be metal energy.

ii) 2 within the Wood Body

The wood body is the most activated and moving of all the energies, and the Earth spirit is amongst the slowest; this makes for a complex and difficult combination. The energy of the wood will stagnate within, and unless this 2 energy decides to move its body a lot, it will cause accumulative type dis-ease. The problem is that many of these men are dissociated from their bodies, as they find sometimes complex connection between what they feel and what they are. This bridge must be crossed and connected through physical movement. Built-up aggressive energy can cause massive problems like cancer particularly, possibly with high blood pressure. This body can make the Earth energy heavy and edgy and violent in bursts, unlike his character. Mental health issues can develop around feeling constant worry and anxiety, which increases in this body structure.

iii) 2 within the Fire Body

This combination is very good, perhaps the best for this energetic type. The fire body heats up and consolidates the Earth spirit within it. It allows this type to feel more and more at home within the physical structure. It is natural for these energies to feel attached to the physical and material, and hence for the spirit to be more material than a wind spirit, for example. The body and how it feels will be important, if they can accept it the way it is. The fire body will strengthen the expression of the person to be himself, and this will allow him to feel more content and less anxious. There can be an increase in appetite for life and for food. The 2 energy is very concerned with food and storage and are often overweight. The fire energy body will consume a lot and will increase the need to eat, so wasting and thirst patterns (diabetic like patterns) issues may arise if they are not careful, and heat accumulation type dis-ease if they do not move. However, the fire moves at a slower rate than the wood body, so there is some balance to be had.

iv) 2 within the Earth Body

This is a very comfortable energetic connection. The 2 energy is too slow to be helpful, but it is comfortable in the Earth body. The Earth structure is luxuriant with a rounder, fatter body shape and thicker flesh. This suits the field/2 spirit; feeling at home and comfortable. The key problems are over-eating and being overweight, due to lack of exercise because this energy moves quite slowly. Men are especially prone due to their more internal worries. In a woman, this combination is very good for pregnancy. The Earth body makes for a very soft and smooth body shape in tune with a more female form, so it again can be more difficult to accept if mental patterns get in the way.

v) 2 within the Metal Body

Again a good combination, the 2 energy is a natural mother, and in this case it mothers the body and creates form from it. These people will be very physically conscious and are entrenched in the earthly energies, so metal and Earth meld together. This formation gives more structure to the spirit and draws essence and direction from it to create aesthetically beautiful forms. These people can be models and actors. They will use the body to create form and shape, and this is important to them as is financial wealth and security. The metal body is tough and defensive and therefore can easily characterise a fatherly energy, whereas the spirit is motherly. The combination makes for a fatherly energy, which is soft but can hold very traditional, conservative values and be rigid in opinion. These people do not have the flexibility of movement that wood energy has, so it is more Earth and structure that form their lives. This can make them colder and cooler, making men of this type dominate, rather than be in a subservient role, which helps them.

vi) 2 within the Water Body

The water body has a heavy bone structure and a lot of internal power from the Ming Men fire which burns brightly within. This energy warms and tonifies the central region where the spirit energy of the digestive system is focused. This powers the digestive system from the opposite direction; as fire energy comes down through the pericardium system, this energy burns from below. The water body is generally too cold exteriorly to promote the radiation of the Earth spirit, for it is a social energy. There are aspects that would help support the digestive energy, but the expressiveness of this body type would be complex for the spirit. This would cause again feelings of melancholic depression, heaviness, and sogginess, not being warm enough to be social, and therefore feeling within a body that protects the exterior and so isolates from others. This is not effective for Earth. This makes for a very internal world and one that can be miserable because of the feelings of isolation.

3.10 The Quality of Water Energy within the Male Body
i) Kan/Water/1/Water Yang

The 1 energy is of water, so one would expect perhaps a female energetic. However, this energy is quite specifically cold sea-salt water. The yang line within the yin shows that there is great and powerful potential within the coldness of the yin, such as a strong current or salt, a potent heating energy. The energy is like a powerful cold sea, and this gives notions of the masculine; Poseidon and water dragons are expressions of this energy (yin water is the lake energy 7). The masculine body takes this on well and forms the loner, the isolated masculine energy, and the wise and yet dark and potentially dangerous energies of the deep. The water is cold, but it has a warm heart, rarely seen but, upon his direction, is seen to shine brightly.

The north is cold and requires substantial will and perseverance to brave it. In the dark and cold winter, it too requires energy to consolidate within and not be lost in the bitter exterior. These men tend to be prominent thinkers and writers, having vast reservoirs of knowledge and often having depth and understanding in them. The need to be true and honest is very important but only to their own internal beliefs about what the truth is, and so this can create problems for others. They always see the negative before the positive and consider the intellect as all-important, looking down on those who are lacking in their means of expression. They tend to shun the exterior as superficial, but all the same they wear things that show their inner fire and often look impressive and powerful, dark but yet with light a powerful combination.

They can be cold and arrogant and often will be those in society that don't seem to fit in. These people may go to the depths of things, journalists who find out the truth, or those who seek it. They use their own power to delve into the dark places of human consciousness—including sex, deep passions, deep mysteries, all things deep. They wish to find its fathomless depth and know it, and this can mean they become dark and dangerous and morally obscure. These are very secretive people; they may know the truth but would not necessarily speak it. They may let you know if they trust you, but there will always be another depth of which you are unaware.

They are often cold and critical, hard and harsh, like the winter. They can be blunt and tough and a stranger even to those who know them. However, there is a spark of kinetic change in them, and this can be as powerful as a tidal wave or as gentle as a ripple along the coastline. Water, in this sense, has great properties of hard and soft. The Taoist ways often express to men, especially in the Tao Te Ching, to how they should deepen their knowledge of the Way of water, for it is profound that water has both the characteristics of strength and softness but pertains to the yin. These people are often long lived. The seeking of the depth, however, can shorten these people's existence, as it can drive them to destruction.

ii) 1 within the Wood Body

The 1 energy is sexually potent, and the wood body is actively aroused. Hence, this energy is a very sexually strong mix. This allows for the water energy not to be so stuck within the wintertime, as the body draws the spirit to the next season of spring, so his ideas can present themselves to the exterior. This is very important so that this person doesn't get lost within his own mind-identity. This spirit is normally very still, but the body of this one pulls him outwards to express. Hence there is a channel for creativity, so the water energy doesn't get too stagnated.

iii) 1 within the Fire Body

The fire body is warm and expressive; the water is cold and still. This is a complex matching. The body wants to move and express; the spirit wants to be still and think. These two are at opposite ends of the spectrum. There is an overall tendency for the water spirit to hold back the body, so it overheats internally and numerous heart problems arise. The water spirit holds the body back so that the spirit will not feel too exposed and helpless. The stubbornness of this energy to go its own way can make life very difficult, especially in such a body. It is likely that the heat can cause psychological disturbance and turbulence, unexpressed emotions, and a withered body, as the heat will burn it. Dehydration is a problem, and kidney related problems could develop.

iv) 1 within the Earth Body

The Earth body is warm and solid and social. The water spirit is none of these things. Fluid, but with a will, it has no need for the slowness and heaviness of the Earth body. Whereas the change of water is downwards, there is no change at all with Earth. It just seems to revolve around itself in an egotistic manner. Earth body energy disturbs the 1 energy, and the body feels like a trap, holding the water or pooling it, and so blocking its descent. It has to rise and meet the people, and of course the spirit rules and will not do this. This causes great tensions, and the Earth body will feel like a thick and unwanted mass. This may cause the body to lose weight and appetite and become thinner than the normal Earth body, as the water spirit drains the controlling effect of the Earth. This can mean that little attention is placed on the body and more is given to attempt to develop mind-identity. However, this body cannot develop this faculty easily, which again causes frustration and a wilful force to keep trying, draining the physicality of this body. A balance must be struck between intellectual pursuits and the physical so that all aspects are nourished and united.

v) 1 within the Metal Body

This is a harmonic combination. The thick and strong skin of the metallic type is a change towards the cooler and more isolated, and this is where the water energy can find some sustenance. The body of this type provides a structure and mothering effect on the water spirit, so it feels at home within its shell-like existence. There is no drive out of the winter, so this combination makes for a very isolated experience as the body and spirit both attempt to go towards stillness and tranquillity. This type can be very good mediators and are associated with thinking or mental activity. The brain and mind play such a role in the water energy due to the kidney being tied to the bones and the marrow within the bones, which in Chinese Medicine is the autonomic nervous system, which includes the brain. (Note that the somatic nervous system is associated with the liver meridian system, i.e., muscular associated nerves:

203

within fragmented modern physiological view) These people will have the dedication to head projects, to understand philosophical problems, and to seek out new solutions.

vi) 1 within the Water Body

This combination is comfortable and connected. However, the stillness of this energy is quite profound. They can be quiet and isolated for very long periods of time. They may not even need to speak, and they will not be able to stay in the company of others for long periods. The deep internal natures of these energies are cold. Within the coldness and toughness is a warm, internal spark, and this keep glowing in the darkest of nights, almost a solitary soul in the darkness. The water energy wants desperately to connect and to join with, often sexually, as this is the sea it swims in, between the physical and the spiritual, in the brain and mind. This is the connection that the water longs to find, but it is often unable to and cannot sustain it for long, for fear of being lost in the ocean and their individuality being lost. This restriction is merely that of being within a structural body, so the water energy, similar to the wood, looks towards the spiritual, whereas fire looks towards her next stage, Earth, and the physical, as she is already spirit.

Part 4

The Female
Body-Spirit Combinations

4.1 The Female Body and the Nine Energies
i) The Female Jing Body

The overall body structure of the female is smaller and lighter relative to the same type in the male. It is naturally a more open structure with less upper body energy and more strength in the legs and hips. The female body is the birthing place and the place of the "blood"/ flesh baby (rather than "energetic"— see male section above), the vessel of the new spiritual formation, differentiated when it emerges. The genitals of a woman are deep within and are hidden from the exterior. This expression of depth and internal workings is part of the complexity of the female form in relation to the male. The masculine energy is much more superficial and exposed. The female is hidden and therefore more difficult to treat. Women's dis-eases associating with the uterus can be connected to emotional patterning and problems of expression and fearfulness they have experienced. In men, problems are associated usually with physical distress to the body due to exterior forces acting on him, and him resisting them or forcing his way through. This shows how internal dis-ease of the emotion and expression are much more associated traditionally with women's bodies, and as a result require a totally different approach medically, often best given by women. In some medical traditions, it is interesting that women only treated women, and men treated men. This actually makes a great deal of sense, but it only sees the physical and not the spiritual connectedness, as there is yin within yang and yang within yin.

The woman's system is much more connected to the cycles of the Earth and lunar changes. In a perfect situation of connection to Nature, ovulation should occur at the new moon and the period during the ecliptic stage. This 28–30 day change is in tune completely with the dark times, the night, which is also associated with the female body, whereas the sun and daytime are more yang and male. The energies of evening and darkness are within the yin, and this is where the female system and interior of the system are regulated and changed. This too is where the sexual energy of women is strongest, whereas the masculine energy is strongest in the daytime.

The female body represents the sacred vessel, the source of life in many traditions, the tampering or destruction of the source severing life for all of human kind. This too is related to the Earth, and so the feminine body is associated very much with the Earth and the matter of the planet and the universe. The change of the Earth affects the body of women more than men, or perhaps it affects men through women. This is an important consideration, as the destruction and poisoning of the Great Mother Earth through a very aggressive mind-identity-masculine way will eventually lead to destruction of all life. It is an external expression of damaging the yin, which connects to Wu.

The Earth and women seem to endure the ravages of mind-identity of man, or the focused and non-broadening principle. The female principle is related to the Earth, and then through the Earth to the stillness outside of the universe, again going back to the diagram on the cover. This shows how the intuition of the female body allows for openness and a broad perspective. Through stillness, it knows and feels, due to inspiration from the exterior. In this way, women become the teachers of the masculine energy. The Tao Te Ching was a book written not for women's physical energy but for men's, to understand and be taught by the female principle rather than attempting to dominate it. Hence women's energies are beholden with the body formation of the yin. In as far as the spirit goes, just as the masculine body exaggerates the masculinity, the feminine body exaggerates the femininity of the spirit. Hence a yang spirit in a male body is much more yang than a male spirit within the female.

There is a complex balance in yang within the female, but generally women accept and become much more what they are, putting up less pretence in pushing to be a social norm if they do not feel it, especially in today's modern society. However, women often feel bombarded to be attractive to masculine mind-identity structures. Of course, many years of suppression and traditional training—again due to narrow masculine fundamentalism of a no tolerance yang within male body perspective—has led yang women who had feeling of expression (yang by spirit nature) to be forced into submission and a sense of slavery by masculinity. This in some part has changed, due to acceptance of the yang as being the "right" way for all, male or female body…, which is still masculine domination, disguised in male or female forms now. However, if there is an aspect that we have not looked at on this planet, it is the yin. This sacred, deep feminine is inherent in the body structures and forms around us, and this stillness is where the cure of most dis-ease is to be found. The yin within the female therefore are the types who hold the key to understanding of the yin, however most have been influenced by the masculine world. The woman's body is called the "accumulative body type". It accumulates blood and energy to form it into a new creation. This can also work towards pathology, the female body gathering dark substances like blood, which can form accumulative based dis-eases, more than men, especially deep within the body.

The sexual act itself is very different for men and women. Whereas men's ejaculate is the release of Jing essence, Jing essence is not release in this same way in women. This means that for a woman's body, the secreted mucus of the vagina is drawn from the blood. Hence, whereas the prenatal energy of a man is used in sex, the blood or mainly post-Heaven energy (i.e. re-attainable energy from food and air etc.) is used up for women. Sex therefore is not the reducing aspect of the process. For men, ejaculation creates heat within the body, as the Jing is a cold fluid directly from the kidneys. Women who drain their blood from having a lot of sex will simply get cooler and reduce their body temperature. Hence sex makes men more yang/hot (li-fire) and women more yin/cool (unless there is pregnancy which then will make her hotter later kan- water).

It is childbirth that drains the pre-Heaven or prenatal energy of women, especially in the last two months of pregnancy. The male physical body, being more expressive of its energy (through physical activity and ejaculation), will drain faster, unless a woman has many pregnancies in her life, which again can drain her system quickly, as it is often too expressive for her body (dependent of physical type and quantity Jing/ power of constitutional strength). There is always a balance between expression and exhaustion—or lack of expression and stagnation—that must be addressed during life processes. This is the way of the female body structure. Recovery for women is often faster, as the stillness of yin is about regeneration and centred in Wu, the origin of all of life. The stiller the energy is, the more potent with yin, which manifests life (yang).

Mind-identity of women is far more pliable and softer focused. There is the possibility for women to easily open their mind-identity to the stillness they embody. However, if they have entrenched mental perceptions, often from body-sensations or continual mental-emotional patterns, this can be very difficult for them to let go of. What Eckhart Tolle describes in his book *The Power of Now* as the "pain-body", is very strong in women, and this is the difficulty of mind-identity they face: mind-identity's relation to the body. Hence with women, there may be much physical pain in letting go of a mental association. For men, it is not felt as physically but is often more about total association with the head and therefore a lack of awareness of anything other than this, making it more dangerous; the pain acts as an anchor for women. Women, however, driven by the male-dominated mind-identity, are becoming more and more male-like in their use of mind-identity. This plays enormous havoc with their emotions and the physical body, which suffers greatly and can become an infertile base from which creation is impossible. Note that creativity of any form has always come from stillness. So when stillness is corrupted by separation and mind-identity, it no longer breeds change or life, mid-identity itself is a form of warped life, which draws all the fuel.

The two different bodies, the male of expression, and the female of collection, create; the upward expanse to hit the ceiling of mind-identity/head/brain for men, and the deep and penetrative holding in emotion, which is within the organs of the female body, which of course mind-identity influences and causes resistance of flow through. Both however are of the same root, which is the non-acceptance (impossible to "do" anything about) to the unity Wu that permeates both entities and connects them as one. These are, of course, big generalizations, and the combinations of body-spirit can create variations in the yinyang qualities within each seeming "individual". This is in addition to everything that has occurred previously in human history and is held within the tissues of the physical form, activated by the spirit and brought into play again by mind-identity. All of these, for each individual, cause billions of seeming "individual" distortions to the basic format of the five-phases and nine energies in their pure textbook form. This shows its limitation in understanding individuation, but it is in relation to the root principles that everything is One, which can help the what seems like an "individual" back to simplicity from the seeming separation of mind-identity.

As the spirit rules within the moving universe, we will denote the female change of numbers 1–9 and follow with physical descriptions of each of the five body types in combination, as before. It is important to note that for women the 12 branches are perhaps slightly more important than for men. As the Earth belongs to the yin, and yin to the female, so the corporeal-quality of the spirit of women will be an important influence in their lives, perhaps more so than men, whose natures are of Heaven, and so their behaviour is more of the cycle of nine. This of course is a broad-brush expression.

208

At the end of this section of chapters, we will look at how the 12 branches influence the five body types, by looking at the stems and branches arrangement and its connection with the body. This influence is therefore impacted upon the spirit energy; I am associating the body more with the corporeal-spirit quality here. Also, please note that I am using masculine first and feminine second (as with the spirit ages also) not because this is a machination of this work, but because it expresses that Heaven is the guiding principle in the moving universe, and the mother yin is outside it. In fact, all come from the mother yin, which is the emptiness. The repletion of this idea in every step is very much foundation to the approach of this book, please bear with it!

4.2 The Quality of Water Energy within the Female Body
i) Kan/Water/1/Water Yang

This is a complex patterning. The female body takes well to the notion of water. It is a *yin* energy and so is associated with a downward, internal, dark change (this is associated more with 7 lake -water). However, it is the power of the sea-water that is difficult for these women to wield if their perception is drawn towards their physical shape or social dogma. These women are strong and have leadership qualities, often leading themselves to difficult and confrontational situations. They have the energy internally to strive for and seek out truth, and they can be powerful philosophers and thinkers. They will be excellent poets and writers, having deep, expressive emotion or courage and will power. They have great ability at being a boss of a business also, and a financial understanding and develop skills in this area. She will look dignified and powerful and like how clothe++≠s can hide and yet uncover in her control. She knows how to use this to further her direction in the world. As with all the water energy even though it looks cool it has a warm heart, this is opposite to the fire energy, which looks fiery and warm but internally is empty and open.

Women are generally much better at accepting the difference of feeling they have in relation to very *yin* female energies. Generally women know that they not the "wall-flower/ girly" type and often accept this. They tend to feel more secure in the company of other men be they of a *yin* or *yang* type expression than with other women especially of the *yin* within *yin* expression. The masculine energetics within them often make them view the world, which is a masculine dominant expression, as something to be overcome, because they feel that men have been trained to view the world in a very *yang* within *yang* way. The emancipation of women was really an expression from these types who saw that women have *yang* within themselves too. The *yin* within *yin* women form very female groups which continue processes of being very feminine and therefore often playing to *yang* within *yang* male desires; these are 2, 4, 7 and 9 women. Energies like 1, 3, 5, 6, and 8 and are much more assertive and see these women as destructive to the role of women in society. This of course is a perspective difference, and both must be accepted as behaviours. However, for the 1 energy, their partner would need to be a female energy type in the male for there to be balance. This energy is dominating and must be in control of their direction. They often need a great deal of time on their own, and so the company of a partner is often restrictive.

Note that the *yin* within *yin* women, and in fact male with *yin* within (to a lesser extent), have the tendency to adapt and bend to the will of a partner who has a *yang* energy, especially in the dominant energetic position of the five phases. The *yin* energies will tend to flex, making people believe they are not strong-willed, or they may take on the will of their partner, as if she is an extension of him. This shows her adaptability and her ability to be moulded by a *yang* power, which is in the nature of things, but it also shows how influential a *yang* power can be, and so can cause great harm or great benefit to the softer energy. Commonly, this will create the social problem of *yin* within *yin* women leaving their friends when they have a strong man present or in their life. This is often seen as a betrayal to the friends, who are often *yang* within *yin* females and see that friendship and relationship should be balanced; male advance does not sway them so easily, and they often take the advance themselves, even though sometimes this is manipulative/ divisive often through tactics of flirtation rather than directly as in the masculine expression of attraction.

The yin within yin energy is heavily influenced by the masculine energy and so defines herself by him. This makes her friends less important when there is clarity about her purpose derived from being with him. This is worse for males with yin within, as they are seen as weak-willed by other yang within yang males, being controlled by a "woman's influence". The yang within yang male's influence is strong and direct, so it is also inflexible and demanding. The opposition to this is yin within yin. These women are "women for men", whereas yin within males are "men for women". The yang within men and women is more desirous by nature and therefore require these yin counterparts to attract and form union. The 1 woman therefore is directive by nature.

The 1-energy women have a natural sexual interest and a need to go deep into things and seek out the depth and dominance. This can lead them to dangerous situations, and they can be consumed by them. However secretive and yet when spoken to truthful, these people have the will power to get through very difficult times and unbearable terrains where others would fall short. They have tendency to be independent and focused but also quite negatively orientated, seeing the difficult prospects before the inspiring. They are likely to see the bitterness, and sarcasm in a situation. They can be blunt and quite pushy, vengeful if aggressed or if they don't get their way. They often can be seen as being quite still and internal, and then people are surprised by their power, which lies beneath for determined force. Highly wilful they are like the sea-goddess, cool and icy but with a fire of life (salt) in her centre. The mind is their landscape and their place of greatest strength, they can draw massive knowledge inwards and may have strong opinions about things often tending towards their particular vision and can be limited in breadth. She tries to avoid exposé of herself, remaining hidden but at the same time longs to come out into the light and take over, which sometimes happens but not for long.

As mothers, they have great patience but often can be too egocentric for the child to cling on to as much as they would like (as they are dependent on the child). They are better as a mother to the yang spirit energy, which tends to become independent quickly, as the masculine principle moves first and displaces the yin. Yang leads through change.

Feminine spirit energy tends to need more connective attention (i.e., is less archetypically independent), and in a male, this can create men who are seen by women as "needing to be mothered" or looking for a "mother figure". Usually these men are simply yin within and so have a more needy/ submissive, less directive nature. Again, instead of judgment, what is needed is acceptance. These types of men would be suitable to be close to yang spirit women, and in fact the emancipation of women also allowed for the emancipation of the more feminine masculine approach that was also hidden within society. This is more acceptable today, outside of the gay community.

ii) 1 within the Wood Body

The 1 energy is dominant and directing, and the wood body is actively aroused; hence this energy is a very strong mix sexually. This allows for the water energy not to be so stuck within the wintertime of 1 energy, and the body draws the spirit to the next season of spring so her ideas can present themselves to the exterior. This is very important so that this person doesn't get lost within her own mind-identity. This spirit is normally very still. However, the body of this one pulls her outwards to express, and so there is a channel for creativity, so the water's energy doesn't get too stagnated.

iii) 1 within the Fire Body

The fire body is warm and expressive, and the water is cold and still; this is a complex matching. The body wants to move and express; the spirit wants to be still and think. These two are at opposite ends of the spectrum. There is an overall tendency for the water spirit to hold back the body, so it overheats and numerous heart-overheating problems arise. Otherwise the spirit will feel too exposed and helpless. It is likely that the heat can cause psychological disturbance and turbulence, unexpressed emotions, and a withered body, as the heat will burn it. Yin fluid loss is a problem, and so kidney related problems could develop. The stubbornness of this energy to go its own way can make life very difficult, especially in such a body.

iv) 1 within the Earth Body

The Earth body is warm and solid and social; the water spirit is none of these things. Fluid with a will, it has no need for the slowness and heaviness of the Earth body. Whereas the change of water is downwards, there is no change at all with Earth. It just seems to revolve around itself in an egotistic manner. These things disturb the 1 energy, and the body feels like a trap, holding the water or pooling it, and so blocking its descent. It has to rise and meet the people, and of course the spirit rules and will not do this. This causes great tensions, and the Earth body will feel like a thick and unwanted mass. This may cause the body to lose weight and appetite and become thinner than the normal Earth body, as the water spirit drains the controlling effect of the Earth. This can mean that little attention is placed on the body and more on an attempt to develop mind-identity. However, this body cannot develop this faculty easily, causing frustration and a wilful force to keep trying again, draining the physicality of this body. A balance must be struck between intellectual pursuits and the physical so that all aspects are nourished and united.

v) 1 within the Metal Body

This is a harmonious combination. The thick and strong skin of the metallic type is a change towards the cooler and more isolated, and this is where the water energy can find some sustenance. The body of this type provides a structure and mothering effect on the water spirit, so there is a homely feeling within its shell-like existence. There is no drive out of the winter, so this combination makes for a very isolated experience, as the body and spirit both attempt to go towards stillness and tranquillity. This type can be very good mediators and are associated with thinking or mental activity having to do with the storage of vast amounts of information. The brain has an important role, especially in the water energy, due to the kidney being tied to the bones and the marrow within the bones, which in Chinese Medicine are the autonomic nervous system (in a fragmented view) that includes the brain. These people will have the dedication to lead projects and understand philosophical problems.

vi) 1 within the Water Body

This is a combination that is comfortable and connected. However, the stillness of this energy is quite profound. These people can be quiet and isolated for very long periods of time. They may not even need to speak, and they will not be able to stay in the company of others for long periods. The deep internal natures of these energies present cold externally, but within the coldness and toughness is a warm internal spark, and this keeps glowing in the darkest of nights, almost a lighthouse in the darkness, 1 is inclusive of both dark and the light. The water energy wants desperately to connect and to join with, often sexually, as this is the sea it swims in, between the physical and the spiritual, in the brain and mind. This is the connection that the water longs to find, but it is often unable to and cannot sustain it for long, for the fear of being lost in the ocean and their individuality being lost. This restriction is merely that of being within a structural body, so the water energy, similar to the wood, looks towards the spiritual, whereas fire looks towards her next stage, Earth, and the physical, as she is already spirit.

4.3 The Quality of Field Energy within the Female Body
i) Kun/Field/2/Earth Yin

The female energy is represented as the three yin lines of Kun within the pre-Heavenly sequence of the trigrams. In the post-Heaven, it too has an important expression of Earth and the yin and the centre and the southwest, all together. These give us a picture of yin within the yin, the pure feminine within the female form. However, Earth is still between yin (metal) and yang (fire) energies of the 5 phases, so the 7 energy within the female is perhaps the most yin of all the energies. The combination of this creates a perspective of a very female energy, where there is a deep acceptance. This structure is most suitable to create life from. It is the right motherly energy that needs love and attention to then be fertilised and offer the fruits of children, which are cared for with patience and tolerance and contentment.

This energy loves to be pregnant and in fact sees most aspects and roles in her life as a pregnancy, wanting to grow everything from her physical body—or what she can do, not what she can say or think, and being very much one with it. They love being pregnant, and this is their ultimate concern. Feeding children and food in general is the focus of life and the source of all goodness. They don't know how to be expressive in a more flamboyant way but are constant and nourishment-oriented. This is a peaceful energy. It moves slowly and gently; it rolls and doesn't force anything. The archetypical female energy, she is gentle and solid yet has the foresight of great intuitive ability in relation to the physical plain of existence. She is practical and down to Earth but intuitively knows at what time she must do something, and her change is very gradual, much like the seasons as the Earth turns. She accumulates material things and likes to do this, she draws many things to her, objects which can be of value and some of no value, but storage is very important to all the earth energies and the 2 is no exception. Initially this energy draws inwards, are frugal by nature and hold onto money and keep account of things very well. They can very often be seen to be greedy or selfish but often this is a process of accumulation without reason, in a sense she doesn't know why she wants so much and very often this pattern is due to not having what she wants as far as direction in life or something to hold onto and draw in towards her, like pregnancy.

She is a (pregnant) mother first and foremost, as are all the Earth energies within the female body. They love to be nurtured. They are sensual and physical, loving touch, and sexual contact. She often endures force from the men that need her softness. She can be alienated for long times and can be abused by men who do not understand her gentleness and what she needs. Women of this nature can often find that men take what they want and leave. This may make them feel that men are entirely a different species, and this is how they respond to them. Often they will gather in groups to socialise, finding strength in numbers, and often their attitude is to find ways to control men through their needs rather than directly. This can make them seem manipulative. However, their motivation or need will often benefit more than just themselves, and they are self-sacrificial in many ways, but with resentment and bitterness, especially if they feel something is being taken away from them physically.

Loved by most people around them, they will be very social, finding the contact of others enhancing and unifying them with their role. They are often workers who are directed by a boss to diplomatically deal with situations, nurses, personnel officers and people involved with other people, secretarial roles, or roles where they are ordered to do a physical job which is their speciality, and they are dedicated to the smallest of tasks. They do not have fast mental expression but learn more by touch and feel and taste than by anything else. They often find themselves in roles where they care for others.

The 2 energy is an example of yin and yin is about depth and simplicity and no expression. Therefore these women tend to be, along with the male energies of this type and the 4,7, and 9 energy less outwardly expressive or they pretend to be outwardly expressive to fit in or because they think they aught to. Very often they are plagued with thoughts they are not enough and need to be more often in self-image and visually expressing themselves, what they wear and how they carry themselves. Very often there will be learned rules applied in these types, which are based on yang ideologies of how they should be or behave or look. Left to their own devices actually the yin is plain and simple. It is not outwardly flamboyant and the yang expressions would call it boring and "not interesting", however this to so overlook the fundamental truth of the expression which is yin, and yin is not flamboyant, it's the truth and the mother, the source. The source is never flamboyant; it just is what it is. Generally females like this are less shocking and more what others would call "prudish" but actually is a protection and true shyness. The only way this energy gets corrupted is through the masculine expressions into the world and how this then corrupts the expression. These types are easily influenced to the benefit of all or to the fall of all.

ii) 2 within the Wood Body

The wood body is the most activated and moving of all the energies, and the Earth spirit is amongst the slowest; this makes for a complex and difficult combination. The energy of the wood will stagnate within, and unless this 2 energy decides to be in their body a lot, it will cause accumulative type dis-ease. The problem is that many of these women are dissociated from their bodies, as they find little connection between what they feel and what they are. This bridge must be crossed and connected through physical movement. In this case, built-up aggressive energy can cause massive problems like cancer and possibly high blood pressure. This body can make the Earth energy heavy and edgy and violent in bursts, unlike her character. Mental health issues can develop around feeling constant worry and anxiety.

iii) 2 within the Fire Body

This combination is very good, perhaps the best for this energetic type. The fire body heats up and consolidates the Earth spirit, allowing her to feel more and more at home within the physical structure. It is natural for these energies to feel attached to the physical and material, and hence for the spirit to be more material than a 4/wind spirit, for example. The body and how it feels is important if they can accept it the way it is. The fire body will strengthen the expression of the person to be herself, and this will allow her to feel more content and less anxious. There can be an increase in appetite for life and for food. The 2 energy is very concerned with food and storage, and they are often overweight. The fire body will consume a lot and will increase the need to eat, so wasting and thirst (diabetic type patterns) issues may arise if they are not careful, and heat accumulation type dis-ease if they do not move. However, the fire moves slower than the wood body, so there is some balance to be had.

iv) 2 within the Earth Body

This is a very comfortable energetic connection; it is a very slow combination of spirit and body. It is comfortable with the Earth energy, and its motion is consistent and revolving at its own speed. The Earth structure is luxuriant with a rounder, fatter body shape and thicker flesh. This suits the field/2 spirit, and it feels at home and comfortable. The key problems are overeating and being overweight due to lack of exercise, because this energy moves quite slowly. In a woman, this combination is very good for pregnancy. The Earth body makes for a very soft and smooth body shape in tune with a more female form.

v) 2 within the Metal Body

This is a good combination for the 2 energy and a natural mother. In this case, it mothers the body and creates form from it. These people will be very physically conscious and are entrenched in the earthly energies, so Metal and Earth meld together. This formation gives more structure to the spirit and draws essence and direction from it to create aesthetically beautiful forms. These people can be models and actors. They will use the body to create form and shape; this is important to them, as is financial wealth and security. The metal body is tough and defensive and therefore can easily characterise a fatherly protective energy, whereas the spirit is motherly, so the combination makes for a stronger/reinforced motherly energy, which is soft but can hold very traditional, conservative values and be rigid in opinion. These people do not have the flexibility of movement that wood energy has, so it is more Earth and structure that forms their lives; this can make them cooler.

vi) 2 within the Water Body

The water body has a heavy bone structure and a lot of internal power from the Ming Men fire which burns brightly within. This energy warms and tonifies the central region where the spirit energy of the digestive system is focused. This powers the digestive system from the opposite direction; as fire energy comes down through the pericardium system, this energy burns from below. The water body is generally too cold exteriorly to promote the radiation of the Earth spirit, for 2 is a social energy. The inexpressiveness of this body type would be complex for the spirit, but the fire within the deep core of the water body type would nurture the inner aspect of the 2 energy generating it. This would cause feelings of melancholic depression, heaviness, and sogginess of the Earth energy, not being warm enough to be social and therefore feeling a body that isolates from others. This is not effective for Earth, but the coolness may reduce the anxiety. In another way, this makes for a very internal world and one that can be miserable because of the isolation feelings.

4.4 The Quality of Thunder Energy within the Female Body
i) Zhen/Thunder/3/Wood Yang

The 3 energy is the most yang of the energies. It is expansive and expressive, and this combination is therefore a complex one. It is impossible for this woman to feel that she must suppress her energy in an attempt to fit into the yin within yin type female approach to life. If this is done, it will in fact curb and destroy the body of this energy, as the impact of holding in such expressive energy is devastating to the physical structure.

Women of this type are striking, often expressively beautiful and admired by many. This is a very important example/ expression of the yang within yin female (1,3,5,6,8) The outer appearance is greatly taken care of and considered, there is much more association with visual and look then the yin within yin female of the 2,4,7, and 9 which generally have much less interest in external appearance as the energy is internal, this is true of the yin males also. The yin in all cases is not exposed, it is hidden. The 9 energy is yin but also tries to move exteriorly so it is in-between, but the 3 has flare and imagination and creative clarity when it comes to design and image. The Yang within yin female will often wear something because it very much an expression of herself, but also for effect and to attract male attention or even to vie with other female expression. Unfortunately the media manipulation of female form means that women feel bound to be something they do not feel which they then resent greatly. The yin within yin female expression tends to be much more modest and less expressive, and as a result doesn't attract attention to herself but are in fact required to be sought-out by male expression. This is very different and non-manipulative. Although many yin within yin female expressions have turned toward yang expression in order to fit in. They are not natural at this.

3 energy hold the attention of others and can be hypnotic in their charm. They are often highly energetic, playful, and gregarious with their expression. This can get them into difficulties with men who believe this expression is a form of flirtation and so attempt to move on it. Women of this energetic type do not like to be in the submissive position; they like to call the shots. For these women to feel they are being themselves, they must be the masters of their own destiny. They are unlikely to wait around in a difficult situation; they can change in a flash and be gone in a moment. Generally they can be self-centered and a magnifier of expression so forming very dramatic and expressive forays or acts of expression, which some might find disconcerting, but they are simply expressing what they felt at that moment. Often they feel misunderstood, and feel like they should be and could be understood if people were able to really see them and realise that their expression is about expounding expression of light and passionate connection towards others, rather than what it is seen to be by the cooler earth and metallic energies, as a cure or her being a source of "problems". They often therefore feel misfits to society and greatly appreciated by few rather than many. Above all needing Calmness and centeredness to still the agitation and anxiety that is how performing in the material-world affects them, these types are sensitive like static charge, they pick up on alot that is missed by others, and as a result are told they are "hyper-sensitive" rather than naturally sensitive.

Unlike the male counterpart, who would be a little more aggressive, these women can be aggressive but are less so. They are usually more assertive and expressional. However, these women are seen often as out of control or without boundaries by the Earth type and metal type; this is often the judgment of society as well (an Earth/metal construction). They can hide internally if they feel shamed and of offence to others turning their energy inward which is highly destructive, or they can outright attack back which is very often through words, not physical expression. Often they are highly sensitive and they can pick up on events and intuit images of events easily. This makes them greatly valued by others. Visionaries, however, are seen as witches, and this has been their persecution, which still occurs. Given a powerful position, they can be either dictatorial or enlivening and energizing, creating bright clarity. The major tendency this type can fall into is an over-association with the exterior world. They are highly attuned and sensitive to the exterior and so highly affected by it, although in the end they will go the way they want to, hell or high-water! Very often there will be a great fluctuation between bursts of their own expression and insight and the total reverse when confronted with an exterior force which questions or criticises their expression. It is likely they will immediately feel lost or utterly winded, thus stopping their expression. This is because they are pungency personified; the energy of spice goes to the exterior and therefore loses centre. They are often involved in expression, and when this expression is questioned or toppled, they are lost. They have no centre to fall back on and can perpetually look for the exterior to provide them with acceptance rather than finding the empty source of their expression again. The key is to live rooted in the Emptiness within. The main issues for this type, male or female, will be anxiety and agitation, this energy tends to affect the digestive organs and sleep patterns together and most of the time this will be the main complaint. The reasons are always mind-identity within emotions.

Generally, as mothers, most do not have the control over their energy to be mothers that are constantly feeding their child's physical needs. However, they have the expansive and attractive energy that makes children love them. They have a great deal of fun with children, as the energy of thunder is associated with wood in the human, which relates to new growth and expansion, much like that of the energy of children. Pregnancy therefore is hard for the thunder energy, as this weighs down on their natural moving energy and confines their freedom. Interestingly the 3, 4 and 6 energies all have some aspect of wood, the wood energy is a natural chef and good at manipulating ingredients with a keen interest and understanding about nutrition although they may not be big eaters or particular interested in consuming food. The earth energies however tend to be linked to food so much as they enjoy the taste and love eating. The difference is clear - one side is the expression of moving the earth energy (wood) the other side is the absorption of food (earth).

Of course it depends on the child, but as far as natural mothering goes, this energy finds it difficult. Although the female body allows this energy to adjust to the physical nature of life, it is truly the expressional, the intellectual, and spiritual connectedness that this energy finds its sea to swim in. This energy constantly needs connection with its body, as it is so easy for such a yang energy to rise to the Heavens. A male body is better to carry it; a female body is more earthbound in a way, so as a vehicle it is more difficult. The body of a woman is soft and physical, and this can be difficult again for this energy, because it requires a tougher structural format, or more exterior energy, whereas a female's is more interior. This means that the body and expression can fight, until acceptance is formed.

ii) 3 within the Wood Body

A harmonious combination, this joining allows for full expression of the liver's physical energy to the exterior and so allows the body to be very healthy in its feel. The wood body creates tension, and the spirit distributes and expresses it, so there is a constant flow met between interior and exterior. This is very useful as it means nothing can stay trapped for long, and expression is easily expressed from the system. This stops stagnation and so stops anger and aggression from accumulating. A healthy system results, and perhaps exhaustion would be the main cause of dis-ease here.

iii) 3 within the Fire Body

Fire and thunder work together to create lightning, so this is a very expressive combination. Very well-expressed thunder energy is much better than held-in thunder, which can be internally or externally explosive. Here, the thunder is drained by the fire, which helps it to move and lose tension. This is useful and creative, and this person would be accompanied by insomnia and exhaustion but little accumulative illness. Psychological mania is possible, as are high anxiety states. This body creates a lot of heat and is fuelled by the wood spirit, so overheating can be a problem too. Sexually expansive, this body will allow for this expression well.

iv) 3 within the Earth Body

The Earth body presents an obstacle to thunder, and the thunder will force through it. The body will block the spirit's flow through the system, not allowing it to express the way it wishes to, which can be very frustrating and annoying. In such circumstances, the thunder will attempt to force the body to do its will. This can create heavy exhaustion patterns and tiredness for the structure, which cannot be easily moved to action. This person is seen to be very hard on herself and often will take this out on others in aggressive feelings or shouting.

v) 3 within the Metal Body

The skin and outside act as a barrier to the thunder energy. Hence trapped emotion, repression, depression, feelings of suicidal rage, or aggression can result. The person, unless allowed to express, will explode or implode, causing massive damage to the body physically and emotionally. Cancer, accumulation illness, strokes, and other nervous/high blood pressure and stress syndromes can result. Blood pressure issues are due to the withholding of the system's energy within. Sexual repression can lead control issues around sex, which can be damaging to this person.

vi) 3 within the Water Body

As with the 4 energy, the water body acts like a springboard for this energy and can also be a firm anchor. Allowing for a more controlled expression than the fire body, this construction allows for expansion without going too crazy. There is great power, so the body can last a great deal longer as a result. Emotionally, it will subdue the yang expression somewhat, but not in a negative, controlled way—more in an anchored way that allows for more body awareness, which is strongly lacking in this type. Sexually, this allows for a less damaging start position, having a strong foundation.

4.5 The Quality of Wind Energy within the Female Body
i) Sun/Wind/4/Wood Yin

Yin wood pertains to the female, so this is a smooth combination. The yin of the wood phase is the cooling wind. This complements the complexity of the 6 energy, which is solid and heavy metal but is the yang of the metal phase and hence is wood in metal. The 4 energy is of the wind and is gentle and soft in its flow. It is a communicator with all aspects of life; the wind or air passes between all aspects of the physical world and so joins aspects together, through itself. It has a cool but caressing quality, which denotes a kind attitude. However, in its expression is defensiveness, so these energies are difficult to pinpoint for the recipient. This energy expresses itself and moves fast but is cool.

Women of this quality are indeed feminine. They are likely to be a little more reserved in their looks, but this energy still can reflect a comfortable air of expression. It is less bright and boisterous than the more yang energies, although still expressive. They usually wear flowing clothes that move, rather than the structuralism of pure metallic quality like the 7 energy. This is a usually serious energy, but they can be quite playful in their expression. They will have an emotive expression and are perhaps the most sensitive/stuck of all. They can be in deep psychological torment and will cry and be deeply affected by things around them. Small things can upset them, and they can be quite intolerant of people who attempt to restrict their space; they are quite claustrophobic in this sense. However they can also be highly detached almost as a protection from the world and go into a space of total non-engagement with the exterior being numb to abuse…this tend to be in the mind-identity expression of wind and is pathological.

The soft quality of this energy can lend itself to helping others through expression such as physiology or therapeutic work, but also academia, music, teaching, and other intellectualised, non-physical activities. They must balance their nature of "taking flight" away from the physical body, as they can easily move into the mental and be locked in. Their expressions are often wrapped in an easy-to-digest package, but the expression itself may be of a very difficult nature, which makes them doctors in the psychological fields. The definers of logical analysis are often of this energy as well. The 7 tends to be practically what the wind is mentally. These energies have insight into mind-identity, thought, and emotion; hence they can swim these waters easily. As mothers and physical beings, they can also do quite well. They have patience or tolerance and can be very caring and take responsibility for things quickly. They are not naturals to the physical aspects of mothering, but to the intellectual development and the psychological aspect, they care very effectively and make excellent teachers for children. Being good at analysis, these people make good diagnosticians, though they are not as insightfully intuitive like thunder energy. However, they can see the objective aspects and be one with the coolness to see clearly, so can become doctors, though they are not practical unless they force themselves.

It is an energy easily accepted in society, and expressions of music or poetry writing (rather than performance) are perhaps their greatest roles, as they are natural at the communication of emotion through this medium. Photography is also another skill which often is common with this type, not so much to find the visual image which they are not concerned with, but to paint a picture of an expression often of melancholic quality they see around them or want to show in a medium other than verbal. There is often hidden meaning to everything they do which is sometimes almost impossible for another person to understand unless they understand this quality about the person.

ii) 4 within the Wood Body

This body would seem correct for the 4 energy because it is so-called a wood energy. However, this is not actually the case. The 4 energy is autumn within spring; hence this body is the most difficult expression of the 4 energy. This body is very powerful and sexual, while the 4 energy is much more gentle and ethereal. The physicality of sex is very draining to the wind energy, and the need for this body to express this energy becomes a drain. Often they can attempt to dissociate from this and become very mental, ignoring the body and the needs it has. Doing so can cause digressional illness and mental problems. Also, they can become angry if heat held within the body is not expressed. They often need to do exercise to prevent problems. This energy is like holding in an explosion, so one must be careful knowing this type, as their energy may be very erratic and seem mentally confused, especially in old age. Blood pressure and blood stagnation problems can ensue, as well as circulation issues.

iii) 4 within the Fire Body

The wind is a cooling energy, but the heat of the fire renders the coolant useless, and so the heat of the fire body, as with wood, overpowers and overheats this temperament. It is too energetic and expressive for the wind energy, which is more self-protective and enclosed. Pressures build up internally as the wind attempts to cool the fire, but this struggle often makes the fire body stagnate and overheat, and the wind energy cannot feel cool and clear enough to express its clarity of feeling and caress the surfaces. Overheating affects the heart, which can cause high blood pressure and issues of this nature.

iv) 4 within the Earth Body

The Earth body actually suits the 4 energy very well; it tonifies the physicality of the metal coolness of the wood as it goes from late summer to autumn. The wind energy is fed by this energy, which provides the earthy coolness from which to express itself. These people will be creative as chefs and attempt to make simple and tasteful foods. The metal energy softly cools the Earth's. They can move towards a slower metabolism and have a tendency to store or put on weight.

v) 4 within the Metal Body

This is also a very ideal physical structure for the wind energy. The Japanese body structure for example, is similar, and the cooling nature of it is very easy for the 4 energy to unite with. The nature of courteousness and non-aggressive action in accordance with others and in social relations works very well with this type, as they have an overall quiet tendency. The wind energy is cool and relates to this temperament and so can feel comfortable and does not need to feel like its expression is being forced or blocked; it can be as it is. These types tend to be very long lived. They can relate to their physicality well in this body, and they tend not to be overly sexual and so conserve their energy and are able to control over-expenditure of it. They can be very good with money, saving it very effectively, but they make very bad investors, as they don't know how to take risks easily (unlike 3 and 6, for example).

vi) 4 within the Water Body

This physical structure is a harmonious one for this energy. This body has more physical power but is steady and has the ability to help the 4 energy express its nature in mental applications. Also, they have a solid bone structure, allowing the wind to feel anchored into the body, not like it is being burned off the surface, as in the fire body. This combination is also very long lived.

4.6 The Quality of Taiji-centre Energy within the Female Body
i) Taiji/Yin Yang/5/Earth Yang

This is a yang energy within the confines of a human being. Therefore, this becomes a complex mix for the female body. This energy is an Earth energy and as such creates the feeling within the female body of power and control but with child-loving and creating form. This is a motherly energy, the female body drawing this energy to be a more powerful and more expressive version of the 2 energy, but far less yin. They have great over confidence in their own self-sufficiently which leads them into massive problems and individualism which can be the biggest difficulty in their lives as they can cut off from others because they believe they can do it themselves, which they can power-through but in the end this doesn't allow them to connect to other people and accept them, which she finds so difficult. If she could see the benefit of being helped by others this would benefit her, and others more as her leadership could be designed for the good of everyone including herself, rather than explosively herself, this later is mind-identity.

This is dominant yang spirit energy, and this energy must be in control. These people are very powerful and dominant controllers of women and men alike. She has the tendency to use her form as a female and draw masculine energy and help by this, which tends to be quite manipulative which they feel very hurt when this is explained to them, because the spell is broken! These energies are queen-like in form; they demand a certain respect and have authority. This may manifest in the role of a housewife or as a manager—they will always demand respect. They can be very manipulative, creating confusion amongst other female counterparts or friends, so that they get what they need/want in a given situation. They can be very self-serving and disliked by others for this. However, they can be enormously generous at times, often giving things they have finished with, but all the same, giving with great expression. One often finds, though, that there is an expected or accepted ownership, that you then belong in some way to her. If you don't do what she wants then you can be made and outcast and never again spoken to or cut off for not behaving the way she wants you to. They have generally strong egos and drive and can happily talk all day. Talking is their means of identifying with how expanded their absorption of the universe into themselves has become.

As mothers, they provide for their children, especially the ones they like the best, and will do what they can to benefit them, as long as it too encourages their own ends. They can have a less traditional approach to things and can notice that they do not behave or wish to behave like yin within yin Earth, which they find too cliquey and submissive. They are very much in control of which they choose as a partner/friend or even family, and they will do what they want, usually to get money or power, although they can be quite considerate if they want to be. Theirs is a more internal control compared with the more extroverted control of the male with this energy, and so it is usually bound to the house and home, making this their castle. They can be exceptionally determined to obtain what they want, and these women love to shop and spend money to obtain things, especially for the home, although they will only buy a bargain, never the full price. They tend to make everything they want "just so".

Generally these women have a keen visual interest and dress to be impressive although will get upset if all the attention is applied to them only because they do not feel in control of it and so it can be overwhelming, they are happy to be the centre of attention when they decide to be, and not a moment sooner! She will tend to have numerous male friends, which is common to all the yang female energies as these people she generally has control over, especially if they are yin masculine. She needs to be desired, as do all the yang energies male or female, if she chooses her own direction and takes it. Yang within yang males, interestingly tend to like male fellowship/ comraderie most, and females are seen more as part of their lives. Yin within yin tend to like the company of other women of the same nature and male company sometimes. Yin within yang males tend to like female company much more as there is less threat from the masculine and they feel they can provide and are wanted.

ii) 5 within the Wood Body

The wood body has a great deal of yang within its liver, and this conflicts in explosive measure with the 5 spirit, the out-of-control energy of the liver, which is spring-like and expressive. It does not give the 5 spiritual control and therefore acts like an irritant to a powerful energy. This simply drives the 5 energy stronger and makes for an irritated or agitated body. They can accumulate anger and become very aggressive physically, expressing this by behaving forcefully towards others. If they do not have a physical outlet for their expression, it can become internally destructive, particularly causing depression. However, serious illness usually doesn't come, as they are body-conscious and behave as one with it. They do not ignore the body as much as the more ethereal energies do, such as 3, 4, and sometimes 9. They can be very demanding and dictatorial in a Napoleonic way and can be frustrated and determined to take control. A force to be reckoned with, they often drive others away.

iii) 5 within the Fire Body

This works well in so far as comfort for the spirit, which feels at ease within an energy form that is feeding it; however, this can simply drive the egoism and power of the 5 energy further and further to become stronger. This is a problem, as the 5 needs to express more, rather than accumulate or benefit from its own role in the scheme of things. This makes a very dominant energy. The fire expression does make this energy stronger though and express it's feeling, but its feeling will often be to demand and be constantly in motion. This energy expands and grows stronger, and often this type will have many children or be a family person. As a mother, she will be a strong one, but one that is in utter control of her children. She can be dictatorial in this regard too. She has a very healthy appetite and adores good food. Often women have problems associated with eating too much, which is a societal dictate, and these women often pay little regard to what they should eat and are driven by their senses, overriding the masculine dictates. Collections of items, usually of financial value, often go along with this type, as well as with the 8 energy, although here they are more numerous and larger collections, and with a miserly approach. This combination can create a very long-lived body, only troubled by high blood pressure and possibly diabetes.

iv) 5 within the Earth Body

A very comfortable combination of energies means this type may become too heavy and motionless. They can be overweight and have a passion for food, as well as being content with themselves, which leads to difficult situations. Earth within Earth is so heavy that it is hard for this energy to become anything different than it is. It is very stubborn and doesn't like change unless it is through its own judgment, and this will be slow. Wasting and thirst pattern (diabetic type patterns) is the main complaint, although this type can have high blood pressure too if overweight. Accumulation of foods and things can be their main difficulty, but they can be softer and more generous than the fire body and less aggressive than the wood body.

v) 5 within the Metal Body

The metal body is the best and most harmonious for this energy. It draws away the excesses of the 5 energy so it doesn't become too strong, and also to see itself. The metal creates form, driven by the Earth spirit, so it makes for a wiser and more pragmatic person. They can be more aware of an exterior to their energy, and they will be more refined in their desires. Physically, they can be very defined and strong. They can be models and the like, showing off their body publicly and being praised for this. They can be egotistical too and colder than the fire and Earth, so concealing themselves sometimes. This can lead to sexual issues, but there is less desire for this type and more of a physical need to connect than spiritual.

vi) 5 within the Water Body

This is a very powerful body for the 5 energy to be in command of, and as such, it suits its purpose. Similar to the 8 energy, this energy is a more interior version to the fire body; hence there is a source of internal power to this body. However, the 5 energy is more expressive and less cool than the water, and this makes for a difficult balance. The water draws inwards, and the 5 can draw inwards also, possibly leading to internal control and depression. It will be less damaging to the exterior, but can be damaging enough to the interior to drive the 5 to deeply isolate herself. In some ways, they will be very self-sufficient people, but usually because they are hard and cold and move away from others rather than towards. The interior heat is key with this body, so high blood pressure can accumulate especially due to the control that the Earth spirit has, causing the body to have internal problems, including possible uterine issues and urinary tract problems.

4.7 The Quality of Heaven Energy within the Female Body
i) Qian/Heaven/6/Metal Yang

This is one of the most complex combinations for the female body. Qian is three yang lines. It is however within a yin zone, being in the late autumn and also being associated with metal. The energy that this creates within the woman is not soft at all; it is quite hard but warm and creative. This energy is firm and forceful in its approach to life, quite unlike its body which is soft and solid and yielding. This energy has its main problems in relationships. Often, this and the 3 energy find that they cannot find a man who is suitable, because men, coming towards these types (which is the wrong approach as they are yang in spirit so often they must move towards a yin man) often find that although they are elegant and beautiful, this is not necessarily a means of attraction and they will not yield a single step in their expression, it is all about what they want, not whether the male energy wants them necessarily, especially if this energy is without mind-identification.

The Heaven energy also has this approach, being very sexually alluring, shapely, and with powerful expression and momentum. They are leaders and are dignified. They walk with an air of authority and can be haughty and demanding. They are admired and desired from afar by many, though few come close enough to be able to care about them, as most are sent away. They use their bodies as decisive tools to create a particular effect or get what they desire in a similar but more refined way than the 5 energy. They are great manipulators and great geniuses in leadership and controlled attack. These people are born courageous and clear in their logic, and they refuse to believe anyone else, even if the answer is obvious and they have been deceiving themselves. While the 5 energy is really too coarse in her expression to be commanding queen, the 6 is the leader and takes command easily.

They must have this attitude, for it is hard for them to be within the male structure, which doesn't often give them the respect they deserve. They often try to climb towards their own goals, only to find that men get in their way and attempt to manipulate them. These women will fight back at every turn and have a plan before you know it, often involving legal procedures in which they are expert. They eventually get their way, even if it is through dangerous waters and off the track of what is exactly sticking to the rules; it seems the rule can bend if needed. Sharp, brilliant, and spatially aware, they make excellent architects, planners, and designers of all sorts, the bigger the project the better, also as artists they tend towards large sculptures and three dimensional or creative images. They also are expert surgeons and medical doctors known for their sarcasm and flamboyance as well as command of what they are doing, as well as dancers and aesthetic form artists. Shape and form involving "high" culture is what they believe is their plateau from which to look down on their people—and this is where they can fall from too. Their warmth and creativity come through the physical, so these activities are often related to yang.

As mothers, they are quite brilliant, having massive patience, care, warmth, and devotion. They can seem more fatherly than motherly, often being judgmental and providing disciplinary measures to make children behave; they can enforce a very regimented family.

These mothers are less likely to be around all the time. They like to be magnanimously there but physically elsewhere. The density of their yang can make them homebound, although this isn't usually the case. As one can see, partners could find this difficult, especially as the masculine must yield to this expression and allow it to be what it is and be led by it. This can be too difficult unless the energy of yin within her partner is supple enough to be himself whatever the odds. Women of this type can be seen as "lesbian" in their stereotypical expression and will either exaggerate the feminine aspects of themselves and their body to compensate, or just shrug off this idea and be who they are, which can be difficult as they often find themselves alone at the top of the profession they are in.

ii) 6 within the Wood Body

The wood body is well suited to the 6 energy because the 6 is actually a spring energy, although found in autumn. The wood body provides ample power and physicality to express the tendencies of this type. This body is active and energetic if used effectively. It also has structure and muscular power, all of which is associated with the 6 energy. This body has high sexual power in the form of blood in ample amounts, so this can make this energy quite forceful or moody. The 6 and the wood body are a perfect match, so there are very few problems, except that they can overdo it and so create high blood pressure problems and overheating of the system. However, if they exercise constantly, which this type tends to do, they will not have these problems. If there are issues it will be with the digestive tract and high tension patterns developing if this person doesn't express themselves resulting normally in digestive upset or bowel issues.

iii) 6 within the Fire Body

The fire body suits the 6 energy also; it is more expressive and energetic of the power of creativity within its expression, and so this is like its outlet valve. They will revel in it, but the raw, physical power of the 6 energy will be difficult for the fire body to sustain. It can drain the fire body easily, if this woman has a number of children, as this is the weakness of this body. Essentially, it is like throwing logs on the fire, so it makes the fire body overwork and over-live, having a short but energetic life. The problems here will be from high blood pressure, the heart, and difficulties of this nature.

iv) 6 within the Earth Body

This is not a harmonious relationship. The Earth body is too static and heavy for the 6 energy. There is not enough movement and creative expression for the energy to expand, and so this body becomes too cumbersome and static for the expression. There can be depression feelings or aggression from this type because they cannot express themselves directly. Issues will be of fatigue, lethargy, and feeling heavy and irritable. They will feel like they are trapped in a heavy and rounded structure, whereas their energy is much more about erecting creative expression in the world. This is a difficult combination.

v) 6 within a Metal Body

Though the 6 energy type is often connected with the autumn and metal, its expression is spring within autumn, so it counteracts this energy. Hence this is probably the most difficult combination for this type. This body cannot withstand the physicality of creative expression of the 6 type. It is too fragile for that and so can easily weaken with massive sexual output. These body structures tend to be cooler, calmer, less expansive, more yielding, and yin. Hence this energy will be highly over-controlled and feel trapped and isolated. Depression is the key problem, but also hypertensive blood pressure, skin problems, and constipation, to name a few, as a result of the tendency to hold in.

vi) 6 within a Water Body

This combination is perhaps the best for the 6 energy. The water cools and controls the passion of the 6 energy and holds it down, giving it a foundation and strength. There is a strong sexual capacity, which means this type will not drain its physical form, and this is beneficial. She will be able to give birth many times without problems. It is slower and steadier than spring, but here the 6 energy can find balance and less harmful expression.

4.8 The Quality of Lake Energy within the Female Body

i) Dui/Lake/7/Metal Yin

The metallic energy of this type is comfortable within the female form. The lake energy is yielding and soft and has a deeply hidden power to it, but especially has expressiveness that is playful, designer oriented, and logical. They are judicious and can be sacrificing if the cause seems worthy enough to them. They are often legally nervous and follow laws and observe regulations meticulously, but sometimes can be rebellious if they feel that they are being told what to do. They like to have their own way because they are insecure and nervous. This can often lead them to make rash moves that can lead to painful consequences. They may hold back emotion and pain, commonly making light of a situation they cannot control or do not want to involve their feelings in. This can lead to dissociation and pathological dissociation from the world around them.

This energy is indirect and therefore often seen as being manipulative although their manipulation is simply indirect ways of communicating due to their lack of sense of inner power. Real manipulation has intent to self-serve or be able to forward themselves or get what they require. The yang in the male (direct) or in the female (indirect) is more likely to be manipulative. There is association of this type to be used sexually by men due to them being flirtatious, however the flirtatiousness of these types is often mistaken. It can be nervousness. Often they have a great natural naiveté and this too makes for an easy target. However, the understanding of their body and sexuality is their base and root. It can be their deep connection to Wu and to the origin, being natural mediators and deeply Still, as they are the most yin of all the energies. They are almost childlike in their approach and can behave as though very innocent of the "reality", which often they are. They may keep going and going, doing things that they feel are morally correct, but that actually damage them, in order to feel okay with themselves. They can become detached from themselves very easily and drift into a controlled way of being, which could be associated with compulsive habitual behaviour patterns. Most things they do are structured and have order and form. They like to neatly package and design things and keep their work places tidy. They can provide structure and format but have little creative spark and are unable to think in absurd or out-of-rhythm/ out of the box, ways. This conformity tends to build a prudish characteristic in later life often. The stubbourness of their expression is profound and they will dig their heels in and not move as a way of protection from a nervous situation for them. They have a distinct internal life, unknown to the exterior, and this can be quite grief stricken. Though they are very refined in taste and know what looks right, they have simplicity of taste generally and they will present themselves well always but have little interest for their exterior really and are more interested in being of service and being of help to a yang expression.

As mothers, they are very good, similar to the 2 energy, but they understand the nature of children and so can often work and be one of them much more easily than the other types (apart from 8, which is even more childlike). The 7 loves to be needed and asked to help out; children can help this as they provide her direction, and other people will draw towards the children, but she may end up feeling left out of the picture. This type is often quite cool of temperament and does not get upset over emotional things easily, but nervousness often is at the base of this and inside they can be like a rabbit caught in headlights. Often they have quite a bit of resilience (but often not as much as they believe they have) till you hit a nerve; then they can respond ferociously for a short time in order to protect themselves. Their main difficulty is dealing with a lack of expression of emotion, because they can be so inwardly drawn and have a surface of bright expression that is really very shallow and not very convincing. Lake is often described as "joyous" in the classics. The "joyous" aspect of lake energy is the fact that that they have a bright, reflective surface hiding what in fact may be deep woes and grief. It is a very female energy, which, if allowed to be free of conflicts, can find deep clarity and stillness within as a natural state, but they must look deeply into themselves rather than fearfully detaching. They must also keep moving or at least placing themselves in situations where yang energy can help them to move or they will stagnate. This is the same with all the yin energies be they male or female; 2,4,7 or 9.

ii) 7 within the Wood Body

This is a complex variation. The mix of the highly potent wood energy and the spirit of the soft and complex lake energy creates a mismatch of types. This time, the body is stronger, and the spirit is softer and cooler. This means that lack of activity can cause this person problems. The calmer, stiller energy of a lake doesn't know how to control the turbulence of strong wood energy, and so a withholding of expression often occurs. There is no reason for the energy of the body to be used, so it can stagnate and cause accumulation based dis-ease and cancers. There is a desperate need of the spiritual energy to control its body, and this brings a sexual element into it. There will possibly be a withholding of the sexual energy and an attempt to control it, or self-control to some degree, which can lead to ovarian cancer and accumulation within the genitals. The wood body needs constant sex for satisfaction and sexual activity to allow the energy to flow. However, the lake energy itself is not about spiritual connection but about form and creating shape. Hence, these two form complexes, and erratic sexual binges or augmented fetishism can arrive if there is not an outlet. Here, the body and spirit have to find acceptance for there to be an allowance of freedom.

iii) 7 within the Fire Body

Again, this is a complex match. The fire body is fast and energetic and warm, and the opposite is true of the lake energy. The lake wants to move towards stillness, the fire towards the Heavens and expanding out. Hence the fire attempts to overcome the spirit. This can mean that the qualities of grounded-ness and subtlety of the 7 can feel burned and bruised by their over-energetic system. They cannot settle, and this can cause them to become anxious. This anxiety often expresses as a frantic nature or a person very busy and without stillness. They can be quite short and sharp and can be hurried and constantly attempting to do things neatly and in an uncluttered manner. They want to be organised and can be judgmental and critical very quickly. The fire almost propels the lake to become expressive, even though it

227

doesn't want to be, and this is what the expression is. There is almost a withholding being forced out. If it is held in, this can form a more forceful outburst when the bubble breaks, and this can be quite explosive.

iv) 7 within the Earth Body

This is a harmonious relationship. The spiritual energy uses the Earth body to form deeper connections to the physical energy, and there is overall bodily comfort. The femininity here will come out with far more motherly attributes. There will be a far greater confidence with this construction, and tasteful food will be presented well and considered. Connoisseur chefs can have this combination. They will feel at ease. This type can live for a long time without any problems, perhaps only lack of movement, but they are often logical enough to do something about this. They can be prone to accumulation dis-ease as well.

v) 7 within the Metal Body

This is a simple combination of body and spirit. The metal body is a very good home for the lake energy. The slow and subtle formations of the metal body, which tend to be smaller and subtler than its fire counterparts, perfectly combine with the lake energy. Subtle hands and feet make for gymnasts, ballet dancers, designers, and form experts of every style, including interior designers and the like. Very long lived, cool, sullen, cat-like in approach, and distant, they are less motherly and can do their own thing, often disconnected from others.

vi) 7 within the Water Body

The water body is heavy, compact, and powerful. It is perhaps too dense for the lake energy to feel in moving, but the density of it means that there is a good weighing-down feeling that allows the lake to feel comfortable. It is a long-lived expression, but the heat of the water body can be a little too much for the lake, whose waters are cool. However, the water body endures, and the lake energy can reside in it without being fired to the exterior like the fire energy type. This does not help the coolness though, and the exterior, at least, can be very cool with this combination, leading people to believe this is a cold person. However, they are often softer than imagined.

4.9 The Quality of Mountain Energy within the Female Body
i) Gen/Mountain/8/Earth Yang

The 7 and 8 energies are perhaps the newest energetic creation. The basic energies of the 1, 2, 3, 4, 5, 6, and 9 have all been there longer than the energies of 7 and 8, which are accumulations that have recently arrived. Mountains have formed in a relatively short time from the 2/5 Earth energy, its mother, and lakes are accumulated rainwater from the sky. These are born of the basic energies and so are newer in form. The 8 is considered a "newborn" being the second youngest in nature, perhaps that of a small child of 8–10 (lake is the newest born, perhaps the age of a seven-year-old girl).

This energy therefore has a stubborn nature, like that of a boy wanting his way and standing firm with it, even in the face of conflict. He wants, and he likes to accumulate and collect. Within the female body, this energy creates a complex mix. The mountain energy has a tough surface and a high head, which is the aspect that men of this type often play to. In the female body, the core of the mountain is more present, which is a soft magma—heated Earth. This energy tends to be very focused motherly types. They really want children and to be mothers, but at the same time do not want attention drawn away from themselves, and this can cause them to give up the idea of having children.

These women are often similar to their 5 counterparts but have less control and a more childish nature. They like to have fun and socialise and be the centre of attention. They are natural performers and do well when allowed to do this in their own way. They are usually very funny and charismatic in a theatrical way. They can be very aggressive if pushed to use force, and when they do, it is very physical and emotional, and they will never forget being wronged, as they see it. They have very fixed attitudes and very manly concepts of what a man's place and a woman's place is, and they will try to control a situation so that the man is made to do their will. They are good leaders in the sense that they have high ambition and can work hard to get there, but social activities often get in the way of actual achievement, so they may not reach the peaks. They find it difficult to look outside of a conservative ideology and can't normally see things from broader views than their own.

Women of this type are quite abrupt and self-centred. They are usually loud and like to have a conversation in which they are the master of ceremonies and tell you their life stories, over and over again. They like being flattered, and it goes a long way with them. They are inherently practical, and anything that they feel is not practical is seen as useless and dismissed, at least until it becomes of value to them again. Once they believe something, it is very difficult for them to be persuaded otherwise. It will take something within them to change, and this may or may not happen.

They love material things and collecting and obtaining. This to some degree makes them excited and they can accumulate a lot. They can be boisterous and shun the Yin female aspect and instead become strong physical workers or compete alongside men in physical industries; they see themselves as "one of the boys/lads" something shared with the 5 energy also. Adventurous and ready to take on a challenge or a test of strength, they can look very feminine, but their attitude can be very male in this respect. Family is most important to them, and being at its head even more so. Children and life within the home—forming the home into a castle that is better than the rest—can become an obsessive pursuit unless they are careful. They can be kind and loving to those around them when their needs are met, and can give generously and feed a family. They need to be balanced by a more yin partner.

This person is practical not verbally orientated, as with all the earth energy male or female there is a temptation with this type to take on the intellectual world because the practical and social world that they navigate easily is seen as being "lesser" by the social media, so they often feel like they are not good enough because they can't force themselves intellectually and are physical kinaesthetic people. This unfortunately causes problems of mental stagnation and analytical thought patterns that are circular and cause problems. They are best being what they are. Very often this type will dress to impress more than most, they are very expressive and use big rather than dainty expressions relative to their shape.

ii) 8 within the Wood Body

To reconcile the situation, physical exercise and movement—use of the body—is very important. The wood type body is the strongest, as it holds much blood and energy; hence the mountain energy has to be able to expand its energy out, rather than blowing its top. This will cause calmness; otherwise, this person may be severely disliked. They may be considered rude, impetuous, angry, and violent in a very physical way. This can lead to isolation and deep depression or alcoholism. Many modern Western people have this energy issue.

iii) 8 within the Fire Body

This can create a much more harmonious relation than the above. In this energy structure, the mountain's energy is allowed to radiate. The mountain feels strong and comfortable in this body structure, and it is most beneficial to the spirit of this energy to grow and prosper. This union can allow the fire body to be used up creatively, and the heart of this person will not be so taxed as a fire within fire type. The key is that the body's energy is harnessing the accumulative and upward structure of the mountain. There can be issues of over-consumption of food, however, and this may express itself as generosity and social kindness. This will be on her terms; she will be confident and charismatic, physical and sensually interested. As food connoisseurs and enthusiasts, they are very interested in volume and quality, but volume is more important to them.

iv) 8 within the Earth Body

This combination is smoothly connected. The body and energy are earthbound; they are completely unified. This person will be very practical, of little words, and not a particularly good analytical thinker, but can devour massive quantities of books and information, which is stored. They love food and eating, and this can lead to obesity and greed. Accumulative disease is their main problem, such as cancer and heart diseases due to food or alcohol and lack of exercise. They move slowly, so unless interested in a sport, they can become too heavy and stagnated. The difficulty with Earth types is that they are body- and sense-oriented and like moving only slowly, so after their youth, they often lose interest in the sports that would be most beneficial to them. Hence, accumulation sets in around the 30s and 40s.

v) 8 within the Metal Body

Here, the ways of the body are benefited by the ways of the spirit. It is again a good balance for the body and spirit. The metal body is also slower and methodical. This leads to a good combination. Achievement comes after long periods, and this body is suited to this kind of process. Whereas the fire body gives a spiritual and lighter connection, the metal body is more sullen and tougher. It can achieve much, but stagnation patterns and habits will be hard to break. This cooler temperament makes for longer life, but this person but must be careful not to hold in emotional expression that results in accumulative illness. Overall, they are more sullen and more introverted, a cooler mountain.

vi) 8 within the Water Body

As we have seen, the 8 energy is associated with Earth in the five phases, but as an energetic formation, it makes up Earth as associated with the cooler period between the beginnings of spring and mid-winter. As a result, the water body conflicts somewhat with the Earth energy, but not entirely. It is in fact quite a harmonious combination. The water body is associated with the Ming Men and the right kidney. This powers the digestive system that is the residence of the spirit energy of the 8 types, just as the fire type resides mostly in the heart of the person. The water body, rather than providing nourishment via the heart fire, provides nourishment via the Ming Men fire. It is the difference between the action of the pericardium, of moving the energy down back to the Ming Men and then up to the digestion, or directly though the kidney yang to the digestion.

The 8 energy therefore thrives on this too, but in a much calmer way, one that is less irradiative. This is the heating energy of deep inside, so it creates a warm base for this energy to grow. It supports the growth in a similar way that fire would but with more power and anchoring, and less exterior heat, expressing as showing off and needing to be number one. This one will become number one through her own hard work—a very hardy opponent and a very powerful body and spirit combination. However, Earth energy is about the southwest region and the centre, and this extenuates the isolation of the mountain, the winter mountain. Hence these types are prone to depression, isolation, and loneliness, especially if they create a massive internal heat, which can turn into a pressure, high blood pressure being a very common one. They can be cold and miserly and withhold; this does little for them socially.

4.10 The Quality of Fire Energy within the Female Body
i) Li/Fire/9/Fire Yin

The 9 energy is a yin energy and thus connects to the female body easily. We can see that that the fire, though bright and radiant, is softer than all forms. It has the power to burn but also is soft, a complicated nature. Water too has the power to freeze and drown in turbulence, yet it is the source of life and yin. Hence, by its very nature, fire and water are a mix of yin and yang. This sometimes means that the 9 is seen as a yang energetic and the water as a yin especially as 1 with the female could be considered more yin, and 9 within the male more yang. In the female however the 9 energy is still yin but with more yang than the other yin energies of 2,4 and 7. The Fire in some ways can seem to have an empty quality, which inevitably is filled by yin/ Wu, and as a result this energy can seem cooler than it first presents. This is opposite to water, which is cold but then has a warm heart.

Within the female, it takes on a bright feminine role. The family position of this energy is the middle daughter, and as such, the middle daughter wants to be noticed. The middle son (water) wants to be hidden, but the daughter comes out into life and is the performer or star of the show. She wants to be first, best, and brightest, and she wants to radiate that to everyone. She can be stubborn in doing so and has a fiery and seeming aggressive stance in some situations in order to protect the emptiness/ yin of her hidden interior. She is often like a star on TV who radiates joy and splendour, but at home is exhausted. These people can be easily drawn into a hedonistic lifestyle of alcohol, drugs, and sex, and it is easy for them to become addicted. The wood types tend to be too sensitive, and the water types will see the falseness in them, and the metallic spirits will believe them to be toxic. However, the fire and Earth's sociability and expression will draw them into this kind of pattern, which often makes them party too long and get very tired very fast. Her energy is that of a partygoer, the life and soul of it. They love this expression; they do not dance for form (which is a metallic expression) but for joy and excitement. They do however get drawn into being a slave to the party rather than being the head of it, and in fact far from what it looks like, she isn't in control of any of the performance but is a puppet usually for another yang energy, a manger, or if she is deeply within herself she can be an oracle, a vessel of light connected to the source itself. These women have great potential for being expressers of truth but it depends what they are "fed" as to what emerges, if they are fed social norms then the result is sickness, but if allowed to connect to her deep sense of emptiness and yin she can be guided by vision and clarity beyond herself, the perfect expression of light.

They are passionate, musical, romantic, and often desirable to men through their extravagance of femininity. They can be the belle of the ball and are also yin, so like a tropical flower; they draw great crowds, often of men, who find their "heat" to be compelling. She is demanding and will lash out in violent rage if she is crossed but it quickly burns out to stillness and softness within. She would do better being led by a strong-willed manager and a cool-tempered person who can direct her to fame. As will many who are in this role, she can look down on people easily. She is proud and can turn a blind eye to things she doesn't not want to see. She can show great awe-inspiring passion and excitement and can literally love a crowd, so they can all feel it. She gives out as much if not more expression than she receives, and as a performer, she can be a virtuoso in her performance.

Their minds are bright, and they are visual and inspired, taking the ideas of the wood energy to the next stage of expression into spirit. However, the direction of it is towards the Earth energy. The intention is to create the physical from spirit, hence they are the point that yang becomes yin again. This opposes the water energy, which does the opposite, turning what is dense and material water into spirit, or vapour towards wood. 9 have a capacity to spread their energy to a crowd and energize and enthuse, although differently than as with the penetration of the 3 energy. They are like wizards as teachers, but they can crash and burn themselves out easily and completely ignore their bodies.

Due to their brightness and passion, if this is directed inwards, it can reverse, causing a deep locking off from the exterior, delusional thinking of being very great when in reality they have been greatly hurt and damaged. Their minds form the weakness as well as the strength and can become manic, anxious, or simply burned out without a clear function. Often these people must let go, or it will literally destroy them. The passion that fire has can be turned so easily to the opposite, just as the peak of summer quickly turns to darkness and the peak of winter quickly spins to the opposite pole. Like a pendulum, winter and summer are the places at the end of the swing, when yin and yang are turning around, and spring and autumn are the middle of the swing, as there is faster change from yin to yang and yang to yin.

The fire energy is yin and can move only through reproducing itself. This energy very much wants to reproduce and is the expression of life itself, so this type wants to give birth and create existence. As mothers, they can be very happy, and are very involved in their children's welfare. They can be passionate mothers. They are highly influenced by the yang in their life and so although they look like a yang energy exteriorly internally actually they are always following and submitting to a yang expression from one direction or another. This energy is of the mid to late teens in its quality, 15–19 years old. They do best to let someone lead and take a secondary position, but often they will think they can do it all themselves and as a result be restricted.

ii) 9 within the Wood Body

The wood body is strong and has a plentiful supply of blood and wood energy. Wood supplies fire, and so the wood body allows fire energy to burn brightly and with a strong drive. The wood body is yang; it supports the spirit as a mother and can be drawn on to provide strength. It can be a very volatile combination, as the wood body has blood that can be hot, and with a spirit that is fiery, this can make for a very angry person or someone who is in a dominant role and will take no other. Likely to be demanding and expressive, this body structure is a pioneering one, and so with the brightness of fire, this is the makeup of a leader or someone who believes they should be a leader and will possibly push themselves to this position. However, due to the nature of being a yin energy, their ability to lead and direct with effect is less firm than the yang energies.

High blood pressure will be the main difficulty, as well as fluid exhaustion problems and alcoholism. Addiction problems can occur, usually with substances that help to relax tension, such as alcohol or marijuana, rather than psychedelics, as they are already quite manic. They can develop depression if they stop moving for any length of time, as the body has stores of heat within the liver. If this is freed up and the aggression is directed towards physical movement, they can be quite spiritual. They need to use their body but will be very intelligent, especially at hunting or seeking out. The liver energy relates to the genital function in women, and so the powerful expression of the wood energy here and the want to express this through the heart from the spiritual energy makes this a very sexual combination and sex-orientated in life.

iii) 9 within the Fire Body

This is fire within fire. This I call a "simple" type, one where the body and energy match. The energy is very fiery, and it is likely that this person will live a very short but very full existence, full of excitement, exploration, and imagination. They will be naturally charismatic and flamboyant, perhaps slightly more accepting of the yin nature of their energy than the previous, but only just. This person can also attempt to be too pushing of their fiery energy outwards and so cause themselves difficulties. They will likely have high blood pressure, and this heat can go into the digestive system and cause secondary diabetes. They can have manic illnesses and psychological anxiety and tension. They can be very fearful when taken to the limit of their anxiety. Heart problems are the key issue, and they are most likely to die of a heart issue if not careful. They have spiritual connection rather than one that is purely physical. They will have very fast minds and be good at music, language, and abstract mathematical approaches. They love being at the centre and in the limelight.

iv) 9 within the Earth Body

These people can gain weight very easily and can become obese fast. They will have a passion for food, as the passion of the spirit is driven into a body that is heavy and Earth-like; hence the spirit will be drawn to satisfy this body. Wasting and thirst patterns (diabetic type patterns) can be a problem, and overheating of the system can occur easily. The fire feeds the body's desire, so this person is likely to be a pleasure seeker, especially of the sexual and touch. They could be highly intellectual, absorbing a lot of books and literature. They may be the life and soul of a gathering, someone who loves social contact and being in the limelight. The ambassadorial role, company boss, or second in command are very good positions for this one.

v) 9 within the Metal Body

This is a complex structural type, meaning the body and spirit are opposite within the system. This has the effect of holding in the spiritual energy, which can cause difficulties. The energy of the body structure is built for longevity, but the spirit is built for speed and expression. This can cause the person to become agitated. The skin doesn't let out heat as fast as the expression comes; this can be a barrier for expression. Skin problems may occur with emotionally related issues. This is very common with Japanese and Chinese people. If the spiritual energy was not in opposition, then the passage of energy would be smoother. They will tend to have bursts of energy to create a kind of manic-depressive tendency, like opening and closing the lid of a pressurised container. Hypertensive problems and cancers may occur. They want to be expressive but don't feel that they can be as easily as a 9 with a fire body. Just following their spiritual path can lead to exhaustion of the body at a slower rate than the fire type, due to the body not being able to respond so fast; this can be frustrating. Acceptance of the unity here of body and spirit is key for this person.

vi) 9 within the Water Body

This is a complex body structure. This is perhaps the most challenging structure for a fire energy type. The legs are heavy and solid, and the whole body is much denser and stronger than fire would wish to be. A fire frame has a much faster build with responsive muscles and developed musculature of the upper body. This is much more heavy boned and slower. It makes the spirit much more held down and heavy, which feels like a handicap. In a sense, it is like an anchor on fire. This can be useful unless the fire energy of the spirit resists its body shape. A light and free mind and spirit feels trapped by this body and overcome by it. This can cause depression, high blood pressure, urinary problems, urethritis, and kidney failure and infections. Also heat-related skin problems can occur due to the want of the spirit energy to move and push the body to action, thereby overheating it.

In the next section, we will look at the 12 animal energies and 10 stems of the Earth's cycle of 12 and how the five body energies are augmented by the interaction with these. The corporeal soul is more related to the animal forms, so we can say that that these energies have a close relation to the physical body, rather than the mix of body and spirit seen in the mix of 9 and 5. The 10 stems are always placed with the 12 branches, so these will be shown together.

Part 5

Relating Cycles of 12 and 10
to the Constitution

5.1 The 12 Animal Forms and 10 Celestial Stem Connections in Relation to the Five Phases of the Jing Body

In order to understand this interaction, we must look into the great works of the classics of Chinese Medicine, the Su Wen and Ling Shu. In several chapters of these classic works, explanations were given of the meridian connections of the body to the stems and branches. This reveals large numbers of connections. The stems and branches have a deep interaction with the physicality of the body. As the 9 energies are not ordered in this way, we see that they are more related to spirit. The following two diagrams express the cycles of 12 and 10 and the meridians associated with each animal, and so we can work out the various interactions.

Note that the 10 celestial stems relate originally to the pre-Heaven universe, so they are basically a reference point for the cycle of 12. Both are coloured with the five phases to make them "suitable for human consumption", so to speak. Hence when we look at the stems and branches, although the 10 celestial stems do relate to Heaven, it is pre-Heaven-Heaven, which within the moving universe doesn't mean much other than as a reference point (still point) for what is occurring within the universe, in this case the earthly branches. Therefore, this is essentially about the connection of pre-Heaven and Earth, whereas post-Heaven is the 9 energy cycle. Be sure to make a clear differentiation between the principles of 9 and 10.

The Branches and Stems constitutions:

There are 60 combinations of constitutions associated with the cycle of the Chinese calendrical system:

(table 5.1)

Associated Element	Stem		Associated Animal
Yang Wood	1. 甲	1. 子	Rat
Yin Wood	2. 乙	2. 丑	Ox
Yang Fire	3. 丙	3. 寅	Tiger
Yin Fire	4. 丁	4. 卯	Rabbit
Yang Earth	5. 戊	5. 辰	Dragon
Yin Earth	6. 己	6. 巳	Snake
Yang Metal	7. 庚	7. 午	Horse
Yin Metal	8. 辛	8. 未	Goat
Yang Water	9. 壬	9. 申	Monkey
Yin Water	10. 癸	10. 酉	Rooster
Yang Wood	1. 甲	11. 戌	Dog
Yin Wood	2. 乙	12. 亥	Pig
Yang Fire	3. 丙	1. 子	Rat
Yin Fire	4. 丁	2. 丑	Ox
Yang Earth	5. 戊	3. 寅	Tiger
Yin Earth	6. 己	4. 卯	Rabbit
Yang Metal	7. 庚	5. 辰	Dragon
Yin Metal	8. 辛	6. 巳	Snake
Yang Water	9. 壬	7. 午	Horse
Yin Water	10. 癸	8. 未	Goat
Yang Wood	1. 甲	9. 申	Monkey
Yin Wood	2. 乙	10. 酉	Rooster
Yang Fire	3. 丙	11. 戌	Dog
Yin Fire	4. 丁	12. 亥	Pig
Yang Earth	5. 戊	1. 子	Rat
Yin Earth	6. 己	2. 丑	Ox
Yang Metal	7. 庚	3. 寅	Tiger
Yin Metal	8. 辛	4. 卯	Rabbit
Yang Water	9. 壬	5. 辰	Dragon

Yin Water	10. 癸	6. 巳	Snake
Yang Wood	1. 甲	7. 午	Horse
Yin Wood	2. 乙	8. 未	Goat
Yang Fire	3. 丙	9. 申	Monkey
Yin Fire	4. 丁	10. 酉	Rooster
Yang Earth	5. 戊	11. 戌	Dog
Yin Earth	6. 己	12. 亥	Pig
Yang Metal	7. 庚	1. 子	Rat
Yin Metal	8. 辛	2. 丑	Ox
Yang Water	9. 壬	3. 寅	Tiger
Yin Water	10. 癸	4. 卯	Rabbit
Yang Wood	1. 甲	5. 辰	Dragon
Yin Wood	2. 乙	6. 巳	Snake
Yang Fire	3. 丙	7. 午	Horse
Yin Fire	4. 丁	8. 未	Goat
Yang Earth	5. 戊	9. 申	Monkey
Yang Metal	7. 庚	11. 戌	Dog
Yin Metal	8. 辛	12. 亥	Pig
Yang Water	9. 壬	1. 子	Rat
Yin Water	10. 癸	2. 丑	Ox
Yang Wood	1. 甲	3. 寅	Tiger
Yin Wood	2. 乙	4. 卯	Rabbit
Yang Fire	3. 丙	5. 辰	Dragon
Yin Fire	4. 丁	6. 巳	Snake
Yang Earth	5. 戊	7. 午	Horse
Yin Earth	6. 己	8. 未	Goat
Yang Metal	7. 庚	9. 申	Monkey
Yin Metal	8. 辛	10. 酉	Rooster
Yang Water	9. 壬	11. 戌	Dog
Yin Water	10. 癸	12. 亥	Pig

From the chart associated with the calendar earlier on page 134 we can understand that the stem and branch is associated with the date of birth which then imbues a beginning of a cyclical expression from the first breath of the baby. The branches and stems can be seen in two ways, one is from the global perceptive of looking at the person as a whole and seeing the strengths and weaknesses from the correspondence of the branches and stems perspective. If we add in the 5 –phase expression we can correlate organs and meridians to this perspective and in many ways doing so with the branches and Stems is a possibility expressed in the Classics. This is one set of correlation, which is connecting the internal 5 –phase with the exterior influences of the movements of the branches. However this also is another imposed order, which is the meridians as part of the exterior, and how they effect the interior. This perspective is seeing in Chapter 41 of the Ling Shu. Hence there is an inside out movement towards the earth energy of 12 and 10, so 5 going towards 12 and 10 and/ or 12/10 being applied to the meridians and so the 5-organs. These are both happening at once. Before we go on lets get a perspective: -

(fig. 5.1)

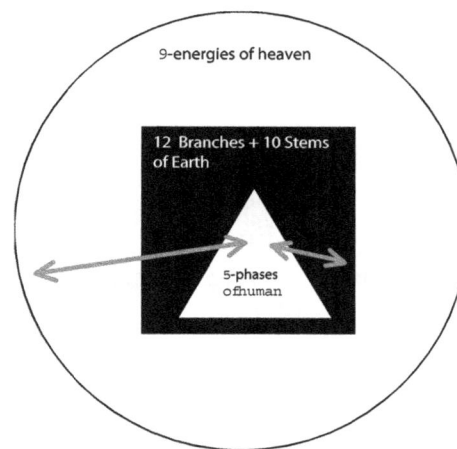

In the above diagram we can see that earth influences the human and heaven does also. We can see that they are all on different levels. The human level is one, the earth another and heaven the last. We can make correspondences of relation to: aspects of no-form within the human associate with 9-energies, aspects of form of the human associate with 12/10….. or we could say 5-phases relates to nine energies and the 5-phases relate to the 12/10. However these are correspondences and they are not actually at the same level of frequency. Hence when we are making relations of microcosm in macrocosm and macrocosm in microcosm we are making a leap of expression, which must not be used rigidly. This is as important for the 9 energies as for the 12/10 branches and stems. At this level there is a looser context. It is best to see the 9 energy influences and the 12/10 influences as a sense of Oneness with the exterior. To do physical medicine based on these expressions is difficult because physical medicine tends to be quite specific. However the correlations to the inner body are very important to understand as they make up the constitution.

Hence the stems and branches above make up the basis for the system of the 4-pillars which corresponds to year, month, day and hour of birth providing a more complex expression than above, however he above is a base and the year is always a base for expression. It is often best to get a broad perspective. The inner correspondence of the 5-phases with the 9 energies provides the idea of the various spirit expression of the body. The correspondence of the 12 branches with the 5-phases provides connection with the organ sand meridians of the body. One is associated with form the other with no-form. This is how we can make total clarity of correspondence. One would not make correspondence of the spirit quality associated with the branches as much and one would not associate with body's physicality as much with the 9 energies.

Remember both the branches and Stems and the 9 energies are all a broader perspective, they are seeing the whole body as a whole not really focused heavily in the specifics of the body but as an overall in relation to other humans. This is why we connect this section with constitution, this is part of Feng Shui so is the expanded view, not internal medicine which is why this view is not expressed in the Nei Jing or other texts however it is associated in the texts of the time associated with Feng Shui and astronomy such as the Huainanzi and the Qing Nang Jing (see annotated bibliography).

Constitutional association of meridians to Branches and Stems in the medical Classics: -

In this section I have combined the inner correspondences of the 5-phase with the outer expression of the associations of stem and branch to meridians. The following are found in the chapters of Ling Shu 41 (meridian association) in collaboration with Su Wen 22 (seasonal association to 10 stems of yearly change) and corroborated in chapters Ling Shu 10, Nan Jing 23, and Nan Jing 7. Originally these were associated with times of the day, which is a 12 bi-hourly cycle, as again Earth change is this. Note that I have changed the PC for Hand-Shao-yin in the descriptions I have used here, as these terms are used interchangeably if we consider the work in Section B of this book.

(Note: - this is the Classical association of meridians to the stems and branches to the meridians from the Nei Jing. The usually offered expression is derived from inferred association to what was called the Zi Wu cycle, this does not belong to Nei Jing expression and is something derived much later in history, here we are just using what is given, as it is, in the Nei Jing and seeing the correspondences and unity of the Nei Jing without add-ons. Please see Appendix 2 for details on this)

Below are the original expressions of body connection through the meridians of the stems and branch patterns. Note that there are 11 meridians accounted for, 10 in the upper body and 12 in the legs bilaterally. This shows the connection of 10 "celestial" stems and 12 "earthly" branches.

(fig. 5.2)

10 stems

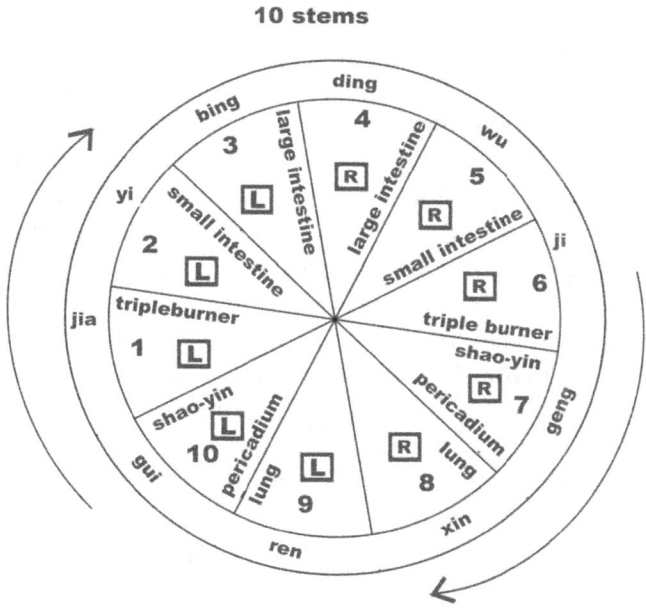

(fig. 5.3)

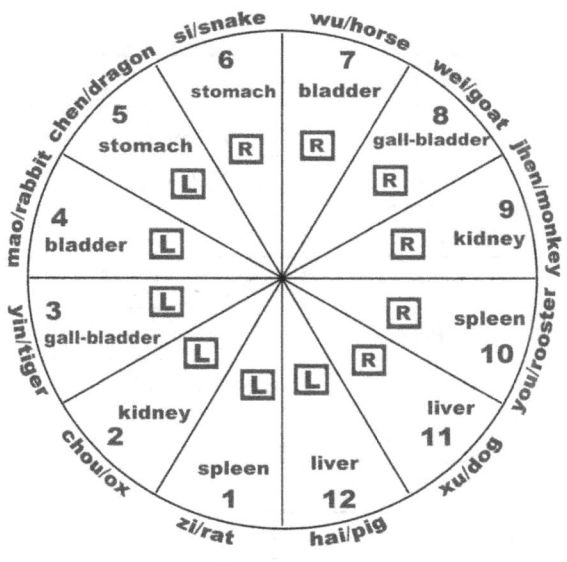

The joined cycle creates a 60-year/month/day cycle (or 60-bi-hourly/5 day, see next section for this

241

cycle) of combinations of stem and branch. In the following table, I have charted the stem and branch combination of 60 and also the 5 different body-types. Hence 60 x 5 is 300 combinations. In each combination, I have given the stem and branch by numbers on the above charts. The branch and stem will have a colour, dependent on which phase they belong to. I also add the meridian combination activated by this combination. This together will give four phases: one branch phase, one stem phase, one stem meridian phase, and one branch meridian phase. These will then allow one to see the strength, and therefore weakness, within a system, which can then be overlaid onto the five physical constitutions, the constitution dominating, and those being within it, having a strengthening or weakening effect on the physical structure. Again, this will of course affect the spirit within this constitution. This below is not an absolute, nor is it a method for point association during treatment, although one could easily be formed by using the points that associated with the meridian's own energetics e.g. Wood point on wood meridian etc (these are often called Horari points in Japan). However this is not the focus of this exercise, as the chapter where this information came from suggests, (chapter 41 Ling Shu) the expression of the system which has the broad outlook of connecting earthly movements with the "individual" body, needs to be kept in broad perspective and not used for the narrowness of specific treatment really, but used as a platform to be aware of the many things affecting us as humans. The ideas below are complex enough, to add more would be to the equivalent of cutting down a rainforest to find the herb that will treat a particular patient.

(table 5.2)

Branch-Stem	Wood Body	Fire Body	Earth Body	Metal Body	Water Body
B1/S1	Wa+Sp/ Wo+TB	Wa+Sp/Wo+TB	Wa+Sp/Wo+TB	Wa+E-/Wo+TB	Wa+Sp/Wo+TB
B2/S2	E-Kid/ Wo-SI	E-Kid/ Wo-SI	E-Kid/ Wo-SI	E-Kid/ Wo-SI	E-Kid/ Wo-SI
B3/S3	Wo-GB/ F+LI	Wo-GB/ F+LI	Wo-GB/ F+LI	Wo-GB/ F+LI	Wo-GB/ F+LI
B4/S4	Wo-BL/ F-LI	Wo-BL/ F-LI	Wo-BL/ F-LI	Wo-BL/ F-LI	Wo-BL/ F-LI
B5/S5	E+ST/ E+SI	E+ST/ E+SI	E+ST/ E+SI	E+ST/ E+SI	E+ST/ E+SI
B6/S6	F-ST/ E-TB	F-ST/ E-TB	F-ST/ E-TB	F-ST/ E-TB	F-ST/ E-TB
B7/S7	F+BL/ M+PC	F+BL/ M+PC	F+BL/ M+PC	F+BL/ M+PC	F+BL/ M+PC
B8/S8	E-GB/ M-LU	E-GB/ M-LU	E-GB/ M-LU	E-GB/ M-LU	E-GB/ M-LU
B9/S9	M+Kid/ Wa+Lu	M+Kid/ Wa+Lu	M+Kid/ Wa+Lu	M+Kid/ Wa+Lu	M+Kid/ Wa+Lu
B10/S10	M-Sp/ Wa-PC	M-Sp/ Wa-PC	M-Sp/ Wa-PC	M-Sp/ Wa-PC	M-Sp/ Wa-PC
B11/S1	E+LIV/ Wo+TB	E+LIV/ Wo+TB	E+LIV/ Wo+TB	E+LIV/ Wo+TB	E+LIV/ Wo+TB
B12/S2	Wa-LIV/ Wo-SI	Wa-LIV/ Wo-SI	Wa-LIV/ Wo-SI	Wa-LIV/ Wo-SI	Wa-LIV/ Wo-SI
B1/S3	Wa+Sp/ F+LI	Wa+Sp/ F+LI	Wa+Sp/ F+LI	Wa+Sp/ F+LI	Wa+Sp/ F+LI
B2/S4	E-Kid/ F-LI	E-Kid/ F-LI	E-Kid/ F-LI	E-Kid/ F-LI	E-Kid/ F-LI

B3/S5	Wo-GB/ E+SI	Wo-GB/ E+SI	Wo-GB/ E+SI	Wo-GB/ E+SI	Wo-GB/ E+SI
B4/S6	Wo-BL/ E-TB	Wo-BL/ E-TB	Wo-BL/ E-TB	Wo-BL/ E-TB	Wo-BL/ E-TB
B5/S7	E+ST/ M+PC	E+ST/ M+PC	E+ST/ M+PC	E+ST/ M+PC	E+ST/ M+PC
B6/S8	F-ST/ M-LU	F-ST/ M-LU	F-ST/ M-LU	F-ST/ M-LU	F-ST/ M-LU
B7/S9	F+BL/ Wa+Lu	F+BL/ Wa+Lu	F+BL/ Wa+Lu	F+BL/ Wa+Lu	F+BL/ Wa+Lu
B8/S10	E-GB/ Wa-PC	E-GB/ Wa-PC	E-GB/ Wa-PC	E-GB/ Wa-PC	E-GB/ Wa-PC
B9/S1	M+Kid/ Wo+TB	M+Kid/ Wo+TB	M+Kid/ Wo+TB	M+Kid/ Wo+TB	M+Kid/ Wo+TB
B10/S2	M-Sp/ Wo-SI	M-Sp/ Wo-SI	M-Sp/ Wo-SI	M-Sp/ Wo-SI	M-Sp/ Wo-SI
B11/S3	E+LIV/ F+LI	E+LIV/ F+LI	E+LIV/ F+LI	E+LIV/ F+LI	E+LIV/ F+LI
B12/S4	Wa-LIV/ F-LI	Wa-LIV/ F-LI	Wa-LIV/ F-LI	Wa-LIV/ F-LI	Wa-LIV/ F-LI
B1/S5	Wa+Sp/ E+SI	Wa+Sp/ E+SI	Wa+Sp/ E+SI	Wa+Sp/ E+SI	Wa+Sp/ E+SI
B2/S6	E-Kid/ E-TB	E-Kid/ E-TB	E-Kid/ E-TB	E-Kid/ E-TB	E-Kid/ E-TB
B3/S7	Wo-GB/ M+PC	Wo-GB/ M+PC	Wo-GB/ M+PC	Wo-GB/ M+PC	Wo-GB/ M+PC
B4/S8	Wo-BL/ M-LU	Wo-BL/ M-LU	Wo-BL/ M-LU	Wo-BL/ M-LU	Wo-BL/ M-LU
B5/S9	E+ST/ Wa+Lu	E+ST/ Wa+Lu	E+ST/ Wa+Lu	E+ST/ Wa+Lu	E+ST/ Wa+Lu
B6/S10	F-ST/ Wa-PC	F-ST/ Wa-PC	F-ST/ Wa-PC	F-ST/ Wa-PC	F-ST/ Wa-PC
B7/S1	F+BL/ Wo+TB	F+BL/ Wo+TB	F+BL/ Wo+TB	F+BL/ Wo+TB	F+BL/ Wo+TB
B8/S2	E-GB/ Wo-SI	E-GB/ Wo-SI	E-GB/ Wo-SI	E-GB/ Wo-SI	E-GB/ Wo-SI
B9/S3	M+Kid/ F+LI	M+Kid/ F+LI	M+Kid/ F+LI	M+Kid/ F+LI	M+Kid/ F+LI
B10/S4	M-Sp/ F-LI	M-Sp/ F-LI	M-Sp/ F-LI	M-Sp/ F-LI	M-Sp/ F-LI
B11/S5	E+LIV/ E+SI	E+LIV/ E+SI	E+LIV/ E+SI	E+LIV/ E+SI	E+LIV/ E+SI
B12/S6	Wa-LIV/ E-TB	Wa-LIV/ E-TB	Wa-LIV/ E-TB	Wa-LIV/ E-TB	Wa-LIV/ E-TB
B1/S7	Wa+Sp/ M+PC	Wa+Sp/ M+PC	Wa+Sp/ M+PC	Wa+Sp/ M+PC	Wa+Sp/ M+PC
B2/S8	E-Kid/ M-LU	E-Kid/ M-LU	E-Kid/ M-LU	E-Kid/ M-LU	E-Kid/ M-LU
B3/S9	Wo-GB/ Wa+Lu	Wo-GB/ Wa+Lu	Wo-GB/ Wa+Lu	Wo-GB/ Wa+Lu	Wo-GB/ Wa+Lu
B4/S10	Wo-BL/ Wa-PC	Wo-BL/ Wa-PC	Wo-BL/ Wa-PC	Wo-BL/ Wa-PC	Wo-BL/ Wa-PC
B5/S1	E+ST/ Wo+TB	E+ST/ Wo+TB	E+ST/ Wo+TB	E+ST/ Wo+TB	E+ST/ Wo+TB
B6/S2	F-ST/ Wo-SI	F-ST/ Wo-SI	F-ST/ Wo-SI	F-ST/ Wo-SI	F-ST/ Wo-SI
B7/S3	F+BL/ F+LI	F+BL/ F+LI	F+BL/ F+LI	F+BL/ F+LI	F+BL/ F+LI

B8/S4	E-GB/ F-LI	E-GB/ F-LI	E-GB/ F-LI	E-GB/ F-LI	E-GB/ F-LI
B9/S5	M+Kid/ E+SI	M+Kid/ E+SI	M+Kid/ E+SI	M+Kid/ E+SI	M+Kid/ E+SI
B10/S6	M-Sp/ E-TB	M-Sp/ E-TB	M-Sp/ E-TB	M-Sp/ E-TB	M-Sp/ E-TB
B11/S7	E+LIV/ M+PC	E+LIV/ M+PC	E+LIV/ M+PC	E+LIV/ M+PC	E+LIV/ M+PC
B12/S8	Wa-LIV/ M-LU	Wa-LIV/ M-LU	Wa-LIV/ M-LU	Wa-LIV/ M-LU	Wa-LIV/ M-LU
B1/S9	Wa+Sp/ Wa+Lu	Wa+Sp/ Wa+Lu	Wa+Sp/ Wa+Lu	Wa+Sp/ Wa+Lu	Wa+Sp/ Wa+Lu
B2/S10	E-Kid/ Wa-PC	E-Kid/ Wa-PC	E-Kid/ Wa-PC	E-Kid/ Wa-PC	E-Kid/ Wa-PC
B3/S1	Wo-GB/ Wo+TB	Wo-GB/ Wo+TB	Wo-GB/ Wo+TB	Wo-GB/ Wo+TB	Wo-GB/ Wo+TB
B4/S2	Wo-BL/ Wo-SI	Wo-BL/ Wo-SI	Wo-BL/ Wo-SI	Wo-BL/ Wo-SI	Wo-BL/ Wo-SI
B5/S3	E+ST/ F+LI	E+ST/ F+LI	E+ST/ F+LI	E+ST/ F+LI	E+ST/ F+LI
B6/S4	F-ST/ F-LI	F-ST/ F-LI	F-ST/ F-LI	F-ST/ F-LI	F-ST/ F-LI
B7/S5	F+BL/ E+SI	F+BL/ E+SI	F+BL/ E+SI	F+BL/ E+SI	F+BL/ E+SI
B8/S6	E-GB/ E-TB	E-GB/ E-TB	E-GB/ E-TB	E-GB/ E-TB	E-GB/ E-TB
B9/S7	M+Kid/ M+PC	M+Kid/ M+PC	M+Kid/ M+PC	M+Kid/ M+PC	M+Kid/ M+PC
B10/S8	M-Sp/ M-LU	M-Sp/ M-LU	M-Sp/ M-LU	M-Sp/ M-LU	M-Sp/ M-LU
B11/S9	E+LIV/ Wa+Lu	E+LIV/ Wa+Lu	E+LIV/ Wa+Lu	E+LIV/ Wa+Lu	E+LIV/ Wa+Lu
B12/S10	Wa-LIV/ Wa-PC	Wa-LIV/ Wa-PC	Wa-LIV/ Wa-PC	Wa-LIV/ Wa-PC	Wa-LIV/ Wa-PC

Key:

    Wo+ = Wood yang = Liver (organ)

    Wo- = Wood yin = GB meridian-(organ), Liver meridian

    F+ = Fire yang = SI meridian (organ)

    F- = Fire yin = Heart (Organ), PC meridian

    E+ = Earth yang = ST (organ)-meridian

    E- = Earth yin = Sp (organ-meridian)

    M+ = Metal yang = Lung meridian, LI (organ)-meridian

    M- = Metal yin = (Lung organ)

    Wa+ = Water yang = R-Kidney (organ), TB meridian, BL (organ)-meridian

    Wa- = Water yin = L-Kidney (organ)-meridian

    GB = Gallbladder meridian

    LIV = Liver meridian

    SI = Small Intestine meridian

    TB = Triple Burner meridian

BL = Bladder meridian

PC = Pericardium meridian of the heart

ST = Stomach meridian

SP = Spleen meridian

LI = Large Intestine meridian

Lu = Lung meridian

Kid = Kidney meridian

How to read and interpret the above chart:

For example:
B6/S2 = F-ST/ Wo-SI

This means:

Branch 6 = Snake/Fire yin (Heart [Organ], PC meridian) and also joins to the Stomach meridian
Stem 2 = Wood yin (GB [organ]-meridian, Liver meridian) and also joins to the Small Intestine meridian

Example:

In terms of how this relates to the CP body types, if we relate this to a wood type, within the wood type CP:

Hence the associations are to the heart (organ), the pericardium meridian, the stomach meridian, the GB (organ)-meridian and the liver meridian. What we can say is that the constitution is tempered by the energy of the stem; it is drawn away by the Snake/ Fire-yin quality of the branch and the SI quality of the stem. It is obstructed slightly by the St (Earth yang) quality of the branch. The overall impact on the system is quite broad, with several phases covered, but for example B7/S3, which is F+BL/ F+LI, has a great deal of fire energy within it, and so has the effect of being more mono-directional.

5.2 The Corporeal and Ethereal Spirit Relations

Shen or Spirit has numerous aspects we will discuss in later sections of the book, but within yin there is yinyang, and within yang there is yinyang, so within spirit there is the more yin aspect which we call P'o, the corporeal or earthly bound spirit, and the more ethereal aspect is the yang within spirit which is called the Hun. The corporeal soul is the aspect of the soul that is earthbound or associated; therefore it associates with body. Earth/Jing and physical formations are a heavier aspect of the ethereal nature of spirit (yin within yang). As such, if we look at the aspects of the stems/branches methodology, these aspects that affect the physical have a more earthbound and therefore corporeal association—especially aspects involving the earth/metal/water categories within stems and branches. Hence aspects associated with the Sp/St, Lu/ GB, Left Kid organs are more associated with this energy, and these centre round the corporeal soul, especially that of the metallic energy.

The fire and wood elements within the 12 branches/10 stems affect the ethereal aspect of the soul more, but because they are within the cycle of 12, which itself is more earthbound, the ethereal association is less so than the 9 energies. Hence they are described as yang within yin. The nine energies are, by nature, more ethereal, therefore affecting the Liv/ LI, Hrt/ SI/ PC/ Right Kid/ TB and BL organs. Hence the yang/spirit associates more with the wood and fire. However, we can see that the yin-more earthy within 9 energy yang:1/2/5/6/7/8 will have more of a movement towards the corporeal, but they are within the Heaven energies (4 tends to be more ethereal due to its nature, but is in fact yin within the ethereal aspect). Hence there is always a spectrum, as usual.

Also, men relate more to the yang and so Heaven and to the nine energies, and women to the 12 branches and 10 stems. This can go on ad infinitum; the important thing is to see how the principle of two, yin and yang, pervades everything. (This is implied by Su Wen 6).

Another important point as we mentioned earlier is that the 12 branches and the 10 stems relate more to the movement of the physical and form and therefore more to the meridians, whereas the spirit energy of the organs is more associated with the form-less dimension of the nine energies. This creates dynamism of form/ corporeal spirit/ body (yin) and exterior/ meridians (yang) with formless/ ethereal spirit (yang) and interior/ organs (yin). This is why in the above section's key, the organ aspects are bracketed off, as these are stems and branches and relates more to the meridians.

The above shows how the branches and stems can join to form the yin aspect of the spirit or the corporeal aspect and how this can augment the physical, and therefore the spirit energy is also affected.

It is important to see how the 12 branches and 10 stems relate to the nine energies and the five phases. The following chart explains this somewhat:

(fig. 5.4)

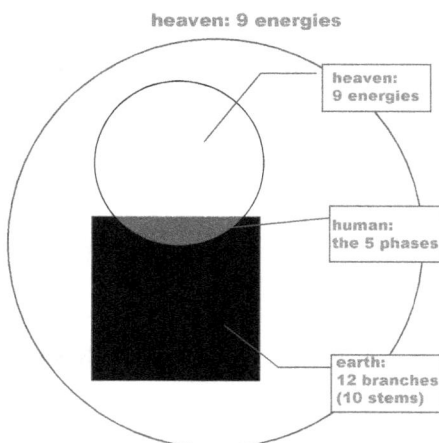

If considered as a whole, the methods of understanding all three aspects of Heaven, Earth, and human can be from human five-phase terms. This makes for a completed understanding rather than focusing on one aspect or other. The 9 energies are the foundation of the 5 –phases the 12 branches influence the 5, therefore there is relative importance in expression.

Part 5

Section A — Endnote

Endnote

This marks the end of the first part of this work on the constitution. In the next section, we look more into the energetics of Jingshen, the combined body-spirit union, in health and in disease and also treatment possibilities. Again, the purpose of this book is for you to have an overview rather than a specific notion. Although the material is dense in some places, the picture it produces should be wide and broad, and you will, I hope, gain more benefit in this way and be able to use the principles yourself when viewing broadly rather than specifically.

As Shakespeare famously wrote, "to be, or not to be"—that is only the question for those of us who question.

For Yoda in Star Wars, it was simple: "Do or do not do, but do not try!"

The point I am making here is probably obvious, but the process of questioning who and what we are always ends in illusion and confinement because it is a trap of the echoes of our mind-identified state. This was understood for millennia in the East, and even within the philosophical tradition of the Modern Western world, several people made the point that the discourse was in it a fallacy, especially Wittgenstein, Nietzsche, and Foucault. It is obvious, if we take a step back from ourselves, that acceptance of who we are and what we are is a moment away. The only aspect that stops us is the fragmentation of our clear mind, ending our ability to unite within ourselves and so within society and universally.

This does not mean the end of change and the end of life; this also does not mean that we all will live without death and destruction, although on a much smaller scale. What it does is allows us to see morality, judgment, and polarizations as products of mental conformity and not of the truth of Nature. The animals (and very young children) can teach us this, as they know how to live through doing and not trying.

In this book, there are many open ends (from a root foundation); these are for you to study yourself, to come up with further developments based on these principles. However, to move away from the principle one will find oneself slipping into individuality very quickly. The problem is that one needs to see things from a perspective that is objective, and as humans detached from the universe (within mind-identity), this is impossible, but to know its impossibility is to know the truth of it (contextualised mind).

All the number patterns we have looked at are products of the human experience denoted for later generations, in order that they not lose track of themselves within their cerebral confusion! A foothold for sanity of Oneness and non-judgmental expression. In general, mind-identity and our thoughts are seen as things needing to be "broken through", when it is the doing of this that causes such fragmentary behaviour in societies and in the world at large. If we are in acceptance of the mind-attached state, which is the whole purpose of that which is "meditative", rather than the away pushing of this and denial of thought and mind-identity, then far from the prison of right and wrong, good and bad, we can find the meaning of all the Natural-people/ sages, who never suggested monasticism, crusades, holy wars, or anything but pointed to …

This, As it is/ Be/ Go with the flow

… in whichever order and with whatever other words were used to make it sound relevant at the time. The essence is the key, the foundations of what one is and the acceptance of this, the Stillness of Wu that permeates all.

In this work, we have seen how, for some people, body and spirit seem to join and become one unit. When there is a brother-sister relationship or a mother-child relationship, there is usually an easier harmony. However, for those that have battles within between yin and yang, male and female, expression and internalisation, these aspects need not be seen as a conflict. "You' are simply perfect the way "you" are; "you" are simply what "you" are, and there is Oneness in this and your connection to everything. The illustrations and drawings show how, if "you" (mind-identify) fight "yourself" (Nature) and create mental tension by doing so, dis-eases result. Those who have husband-wife aspects within the bodies and spirits have need to accept uniting opposite energies. Non-doing this; strength and freedom result. To contextualize what is mind-identity/separation and cognition is to find the feeling of what you are, and these road maps will, for a short period, help you to do so, till it is time for you to fly yourself, which can only happen "Now", as it is. It is as difficult as being a bird that doesn't sing because he/she has an answer, but because there is a song.

Section B
Energetic Anatomy and Physiology

Observe things as they are and don't pay attention to the [imaginings of] people.
- Huang-Po

A man learns by keeping his gaze on Unity
- Plato

Introduction to Energetic Anatomy and Physiology

The next section moves from the realm of the constitution seen from the overall-broader perspective into the inner constitution; this is called energetic anatomy and physiology. We will now move into the realm of the inner body. We have seen how it has been constructed in physical formations of the male and female Jing energy and the union of this with Shen of heaven, the body-spirit formation. Here we look into how this system works, how the microcosm of the body is actually the same expression as the macrocosm of the universe in miniature (depending on perspective). Noting all that was said earlier, with the context in place, we can engage now with the detail. It is important to read the first Section of the book before attempting to look at this part, or it will be utterly confusing. The first part also puts in context the various concepts or principles within Chinese philosophy: Wu, yinyang, principles of 3, 4, 5, 8, 9, 10, and 12. These principles govern everything in the universe, as far as ancient philosophy can reach to; they are simply rhythms or cycles, all happening together with interconnections and rhythms that associate with Heaven, Earth, and humans, founded in the Stillness of Wu.

Please note that anatomy and physiology are still really part of constitution. We are not yet in the realm of medicine, which is an understanding of the body in sickness. In the following section, we focus on the system in health, how the body normally and naturally functions if all is well; this therefore is an ideal model. When we look at aetiology in section C, we will talk about energetic medicine, which is the treatment of the body in dis-ease.

This part of the work is more about the more practical activities of the energy flows within the body and how these can be understood more effectively. This deals with the principle of 5; the union of Heaven and Earth, Jing and Shen to form humans—Jingshen. All of the principles above are explained through the principle of 5 because 5 is the looking glass of the human mind; we can only view things in 5 dimensions, as it were. The universal/heavenly/yang moves in sequences of 9, and Earth is in sequences of 12, but for humans, this often needs to be translated into sequences of 5 so that we can relate it to our bodies and our being here on Earth. The Jing/body and Shen/spirit therefore combine with the structure of the 5 principle. Hence it is the study of the 5 principle that gives us access to understanding the Wu-based body-spirit constitution and defining its physiological processes.

At this point, I would like to express that what I am doing in this work is a following through of logical sense and additional explorations of Ikeda Masakazu's works in English, which are:

Traditional Japanese Acupuncture: Fundamentals of Meridian Therapy, the Society of Traditional Japanese Medicine (Complementary Medicine Press, 2003), ISBN 0-9673034-4-3.

The Practise of Japanese Acupuncture and Moxibustion: Classic Principles in Action, translated by Edward Obaidey (Eastland Press, 2005), ISBN 0-939616-43-2.

Integration of Acupuncture and Herbal Medicine: Theory and Practice. (International Acupuncture Network, 2010), ISBN 978-0-578-06534-2

Ikeda Masakazu is perhaps the most important teacher of Chinese Medicine alive today. The director of education at the Society of Traditional Japanese Medicine (previously the "Meridian Therapy" organisation), Ikeda Masakazu has enigmatically - through clear and deep perception into the classical material - been able to re-unite as they originally were: acupuncture and herbal understanding and give us a way of investigating the classics as combined works, not as fragmented parts seen through fragmented mind-identities. I have a deep acknowledgement and gratitude to this man and his lifetime effort to recreate the ancient lineage of medicine in the modern day. The clarity of his teachings is now available in English through Edward Obaidey, a profound practitioner also. Through these two men and the work they have achieved and continue to do, great Truth has been brought back to Chinese Medicine again, unifying all aspects.

There are many other works in Japanese that can be found along with the works mentioned above written by Ikeda Masakazu. The above works are, in the English language, simply the best clinical guides available for the student of Chinese Medicine until post-graduate level and are sufficient if truly and deeply studied. In my explorations into these works, and through study and training, I feel I have been able to glean more than is explained directly and so feel that the additions I have to make are a worthwhile supplement. I hope that this is in accordance with my teachers, and if it is not, I apologise for any error of explanation I am making. I believe this work to be as rigorous as possible in its explanation.

Part 1:

Understanding Energetics

Part 1: Understanding Energetics

The Jingshen of the five phases is the union of body and spirit. The meridians therefore are the conduits of this union. When dealing with the next section about the meridians/physiological nature of the body and its pathogenesis, we are looking at the union of body and spirit - of the spirit within the physical structure and their total unity (Jingshen = yinyang); they merely take different forms through life- death cycles.

We constantly emanate qi throughout our lives, radiating it. The radiance is life and is also the expression of the changing/transforming of the body to spirit and spirit to body, as yin and yang intertwine and mix, much like the continuum of Einstein's mass and energy spectrum ($E=mc^2$), except more inclusive. Death is simply another/further transformation as the process continues—there is no reason that it would not.

Understanding that there is union of Jingshen within the meridian flows means that we can affect the more physical to affect the more energetic spirit. This I have called Jing Medicine. It is associated with any physically related discipline/practise using tools or materials—acupuncture, moxibustion, herbs, physical manipulations, etc. The alternative is that we can affect the spirit to affect the more physical. This is any form of energetic based medicine, often using no materials—for example, Qi gong healing, Feng Shui, meditation, etc. Qi gong healing is an approach that lies between the physical and energetic, but is energetically based. We will look at all these later, but for now it is important to note that Jingshen IS the meridian energy—blood-qi or yinyang, it's all One. The energy mix, therefore, is human energy; between Heaven (9) and Earth (12) is human (5).

1.1 The Structure of the Organ Meridian System: The 11 Organs, the 10 Celestial (arm/hand) Stems and 12 (leg/foot) Earthly Branches

If we consider that the body of the male and female are not exactly the same but have general similarity, and if we consider that we are now looking into anatomy and physiology, then we have to regard the following understanding as a "healthy"/liberated and natural-state body structure. The next section doesn't explore the different constitutions of the energetic system, but simply the general organisation of all of them. The following charts the organ meridian complex and its formation, as explored initially by the Moxibustion Classic, then fully expounded in the Nei Jing; Su Wen and the Ling Shu.

i) Daily Cycles

The energy of the body in the morning is at the exterior. The defensive exterior Wei chi/ eki (yang) is strong at this time; the nutritive internal ying chi/ eiki (yin) is pushed to the surfaces. This increases to a maximum in the afternoon, yang, and then converts back to moving into the yin in the afternoon and evening. The Wei energy directs inside, and the ying energy is at its highest strength in the evening. The energy generally therefore circulates in the following way:

- Wei chi emerging, ying ki reducing (yang within yin)
- Wei waxing, ying chi waning (yang within Yang)
- Ying chi emerging, Wei chi reducing (Yin within yang)
- Ying chi waxing, ying chi waning (yang within yin)

This can be explained as:
Morning
Midday
Afternoon
Night

Or
Spring
Summer
Autumn
Winter

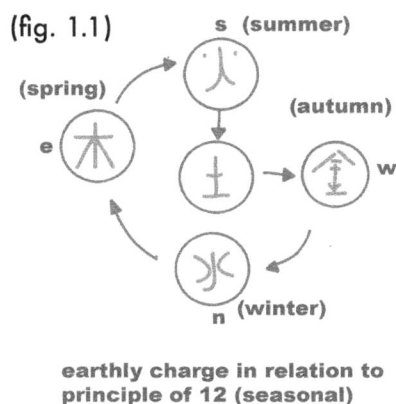

(fig. 1.1)

earthly charge in relation to
principle of 12 (seasonal)

Hence we can see that the flow of the energetics of the meridians is controlled by the earthly/lunisolar cycles. Notice that at the same time as this is happening with the energy, the actual place where the ki is, is different:

Yang surfaces and upper body
Yang surfaces and head
Yin surfaces and lower body
Yin surfaces and feet

Also the depth is different:

Skin
Above the skin
In the flesh and muscles
In the bones

This is how the cycles of the qi flow within the continuous connected loop. From the Ling Shu 10, the cycle is:

LU

LI

ST

SP

(HRT)

SI

BL

KID

PC

TB

GB

LIV … then to LU again

Please be sure to note that this is the continuous loop itself; it is NOT how the chi necessarily flows through the loop. One can start this loop at any point, and it will always connect back on itself. The loop doesn't indicate the flow of chi, it simply indicates the loop. Other forces such as the Earth lunar-solar change/interaction and the five-phase internal change of the organs will affect the chi cycle. The meridians are in-between the inner circulation of the 5 –phases and the outer circulation of the heavens associated with 9, 12 and 10. The 11 meridians are between, their flow is depended on the interaction of the other 2 qualities of circulation,, of course they are all one unit in health and in dis-ese they are out of sync. Classical medicine associated with the body treatment methods always focused on the Inner 5-phases as the root to synchronize again with the exterior. The method is always inside out form contraction of mind-identity to Oneness with everything.

The loop was organized into the cycle of the 24-hour clock assoiled with acupuncture in around AD 1153 (Liu,1999) -more than 1,300 years after the Ling Shu -and called the ZiWu cycle/ Shi Go Ruchu/ 子午流注, although NO evidence of this cycle in connection with designation of the meridians to specific times is explored anywhere in the Su Wen or Ling Shu. Hence we will not be using the ZiWu cycle, as it is a stylistic representation of the meridians and their attachment to time. The meridians are attuned to time in other chapters, see Ling Shu 41 (See appendix 2 for a discussion on the ZiWu or Stems and Branches style of acupuncture).

Another important point is that the meridians are yang and exterior, relative to the organs, which are yin and interior. The Fu organs are yang organs, so they have a closer connection with their conjoining meridians than their organs. The Zang are yin organs, so the Zang are more associated with the interior and the yin; they are true organs, so to speak. This will become important when we look at treatment and the reason the root treatment points are always on the yin meridians (closest to the Zang) and the reason all issues arise from weakness of the interior. The exterior Fu and meridians derive all their energy from that of the Zang; the Zang are the foundations.

We will look therefore at the generator of the energy in this loop of energy, the five-phase system, and, as is traditional, we will start with the spring, the wood energy. I will not indicate the pathways of any of the meridians in this text, as this information has many sources; my interest is to understand the energetics of this energy form.

Before we start, this is a chart that expresses the energetics of the 5 phases: -

(table 1.1)

	Wood	Fire	Earth	Metal	Water
Season	Spring	Summer	Late Summer	Autumn	Winter
Pictorial					
Shape	Cylindrical	Triangular-pyramid/Pointed	Square/Flat	Dome/Arch	Irregular
Time of Day	Morning	Midday	Afternoon	Early Evening	Midnight
Daily Rhythms	Male erection; female clitoral activation, period flow; bowels open exercise, empties abdomen and chest of heat; movement is expressive; eyes are dryer; focusing of eyes is better	Food consumed, peak of energy, heartfelt, expressive, sexually expressive, in-breath is most powerful	Food digested, mood is steadier, talkative, gently expressive, mouth has more saliva, lips are wetter	Cooling, draw in heat from food, wear warmer clothes, do less, calming, meditative, eyes can take in light very well, lungs breathe out deeply	Cold, silent, dark, hidden, sleep deeply covered, heat deep within, hearing is acute
Temperature	Warming	Hot	Mild	Cool	Cold
Zang Organ	Liver (Ming Men, Right Kidney)	Ming Men, Right Kidney (Heart)	Spleen	Lung	Left Kidney
Flavour Associated to Tonify Associated Zang Organ	Pungent	Salty (and pungent)	Sweet	Sour (no lung essences exist, GB being surrogate Zang organ)	Bitter
Yin Meridian Associated	Lung	Pericardium	Spleen	Liver	Kidney

Tissue Type (Zang Ki = Tissue Ki)	Muscles (red part)	Blood Vessels	Fatty Tissue	Skin and Skin-like Tissue, Membranes and Tendon/ Facia	Bone
Entry Point of Flavour	Tendons/Skin	Bones	Fatty Tissue	Muscles (and surface)	Blood Vessels
Flavour Goes Into (expressed in)	Muscles (and surface)	Blood Vessels	Fatty Tissue	Tendon/ Skin	Bone
Fu Organ-Meridian Associated Through internal-external Pairing	Large Intestine	Triple burner/ Small Intestine	Stomach	Gallbladder	Bladder
Fu organ-meridian associated though energetic similarity	Large Intestine	Triple burner/ Small Intestine and Bladder	Stomach	Gallbladder	-
Flavour Associated to Tonify energetic Associated Fu Organ/ Meridian	Pungent	Salty-Pungent	Sweet	Sour	-
Depth of Energy	Deep	Both, between Earth and Wood (Pericardium and most superficial (Triple Burner/ Bladder/Small Intestine)	Below the Metal Layer	Superficial	Deepest
Spirit Quality	Holds Ethereal Soul	Houses the heavenly Life Spark/Yang	Holds the Intuition, Inclination/ Impulse to Follow the Heart	Holds the Earth-bound Soul	Holds the Wilful Action of the Heart
Spiritual Expression	Assertiveness	Joyfulness	Sense of Feeling	Consolidating	Wilful-fearless
Pathological Emotion	Anger	Anxiety	Muddled-ness	Grief-struck	Terror
Colour	Green	Red	Yellow	White	Blue/Black

Governed Region of the Body	Sides of the Body (inner and outer)	Face	4 Limbs and Abdomen	Chest	Lower Back
Sense Associated with Zang	Sight/ Image (note this expresses out—yang—as well as letting light in—yin)	Expression Organ (voice; no sense, and no exterior contact)	Taste	Smell/Touch	Hearing
Sense Organ Associated with Zang	Eyes (inner eye is yang expressive of Shen/liver organ, outer eye is yin absorptive of light Gallbladder/ liver meridians)	Expression Organ (Voice; no sense, and no exterior contact, expresses in the tongue-body)	Tongue and Mouth and Lips (mouth and tongue surface is spleen, lips are more associated with stomach)	Nose/Skin, (the open hole on nose and pores represents the lung organ, the energy of warmth is the lung meridian)	Ears
Direction	East (left side of body)	South (front of body)	Southwest and Centre (centre of body)	West (right side of body)	North (back of body)
Age of Human Life Phase	0–18	18–32	32–45	45–65	65+

ii) The Five Phases of Flavour Energetics

(Please see Su Wen 3, 23 and Ling Shu 63 for expression and differentiation of the five flavours and the five "enterings", or what I call Five Desire-flavours.)

I will now describe the most important aspect of body construction, the five flavours and the dynamics of this energy within the body. A flavour is not only something you can taste, it is also the energy or quality of the five energies within and exterior to the body. Expressing this as a flavour gives great perception of the effects of the five phases through our sense of taste. It is simply another way of describing the five-phase energies. Flavour may be different in an organ versus a meridian even in the same phase relation, but the overall understanding of the flavours will always follow these energetic principles, these are utterly key the to understanding of Chinese medicine, perhaps the most important part of this book:

Pungent

Pungent flavour is the energy of spring. Its energy comes from the inside to the exterior strongly and quickly; it spreads outwards from inside. The flavour is also called acrid or metallic tasting, and this also is a form of spiciness. The acrid type / metallic pungency is deep pungency affecting the deep levels of the body. This is like the taste of blood, hot spices like chilli, and numerous herbs and other spices. The superficial pungency is like cinnamon or cardamom, strongly smelling spices with a warming and opening quality. This flavour is strong and buoyant; it opens out and has the energy to force open and push through.

It moves to the surface from the insides. Pungent flavour is a post-Heaven associated flavour, found in the Lu meridian and the LI meridian organ and the Liv organs of the body. It is also the "initiator" of movement and associates with the Ming Men's ignition energy to begin to warm the body from birth. It is drawn from air-qi and from the pungent flavour of food. This action draws fluid like those of the kidney yin with it, and so it seems to moisten external regions of the body. However, it actually has a more drying effect, as it often releases fluids from within tissues. The pungent flavour has the effect of opening.

Salty

Salty flavour is the hottest of all. It is the flavour of the summer heat. If one thinks of the coldest situation—ice—just a sprinkling of salt will liquefy ice! It is fire in crystal form, a very purified substance. It kills infection through drying out with its heating effect. It draws away water and fluid from the inside and sends them outwards, like squeezing out a sponge. The salt within the seawater is yang within yin and is the mix needed for life. Without the salt, there is no potential yang energy to create life. The salty flavour is a post-Heaven associated flavour, igniting at the first in-breath at birth. The flavour is drawn from food and is found in the R-kidney and heart complex (see the Hrt organ and PC meridian part of this section B), as well as the bladder and SI meridian organs. Whereas pungent flavour heats and expresses from the surfaces, salty flavour heats from within the body. The salty flavour has the effect of softening icy coldness.

Sweet

Sweet energy has a gentle and warming effect. It has the effect of moistening and adding warm fluid to the body and encourages the bulking out of tissues. It is storage of energy to do long-term work and the beginning of the autumn storage for wintertime. The sweet flavour is a post-Heaven flavour and can be obtained from food. It is also found in the spleen and stomach meridian organ networks. It is the relaxing of tightness in the tissues due to cold, and it helps the body loosen in this way. The sweet flavour has the effect of loosening.

Sour

Sour is autumn within the system. The energy starts to move inwards and is drawing in. The autumn flavour makes one's face crinkle up; it's hard to smile in autumn. The accumulation of the energy draws inwards and sucks the juices and sap of the body, to be harboured internally till spring comes again. It comes with the breath out and the process for grieving for the summer's loss. The sour flavour is a pre-natal/Heaven flavour and is associated with the liver meridian and GB meridian-organ. It is difficult to ingest but can be drawn from food. Please note that the sour flavour does not act in its cooling, as cooling is yin; it is actually associated with gathering, due to being able to be still (lack of yang). Hence the sour flavour never acts itself but is always acted upon by another yang flavour. The sour flavour has the effect of gathering.

Bitter

This truly is the bitterness of winter. Winter energy goes downwards. The coldness of the winter affects the mood, and the bitter flavour is very abundant in the nature of winter. This energy is heavy and non-life promoting, many cold poisons being bitter. For survival in this bitterness the body retreats to the darkness where it holds some yang till the spring returns again. The bitter flavour dries too, by making everything turgid with water and freezing it; this is a dryness of cold. Again the bitter flavour is cooling, so it doesn't act on anything; it is drawn on by other yang flavours. The bitter flavour is a pre-Heaven/natal flavour and once lost is very difficult to replace in the system. It is associated with the L-kidney organ meridian only and is almost impossible to draw from food, except in very small quantities that can simply supplement the immediate situation and slow down loss of yin. The nature of yin is to be stiller than yang, hence through stillness yin is formed. The bitter flavour has the effect of firming by freezing. (This associates with the stillness before yin and yang emerge, but of course this is only an association, as Wu is pre/ encompassing yin and yang. Notice this too is why yin is associated with deficiency.)

These five flavours and their association with the seasonal energy are perhaps the key understanding within the work to explore the difference and mechanics of the organ meridian system in a full and complete way. They are key for drawing together the treatment methodology of acupuncture and herbal medicine into one flowing and refined completeness. This is the masterwork of Ikeda Masakazu, whose work I must refer to so that you are able to get a fuller picture of the principles I am exploring here.

(fig 1.2)

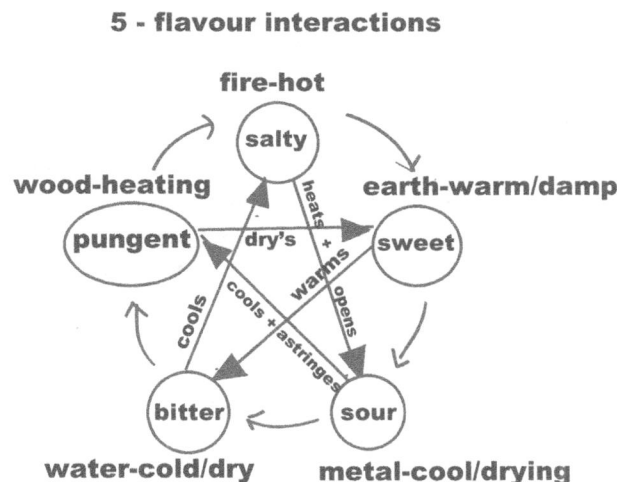

5 - flavour interactions

The image above needs to be explored. The following can be understood easily: pungency tonifying fire would seem understandable, saltiness tonifying sweetness also. Sourness tonifying bitterness also seems logical. However, the difficulty is when the yang flavours join the yin, and yin join the yang. We need to explain these aspects. Note that the five phases simply expresses an overall picture, not the specifics of the energetics in terms of the physiological energetics. Hence, the explanations that follows.

Sweet tonifies sourness through the introduction of fluid energy. As the heat of the summer dies away, so the temperature drops a little and condensation forms. This adds mass and weight and eventually draws down into the sourness of autumn. In the body however, the sourness and bitterness are pre-Heaven energy; the post-Heaven is the pungent, sweet, and salty. Hence one can say that the spleen Zang helps to add to pre-Heaven fluids. However, the intake of a lot of sweet food tends to damage the sour and bitter energy, but taken with fluid it can prevent pre-Heaven fluid being overused.

Bitterness tonifies the pungency? This is again complex bitterness which is very cold, and yet pungency is very warm and radiating to the surfaces. The anchoring of the kidney yin supports the whole life process, yet it isn't really the yin that moves but the yang that begins again. Hence we can say that pungency DRAWS from the bitterness, rather than bitterness actually acting to create pungency. If we think of the liver blood and the lung meridian, both express outwards and move fluids. This is the action of drawing out of the kidney yin. One could say the bitterness is a mother who is ready for her child to suckle, and when he does, spring occurs.

Another issue is pungency controlling sweetness. Again this is about the fluid aspect, and the pungency will dry out the fluid nature of the sweet, hence controlling its energy, weakening the post-Heaven yin that is present in Earth. The other control cycle aspects are easy to explain. Note again that the yin energies of sour and bitter do not truly act to take over the yang, but again they are drawn by its presence, whereas the yang flavours and energy, in general, will always directly approach the yin.

Finally, sweetness is tonified by saltiness. This is associating the process of the heat of the midday sun drawing up all the fluids into the atmosphere. Just after the peak of the midday sun, as the sun sinks in the sky, it is the time of highest humidity and warm heat. The combination is that of spleen energy. Whenever we look at the spleen, we will also be implicitly looking at the stomach, which is essentially the heat of the spleen. Hence the hotter the fire in the body, the stronger the thirst and appetite and the stronger the draw of yin fluids from the kidney; all this forms Earth energy, a combination of dampness and warmth. Notice that Earth is associated always with yin and yang. If we look at the derivation of five phases from nine energies, we can see that the archetype for Earth is the trigram Kun. The other aspects of Earth, the centre and trigram Gen, are secondary to this as they represent yang, but note that they are together within the phase of Earth. However, the focus is the spleen and fluid.

If the above information is too complex please look on to further chapters discussing the energetic anatomy and physiology of each of the aspects of the system and them come back and it should fit into place.

The Functions of the Zang in Relation to the Flavours

This will be discussed in more detail in each section, but I will touch on it here for a general idea of flow, starting with the pungency of wood. Pungency is associated with the vigour of the blood of the liver and of the Ming Men R-kidney ignition of heat that powers the heart. This pungency comes up to tonify the heart function and spreads the energy all over the body up to the skin and pores. The salty flavour is that of the heart, the full expression of the pungency of the wood energy and the power of the liver's blood and heat from the Ming Men fire/ R-kidney. This energy in the body warms the digestive system and draws fluids from the kidney yin to a higher level to be used in digestion, as well as drawing fluid from the exterior; this is the function of the spleen (and stomach). Then the fluid of the spleen being transported all over the body is absorbed into the deep tissues again as day turns towards afternoon-night and the energy of the body goes in and cools down. The tendons draw in the fluid which is the power of the sour flavour associated with the GB organ-meridian and the liver median (the yang or Fu organ within the yin). This process of cooling and condensing draws into the bones and the left kidney where yin is restored and recovers, only to be drawn again by the pungency of the liver-blood and Ming Men again at the beginning of a new day.

This process will become clearer later when we look at each meridian organ in more detail. However, the above is how the energy circulates within the Zang as a basis for the function of the whole body. The Fu and meridians are simply another "coating" on this Zang foundation.

The flavours in this expression relate directly to the change of 5 within the system, which is associated with the five yin organs. The flavours follow these as expressed in the previous chart. These flavours DO NOT express the energy of the time cycles of the 11 meridians, as the cycles of these are akin to Earth energy (we will look into this next).

This is a vital point: The yin organs associate with humans—cycles of five. The meridians associate with cycles of 10 and 12 stems and branches, which are associated primarily with the Earth's circulation in relation to the pre-Heaven universe (10 stems). The organs are truly between Heaven and Earth, whereas the meridians are associated with the Earth more, while the spirit is associated with post-Heaven 9 more. Be careful on this point:

9 energies of Heaven

5 energies of man (represented in the organs)

12 energies of Earth (stems and branches of earth)

Hence the root of man is between Heaven and Earth.

iii) Left and Right Construction of the Body

The sage faces the southern sky and looks out at the horizon, expressing his energy as an emblem of the microcosm in the macrocosm. His face is round and yang, as is the sun, heating at the zenith of the sky. His square back is tough and strong; it faces the coldness of the north at his back and lower legs, and his feet represent the nadir of the sky beneath the visible horizon. To his left side, the sun has risen, so this becomes the yang side of his body. To his right side, the sun sinks; this becomes his yin side.

(fig. 1.3)

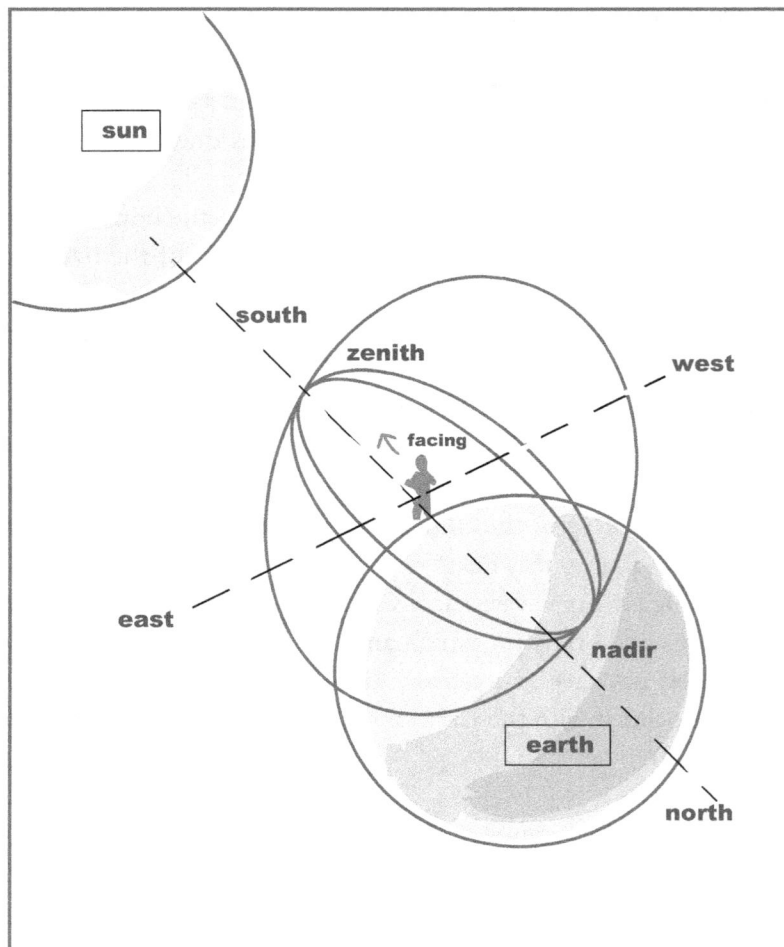

However, in the body there is balance. The heating and cooling of the fundamental energy is governed by the kidneys. The kidneys balance this left and right side so that too much yang is not placed on one side and too much yin on the other. Hence the eyes of the Taiji symbol within the human are the left and right kidneys, with the left side holding the yin of the kidney and the right side holding the yang. It becomes easy to see then why the pulses we use in diagnosis for the yin associated meridians are on the left, and the right associates with the more yang type pulses. Left is prenatal, and right is post-natal; this is another way to look at the same thing. Also, in the dis-eases section, we will look at how the symptoms of dis-ease can appear on the different sides of the body dependent on what the condition of the kidney is.

iv) General Construction of the Meridian-organ System

There are several ways to go through the meridian-organ networks. In Ling Shu 10, there is an expression that goes through the meridians as an interconnected system of meridians, a continuous loop. The other way is to look at the energy circulation generator of the five phases. I believe the best way to look at the interconnections is through this generating energy/ energy producer, this is the five phases within the system. Hence the following parts of this book are ordered in this way.

If the yin organs are called the "roots", the yang organs are "branches" from these "roots", and the meridians are "shoots" from these "branches". If we again use concentric circles, we can diagrammatically express this. Notice that the meridians form the exterior in exactly the same way as the organs form the interior layers. The body is formed from the roots creating branches and then shoots/ meridians. It must be remembered that when I use the term "physical", it simply means more physical or yin, and "energetic" means more energetic or yang. Hence the formation of the body comes about through the yin energy sprouting the yang meridians. As it is with all things, yang emerges from the yin, but it LEADS the change. Commonly in Chinese Medicine, the concept is that blood follows Qi, and Qi follows spirit, which means that Qi of physical energy follows ethereal energy, and this moves the material or the material moves with it. All formations therefore have an energetic base or are governed by Heaven/spirit.

(fig. 1.4)

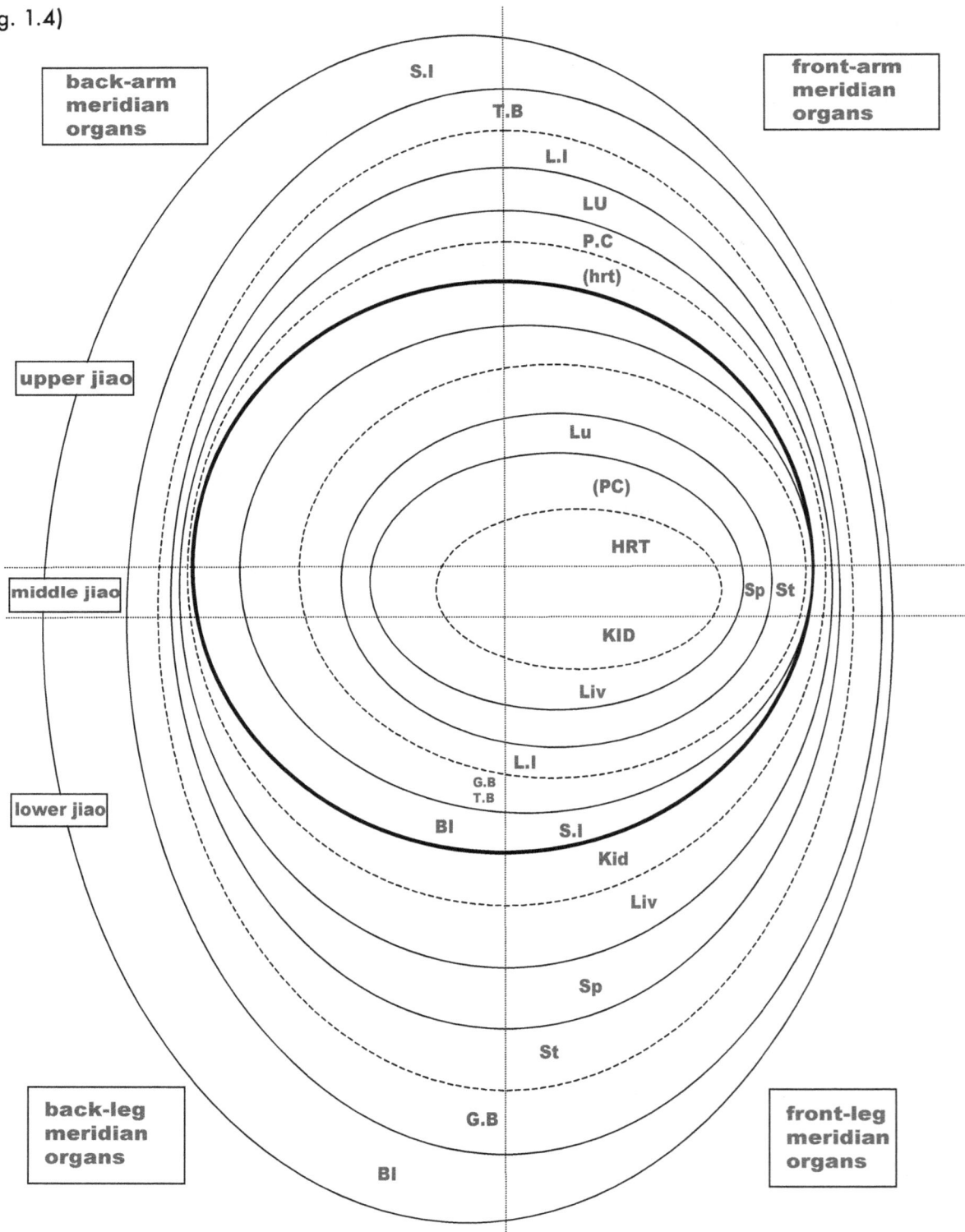

back-arm
meridian
organs

front-arm
meridian
organs

S.I

T.B

L.I

LU

P.C

(hrt)

upper jiao

Lu

(PC)

HRT

middle jiao

Sp | St

KID

Liv

L.I

G.B
T.B

lower jiao

Bl

S.I

Kid

Liv

Sp

St

back-leg
meridian
organs

G.B

front-leg
meridian
organs

Bl

266

On the surfaces of the body, the areas covered by the meridians are expressed to the surface here. Note that the meridians will be at a different depth, as noted above, representing the various tissues of the organ-meridian systems that are associated. Hence the below illustrations cross over significantly and do not have absolute areas of influence.

(fig. 1.5)

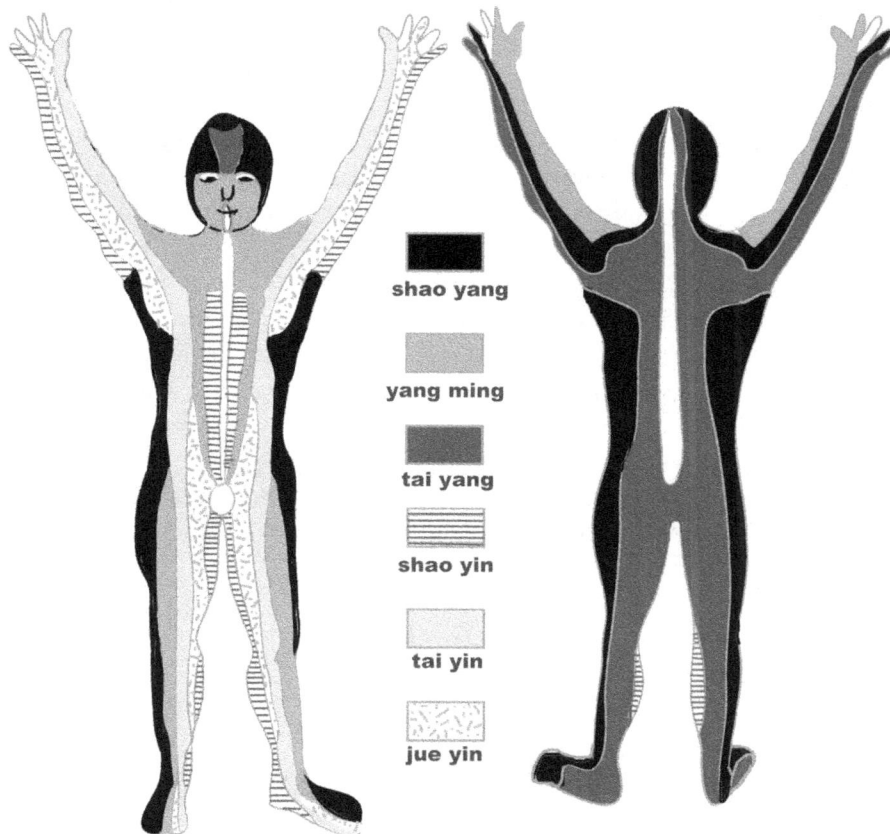

shao yang

yang ming

tai yang

shao yin

tai yin

jue yin

In total, there are 11 true organs and 11 true meridians in the body. This will be discussed at length in the section exploring the heart meridian complex, which for the first time (at least in the Modern Western world) explores the true nature of the heart and its meridian.

If we look further into this general construction, the following picture is explored: There are five meridians in each hand. This consists of three yang meridians, the TB, LI, and SI and two yin meridians, Lu and PC, which we can call the Jue Yin/ Ketsuin/ 厥陰 meridian of the heart also. Notice that fire and metal are at the top and balance yin and yang, but overall the upper body is more yang, and so there is one more yang meridian. The total number of meridians, bilaterally in the upper body, is 10. These are described as the 10 celestial stems, being above and uppermost in the body, and are prescribed time and date organisation as discussed earlier, in relation to the calendar, and this too can be tied up with the 24-

hour clock into a 120-hour, 5-day cycle of meridians (see below).

There are six meridians in the legs, three yin and three yang to balance. Notice therefore that the body is overall one yang meridian over, so it needs balance in the upper body provided by the empty "Shao Yin meridian" (see explanation on heart complex below). The three yang are GB, BL, and ST, and the three yin are SP, LIV, and KID. This is 12 bilaterally. Hence these are known as the 12 earthly branches. There are 22 meridians in all bilaterally. This is described in the following diagram, stems connecting with Heaven (of the pre-Heaven) being above rather than below and branches with Earth, the organs between. This is how the meridians of the body connect to the change of the Earth and the lunar calendar on the macrocosmic perspective (see part 1). This image is drawn in the anatomical image of Chinese Medicine with hands up to show the downward movement of all the yang meridians and the upward movement of all the yin meridians. Notice how the yang within the yin is the upward change of yin, and the downward change of yang is the yin within the yang, and describes the power of yin and the power of yang, when combined with its opposite in movement.

Notice that in the interconnection of the cycle of meridians, for example LU goes into LI, large intestine goes to stomach, etc. The yin meridians in the arms feed the yang meridians, whereas in the legs, the yang meridians cool and condense into the yin of the lower body. This indicates two things: that the yin of the arm meridians expand towards the yang like an upward and outward growth, while the meridians of the legs are focused on drawing energy back into the yin, stomach feeding spleen, GB feeding Liv, and BL feeding Kid. This is important as all the body's yin is found in the legs, and the yang in the upper body. This is the process of transformation of energy, converting it from an ethereal potential to a more dense physicality and round again. That is the cycle of transformation of yin and yang energies in the body.

(fig. 1.6)

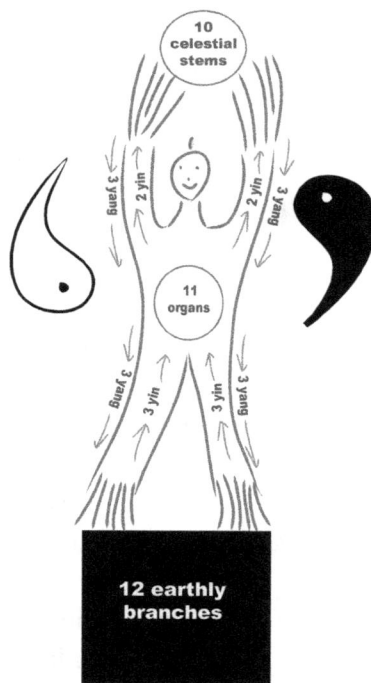

Notice that the foundation of the 11 organs within is the five Zang. The five Zang represent the inside, which is ruled by the nine energies of Heaven. The meridians are connected to the Earth and thus are dictated by its change. The combination of internal and external, Heaven and Earth, is what is associated with the movement of the whole body constitution as one.

The 120-hour (5-day) Meridian Cycle:
(table 1.2)

Branch-Stem	Bi-hourly 24-Hour Time Clock	Organ-meridian Associations Corresponded from Ling shu 41
B1/S1	23:00 to 1:00	Wa+Sp/ Wo+TB
B2/S2	1:00 to 3:00	E-Kid/ Wo-SI
B3/S3	3:00 to 5:00	Wo-GB/ F+LI
B4/S4	5:00 to 7:00	Wo-BL/ F-LI
B5/S5	7:00 to 9:00	E+ST/ E+SI
B6/S6	9:00 to 11:00	F-ST/ E-TB
B7/S7	11:00 to 13:00	F+BL/ M+PC
B8/S8	13:00 to 15:00	E-GB/ M-LU
B9/S9	15:00 to 17:00	M+Kid/ Wa+Lu
B10/S10	17:00 to 19:00	M-Sp/ Wa-PC
B11/S1	19:00 to 21:00	E+LIV/ Wo+TB
B12/S2	21:00 to 23:00	Wa-LIV/ Wo-SI
B1/S3	23:00 to 1:00	Wa+Sp/ F+LI
B2/S4	1:00 to 3:00	E-Kid/ F-LI
B3/S5	3:00 to 5:00	Wo-GB/ E+SI
B4/S6	5:00 to 7:00	Wo-BL/ E-TB
B5/S7	7:00 to 9:00	E+ST/ M+PC
B6/S8	9:00 to 11:00	F-ST/ M-LU
B7/S9	11:00 to 13:00	F+BL/ Wa+Lu
B8/S10	13:00 to 15:00	E-GB/ Wa-PC
B9/S1	15:00 to 17:00	M+Kid/ Wo+TB

B10/S2	17:00 to 19:00	M-Sp/ Wo-SI
B11/S3	19:00 to 21:00	E+LIV/ F+LI
B12/S4	21:00 to 23:00	Wa-LIV/ F-LI
B1/S5	23:00 to 1:00	Wa+Sp/ E+SI
B2/S6	1:00 to 3:00	E-Kid/ E-TB
B3/S7	3:00 to 5:00	Wo-GB/ M+PC
B4/S8	5:00 to 7:00	Wo-BL/ M-LU
B5/S9	7:00 to 9:00	E+ST/ Wa+Lu
B6/S10	9:00 to 11:00	F-ST/ Wa-PC
B7/S1	11:00 to 13:00	F+BL/ Wo+TB
B8/S2	13:00 to 15:00	E-GB/ Wo-SI
B9/S3	15:00 to 17:00	M+Kid/ F+LI
B10/S4	17:00 to 19:00	M-Sp/ F-LI
B11/S5	19:00 to 21:00	E+LIV/ E+SI
B12/S6	21:00 to 23:00	Wa-LIV/ E-TB
B1/S7	23:00 to 1:00	Wa+Sp/ M+PC
B2/S8	1:00 to 3:00	E-Kid/ M-LU
B3/S9	3:00 to 5:00	Wo-GB/ Wa+Lu
B4/S10	5:00 to 7:00	Wo-BL/ Wa-PC
B5/S1	7:00 to 9:00	E+ST/ Wo+TB
B6/S2	9:00 to 11:00	F-ST/ Wo-SI
B7/S3	11:00 to 13:00	F+BL/ F+LI
B8/S4	13:00 to 15:00	E-GB/ F-LI
B9/S5	15:00 to 17:00	M+Kid/ E+SI
B10/S6	17:00 to 19:00	M-Sp/ E-TB
B11/S7	19:00 to 21:00	E+LIV/ M+PC
B12/S8	21:00 to 23:00	Wa-LIV/ M-LU
B1/S9	23:00 to 1:00	Wa+Sp/ Wa+Lu
B2/S10	1:00 to 3:00	E-Kid/ Wa-PC

B3/S1	3:00 to 5:00	Wo-GB/ Wo+TB
B4/S2	5:00 to 7:00	Wo-BL/ Wo-SI
B5/S3	7:00 to 9:00	E+ST/ F+LI
B6/S4	9:00 to 11:00	F-ST/ F-LI
B7/S5	11:00 to 13:00	F+BL/ E+SI
B8/S6	13:00 to 15:00	E-GB/ E-TB
B9/S7	15:00 to 17:00	M+Kid/ M+PC
B10/S8	17:00 to 19:00	M-Sp/ M-LU
B11/S9	19:00 to 21:00	E+LIV/ Wa+Lu
B12/S10	21:00 to 23:00	Wa-LIV/ Wa-PC

Key:

Wo+ = Wood yang = GB organ
Wo- = Wood yin = LIV Organ
F+ = Fire yang = SI Organ
F- = Fire yin = Right KID (HRT) Organ
E+ = Earth yang = ST Organ
E- = Earth yin = SP Organ
M+ = Metal yang = LI Organ
M- = Metal yin = LU Organ
Wa+ = Water yang = BL Organ
Wa- = Water yin = Left KID Organ

GB = Gallbladder meridian
LIV = Liver Meridian
SI = Small Intestine meridian
TB = Triple Burner meridian
BL = Bladder meridian
PC = Pericardium meridian of the heart
ST = Stomach meridian
SP = Spleen meridian
LI = Large Intestine meridian
Lu = Lung meridian
Kid = Kidney meridian

Each meridian has a divergent pathway, which is a superficial pathway, part of the main meridian, but is a tributary of the main meridian on the surface, as are the Luo vessels (see below). Please not the above could be turned into an expression a acupuncture oriented on the timing of point known as Stems and Brach acupuncture. Although this was assorted with Ling Shu 41 the attention focused on these ideas in the Classics is minimal and as such I will not further them here (please see appendix 2 page 820 for details)

v) The Luo (Network) Vessels

The Luo vessels are tributaries of the main meridian networks and are vessels that run between superficial and deep layers of the tissue and also horizontally across the vertical meridian lines on the body surface. This creates an internal and external network of energy; if the main meridians are the warps of the energetic body, the Luo vessels are the weft. This 3-D structure goes deep into the lower levels of the extraordinary vessels layer, which is deep to the main meridians. Hence the Luo vessels meridian between the very surface and the depth of the extra vessels:

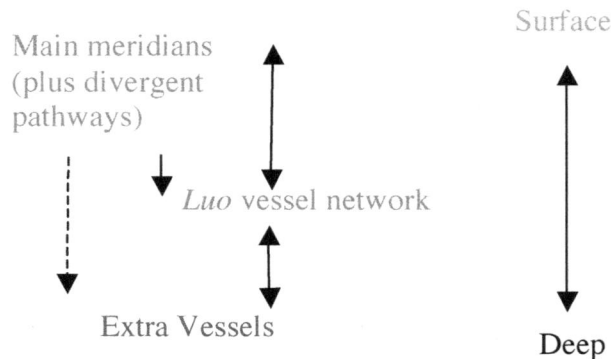

The Luo vessels are part of the main meridian, but a tributary, and so are controlled by the main meridian. Hence they are always secondary in treatment and in focus of attention to training. There are 15 Luo or "network" vessels, which is one for all of the 12 main meridians, one each for Ren and Du mai of the extra vessels, and one other vessel which is a secondary tributary of the spleen (the great Luo of the spleen) which has two Luo vessels attached. The great Luo connects to all the other Luo vessels of the whole body, so Luo vessels can be considered closer to the Earth energy, whereas, for example, extra vessels are closer to the water energy, and perhaps the main meridians are closer to fire energy. The depth of the meridian throughout the system is what is being expressed, so always look at the big picture.

vi) The Eight Extraordinary Vessels

So far we have not explained much about the extraordinary vessels of which there are two with their own points, Conception Vessel and Governing Vessel, creating a central line meridian, front and back respectively. Using points from other channels and linking them together at a deep level to the superficial vessels of the main meridians are the Dai Mai, the Chong Mai the yin Wei Mai, the yang Wei Mai, the yin Qiao Mai, and the yang Qiao Mai, making a total of eight meridians. These meridians are governed by the left kidney and are seen in all ancient Han Dynasty texts as of little importance in treatment. They are seen akin to the Luo vessels on the surface, which are treated by draining the meridians when they become full only (please see Nan Jing chapter 27). Otherwise they are recycled and controlled by the main meridians' flows of the body. They are however more closely associated with the kidney yin and so are not part of the main meridians in the same way as the Luo vessels; they are reservoirs, whereas the Luo are tributaries. This is because they themselves have very little movement. They are yin and reservoirs of surplus energy generally of the fluid nature. They collect and are drawn from. They also hold pathogenic ki and act as reservoirs to prevent this energy from getting to the main meridian flow by holding it. This likens them to the lymphatic system in Modern Western physiology, although again this is too fragmented a model. A reservoir is a very still entity, and its flow is associated with that which draws from it and that which accumulates in it. Hence this again expresses the close relationship with the kidney yin and the association of Stillness and the bitter-cooling energy throughout the system.

The reservoirs of the extra vessels are dictated in their movement by the flows of the main meridians. The main meridians draw and regulate these channels themselves, so intervention is unnecessary, as it is more important to correct the flows themselves. The eight extra vessels have been considered to be close to the eight trigrams of the pre-Heaven sequence, as they are structural, forming an octagonal formation within the body structure, as explained by Yoshio Manaka. This would seem accurate, as the kidney energy of the kidney yin is the pre-heavenly essence and the cool part of the system associated with the root of the yin within the person (see chapter 6 of this section). However the treatment of these would be akin to treating the symptoms, not the root, of a dis-ease, as they have no flow of their own. Due to this, we will not mention them further in this text. If one studies further, one will understand their use as symptom-managing regions in relation to the meridian they are associated with and the root pattern being treated, but very little else.

As far as depth goes within the body structure, as one gets closer to the bone and to the yin, one gets closer to the extra-vessel energy wherever you are on the body. Hence this again shows the close relation to the meridian energy of the kidney and the extra vessel energy; they are one and the same.

Please note the sinew meridians are also a commonly indicated form of meridian however these are simply the association of a meridian network to the energy of the GB. As such The sinew meridians are not really discussed either as it would be like focusing only on the network of the GB system would not be inclusive of any other aspect of the system. They are more exclusive than the Extra-Vessels or Luo vessels to the energy of the GB, where as the Luo and Extra-vessels are involved in more than this. However, the Extra Vessels are the expression of the Kidney yin and the sinew meridians are associated with the Liver-yin and GB energy, hence they are similar in association with yin. The sinew meridians mostly follow the basic line of each of the main meridians of the body and associate each meridians with the primordial tedious like tissue, which is the origin of all the body structures. However treatment focused here is to discount the rest of the body and this would be to not see the living aspect of the flows of the 12 meridians. Again the sinew meridians are treated as part of branch treatment, when the Liv/GB is involved mainly.

273

In the next chapter, we start to look at the meridian-organ system one by one, starting within the wood phase. Before we go on and look into the organs and meridians of each phase, we should first have a look at the language we are using to describe the organs and meridians. It is common to use Modern Western names for the organs and their associated meridians without differentiating clearly what is actually meant. This often gives us a very distorted view of what we are investigating. Hence we will first look into the vocabulary of physio-anatomy within the ancient context, how perhaps the ancients viewed the various energy systems and how their naming of the organs meant something very different from what we understand today.

1.2 Vocabulary of the Physio-anatomy of the Energetic Body
i) Meridian and Acupuncture Point

Let us first start with the meaning of the word "meridian" and "acupuncture point":

(fig. 1.6)

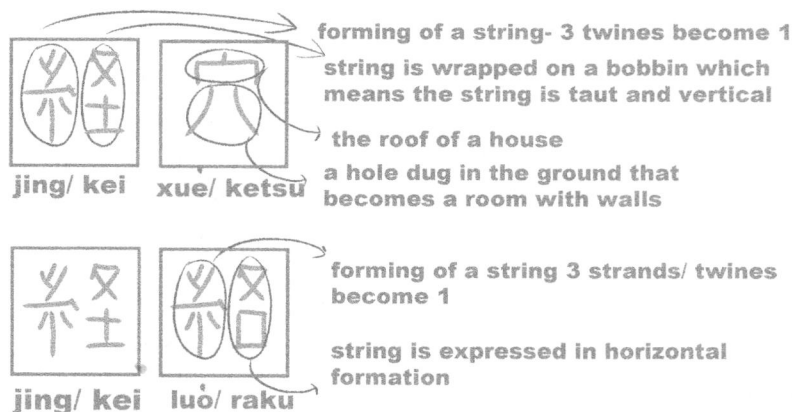

jing/ kei   xue/ ketsu

forming of a string- 3 twines become 1
string is wrapped on a bobbin which means the string is taut and vertical
the roof of a house
a hole dug in the ground that becomes a room with walls

jing/ kei   luo/ raku

forming of a string 3 strands/ twines become 1
string is expressed in horizontal formation

The top set of characters, Jing xue in Chinese or Kei Ketsu in Japanese, is what we call in the West a meridian-point or acu-point. The word Jing/Kei is also used in the term JingLuo or KeiRaku (Japanese), which literally means main-meridian and Luo-meridian in English.

Let us look at the word Jing/kei. The picture expresses on the Left; twine being drawn together to form string; at least three strands form this string. The string is then turned onto a bobbin or reel on the right side, which is held so the twine is vertical and is running down into the reel. If we look at Luo/ Raku we can see the same string being formed on the left side, but this time the string is horizontal. The joining of horizontal and vertical creates a net structure. This forms the meridians of the body, a network of main-meridians and Luo vessels, which are tributaries of the main meridians coming off at horizontal angles. This network creates an energetic "cloth", the fabric from which the body-energy is made. We will look at this later in chapter 8 again.

When Jing/kei is joined with xue/ketsu, we get the idea of a point along this meridian, but what actually is a point? In the word xue/ketsu, we get the idea of a hole being dug out of the Earth below to form a hollow space, and a roof on top of this space. This is along a vertical string, or meridian, hence it literally means hollow spaces along the string of the meridian. The roof over this space shows that it is covered, and also it houses something, perhaps energy, a space for energy to be housed along the meridian. If we consider the meridians to be like movement of water as expressed in the classics, then these regions are like eddies in a current, pools of energy within the general flow.

Now let's look at the organs:

ii) Zang and Fu

(fig 1.7)

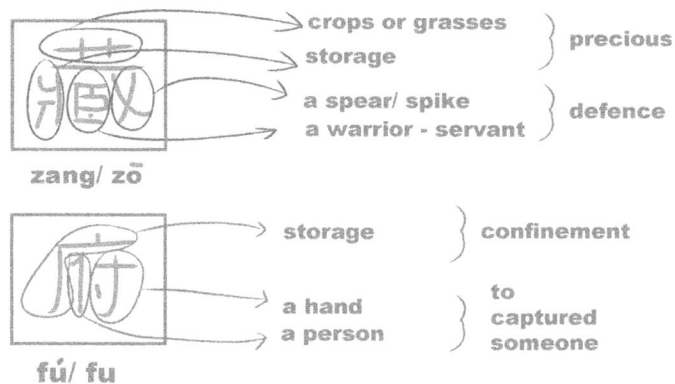

Zang/Zo is a complex character. At the very top of the character, it shows grasses or crops. These are stored in the structure below it, a granary or store house. A warrior-servant or guard protects this granary, indicating that it holds vitally important substance, the food of life.

Fu is also interesting. Again there is a storage capacity or house structure, but this time a person is captured and stored within the structure. This comes close to the notion of a prison or a confining structure. In relation to the Fu organs, we can understand that the organs, although innately empty, capture ingredients and actually change them, rotting them or ripening them—in this case, the idea of capturing a person, perhaps a prisoner or someone who has committed a bad deed. Imprisoning them for a while allows them to mature and ripen, or to become useful to society before they are let out into the world again.

Now we will look at the Zang–Fu organs themselves and attempt to understand the meaning of what we accept as Modern Western stylized organs:

275

(fig.1.8)

a form of flesh/ meat

shield

shield- like
flesh
formation

gan/ kan

This is the character for what we describe as "liver". However, as you can see, this is not as specific an explanation. The radical on the left is common to all the organs except for the heart and the triple burner. These two organs are associated with pure yang and therefore do not relate to form. However, the left side of Gan relates to a shield. Together these form a shield-like flesh formation. This associates strongly with the musculature of the system. Hence the liver is seen as the entirety of this formation of flesh throughout the body, very different from just the liver organ. Now consider the following:

(fig. 1.9)

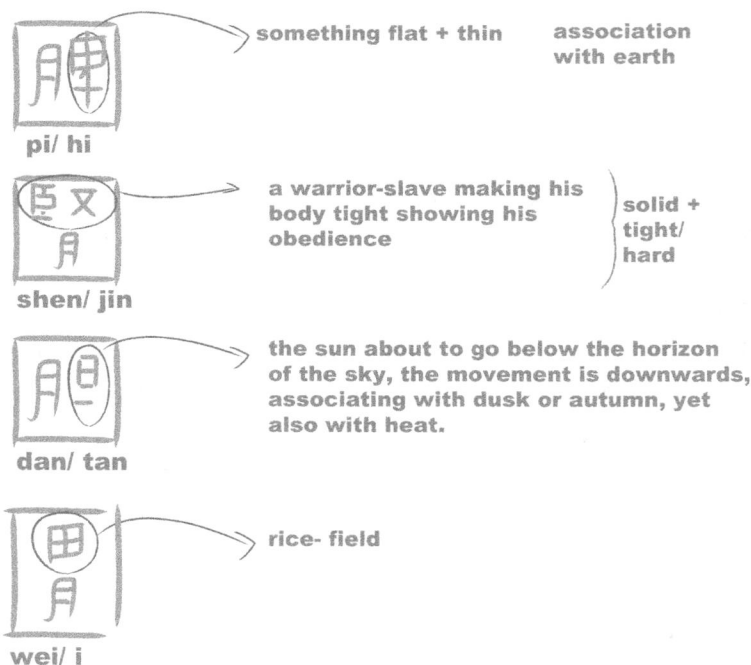

something flat + thin

association
with earth

pi/ hi

a warrior-slave making his
body tight showing his
obedience

solid +
tight/
hard

shen/ jin

the sun about to go below the horizon
of the sky, the movement is downwards,
associating with dusk or autumn, yet
also with heat.

dan/ tan

rice- field

wei/ i

(fig 1.10)

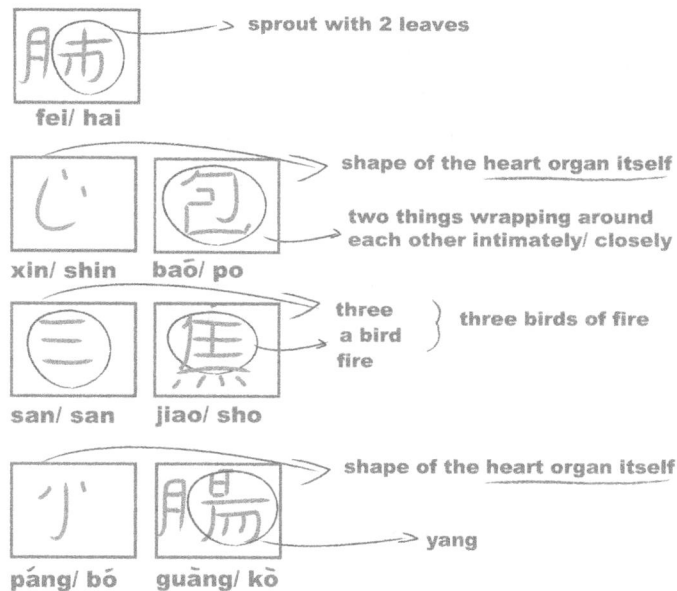

I have drawn a chart below to describe the nature of the labels we use to describe the organs of the body and how wary we should be in insisting they are simply organs. In this art of understanding, we must see a far less specific view of the body and start observing energetics rather than absolute concepts in our investigation from here on in.

((table 1.3)

Name of Organ-meridian in Modern Western Terms	Understanding or organ derived from the ancient characters
Gallbladder	A formation of flesh/tissue that is full and dense and has the tendency to rise upwards to the surface
Liver	Flesh that has a shield-like formation
Small Intestine	Flesh that is of smaller size relative to the large intestine, and is yang in quality
Triple burner (Right-kidney, has no form)	Has no form of flesh associated with it, but relates to 3 fire birds within the body, three levels, one at each level but all fire rising/ flying upwards towards the heart
Heart (associates with Shen, so also has questionable actual form)	The simple shape of the heart itself; given no flesh radical indicating that it has an association with spirit and yang, not form and yin

(Pericardium—this is a meridian only, not an organ)	Again no flesh indication; this is a conduit of the heart that wraps it, considered to be simply the meridian of the heart's energy (see chapter 3 of this section)
Stomach	The nourishment form of flesh, or that flesh which carries nourishment
Spleen (sometimes called Spleen-pancreas)	The Earth-like flesh formation of the body
Large Intestine	The larger formation of yang flesh compared to the small intestine
Lung	The formation of flesh that has two leaves, or lobes of the lungs
Bladder	The formation of flesh which associates with a broad expanse over the exterior and that is wide and big, associating with the meridian energy
Kidney (left)	The dense and hard formations of flesh in the body, bones being one

Note that the yang associates more with the meridian energy, and the yin associates more with the organ energy that we know of as "an organ" of the body. When we are looking at the organs and meridians and attempting to acknowledge them fully, we must attempt to see the formations of energy rather than isolated structures when considering life and natural medicine. With Modern Western mind-fragments, in the language and in the way of expression, we must be constantly aware of this and give attention to words not as reality, but only as shadows of meaning.

iii) The Five Main Tissues the Body Is Made From
(fig. 1.11)

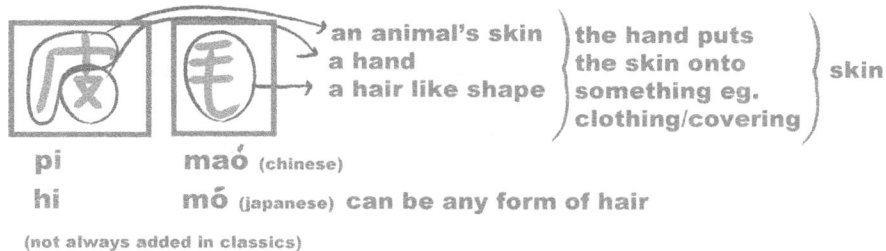

an animal's skin
a hand
a hair like shape
} the hand puts the skin onto something eg. clothing/covering } skin

pi **maó** (chinese)
hi **mó** (japanese) can be any form of hair

(not always added in classics)

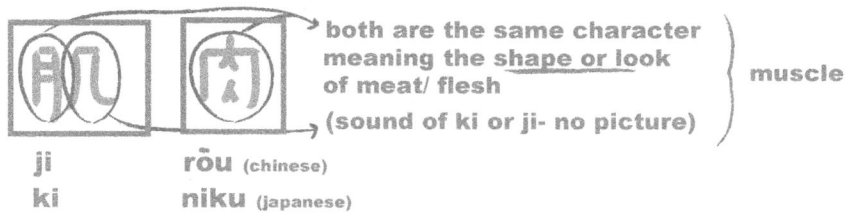

both are the same character meaning the <u>shape or look</u> of meat/ flesh } muscle

(sound of ki or ji- no picture)

ji **rōu** (chinese)
ki **niku** (japanese)

bamboo
power
flesh
} this depicts the form of muscular flesh; when tensed, muscle looks bamboo -like strong +flexible- powerful flesh } tendon

jin/ mó

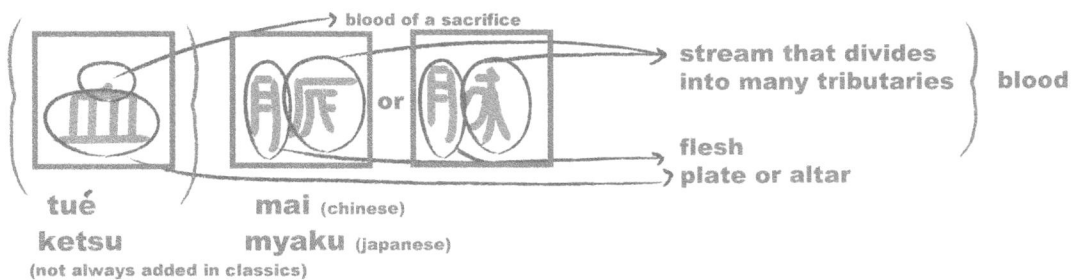

blood of a sacrifice

stream that divides into many tributaries } blood

flesh
plate or altar

tué **mai** (chinese)
ketsu **myaku** (japanese)
(not always added in classics)

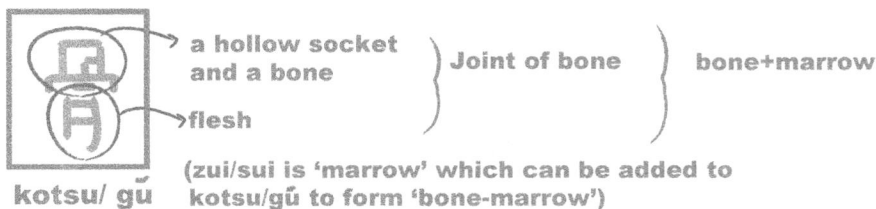

a hollow socket and a bone } Joint of bone } bone+marrow

flesh

(zui/sui is 'marrow' which can be added to
kotsu/ gǔ kotsu/gǔ to form 'bone-marrow')

279

(table 1.4)

Name: Japanese/Chinese	Energetic Relation to Zang	Possible English Meaning
Himo/Pimao	Lu	Skin and body hair
Kiniku/Jirou	SP	Flesh, the rounded look or shape of flesh, associated with fatty tissue
Jin/Mo	Liv	Tendon-muscle tissue
Ketsu Myaku/ Xue Mai	Hrt/ R-kidney	Blood vessels
Kotsu/ Gu	L-Kidney	Bone

Note again in the descriptions of each character above that there is significant difference in what the possible meaning in English could be. This is especially true of Ki/niku which is a very visual expression indicating the shape of something. Considering all the other tissues involved in the body, the only aspect left out is the fatty tissue, which bulks out and makes things more "fleshy". This is why we relate Jin/Mo with the muscle aspect of the body and the flesh-fatty aspect with Ki/niku, which I have described as fatty tissue. Note that body hair has no flesh radical involved; this, as with the heart, indicates a connotation that is not part of the physicality of the body, or as we will see later, may not have relation to the body (yin) or its energy (yang) at all but is part of the exterior.

Part 2

The Wood Phase

The Wood Phase

Please note a key point of difference from the Traditional Chinese Medicine concepts, in the classical approach as explored and extrapolated by Ikeda Masakazu: the meridian and organ have different energetic expressions. This will be explained throughout the text.

Before we look into the wood system, let us clarify the true nature of the wood and metal systems, as they can be confusing. The liver organ is yang, and the lung meridian is yang; the liver meridian is yin, and the lung organ is yin. The GB organ-meridian is yin, and the LI organ-meridian is yang. This forms a total balance between spring and autumn within the body:

(table 2.1)

Yang	Yin
Spring	Autumn
Pungent	Sour
Liver Organ (+ full of air lung organ + partially the function of R-Kid/ Ming Men)	Liver Meridian
Lung Meridian	Empty or air Lung Organ (+ empty liver organ)
LI Organ-meridian	Gallbladder Organ-meridian

The liver organ and lung meridian represent pungent flavour in the surface yin and the depth yin. The LI represents the yang. The liver meridian and lung organ represent sour flavour in the surface and the depth yin. The GB represents sour in the yang. The fullness (of air/ qi) of the lung organ is in coordination with the fullness (of blood) of the liver organ and vice versa. Also, when there is fullness of the organ, the meridians are empty; when there is fullness of the meridian activity, the organs are empty. There is constant balance here. This is the full explanation of the connectedness of these systems. When the lung organ fills up with air/qi the diaphragm presses on the liver organ this gets ready for release of blood. With the lung's out breath the air is now inside the body and invigorating the blood and also the outward pressure courses the blood and air (qi) through the body.

Spring and autumn are the times of fastest change from yin to yang and yang to yin. Notice that the natures of the fire phase network and the kidney yin and the spleen and stomach all have the same organ and meridian function for their respective organs and meridians. This can be expressed in the following diagrams. The three burning spaces or three Jiao are represented here and show that the movement is fixed. I believe this is the Spiritual Axis or pivot, which is the title of the second canon of the Nei Jing, The Yellow Emperor's Classic. It is obviously of vital importance that this axis is kept strong and firm for good health, and the metal and wood energy flex around this movement, which can be shown best like a pendulum's change:

(fig. 2.1/ fig. 2.2)

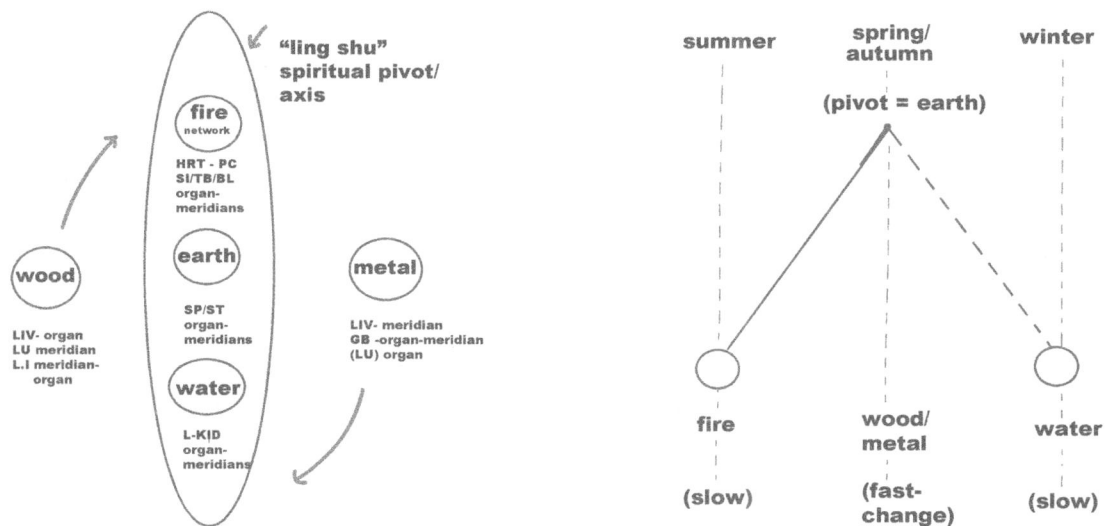

This being the case, we can now investigate the true spring-wood phase in the system, which includes the liver organ, lung meridian, and large intestine meridian-organ.

2.1 The Liver Organ System

The liver is a complex system. Its meridian is sour and cooling, but its organ and blood is expansive, pungent, and hot. This duality creates the pressure of the blood around the system in the tissues of the muscles and tendons of the body. The liver organ is wrapped by an energetic blanket of cool yin—the liver meridian. The meridian wraps the hot liver, bathing it in a cool blanket to stop it from overheating and to calm it down. Also, it provides a cylinder for the blood to gather energy and have a springy effect (i.e., slightly held back by the cool meridian wrapping it). The liver is hot due to the blood it carries, and it is said to store the blood. The combination of flavours that create the blood form life, which in itself is a yang phase and has an overall pungent expression. Hence liver blood is pungent. However, it is internally pungent, whereas the meridian of the lung meridian is pungent on the exterior. These two pungent flavours express the "spring" energy within the body. In the blood, the ethereal soul can anchor itself, and so the liver becomes a good place for the soul to hold to the body. It is the pungency within the blood that gives the blood its pressure.

The root of all yang in the body is the Ming Men, which, as we will acknowledge in the next chapter, has association with the pungent and salty flavours, as it is both the warmer and the initiator of life. Hence we can say that the liver organ is really an extension of the power of the kidney yang (just as we can say that the liver meridian is an extension of the kidney yin; we will look at this on page 328 of this section). The root of the yin flavours are the kidney yin, and the root of the yang are the kidney yang.

The liver meridian is between the deep depths of the kidney meridian in the deep inner, and the Tai Yin/Taiin/ 太陰 meridian of the spleen which, with the lung (arm Tai Yin), are the most exposed yin meridians. Hence liver is between. Between upper and lower, it lies under the diaphragm, and it requires the movement of the lung to express its energy. With every out-breath, the diaphragm goes from a pressured effect on the liver to a relaxed effect. The relaxedness of the out-breath allows the liver to open, and tension is reduced. Hence meditation practises use this to release and relax tensions, which are often to do with the liver system. The liver organ is a large muscle, and so the liver energy encompasses all inner muscles of the body and all interior yin associated muscle. The liver meridian dominates the somatic nervous system within the skeletal muscles, from a modern fragmented perspective, this is the Liver yin energy. The tendons and fasciae can be considered more of the liver meridian, whereas the muscle bellies are more associated with the liver blood and redness, and are associated with the organ. There are two parts to the blood: the white yin aspect, considered to be the liver yin or the liver's meridian energy, and the yang part of the blood, the red aspect, which is the liver yang, or liver organ aspect. This is how the liver relates most to the blood.

The blood of the liver must be considered, in its entirety, pungent in flavour. However, the pungent flavour rises easily to the surface. Hence blood is an anchored pungent flavour. Blood is created in mixes of pungent, sweet, and salty, and bitter and sour to hold it down and in; these flavours make up blood. Combined, these create a heated and powerful but liquid form of pungency that occurs within the body. The pungent flavour of the surface of the body, expressed in the lung meridian, is much lighter and the more ethereal aspect of the pungent flavour. It is key to understand that the expressive energy of the inside and exterior are connected in their energetic quality and general function—circulation and radiation. The liver blood is activated and moving.

As the fundamental flavour of the liver organ is pungency, this draws to it the sour flavour, just as the fire network draws the yin of the bitterness of the kidney towards it. These yin flavours are non-active and yin; therefore the yang movement of pungency pulls the sourness towards it in opposite attraction, and at the same time draws up kidney yin fluids (mother-child). Hence the liver meridian and GB organ meridian act as opposition for the liver organ. The liver yin/GB yin tonifies the kidney yin (mother-child).

The male and female genitals (penis and clitoral-internal genital functions) and the uterus (although ovaries associate more with water [origin]) the liver, and so the sexual arousal energy is determined by the liver's power of blood (pungency) and blood pressure. Note that the left and right kidneys are the source of the power of the liver's yin, and they help to create the red blood (yang) respectively. They are the origin, but the liver is the expression of sexual energy, the kinaesthetic of sex, while the kidneys are the potential. The anus and all muscles throughout the peristaltic systems of the body are liver dominated as well. However, the energy for the peristalsis itself is related to the stomach energy. As with blood vessels and everything else in the body, the liver layer is that of muscle tissue, so it relates to all this within every aspect; this is wood in Earth, or wood in fire, etc. The meridian itself is just a description of the connection of the main areas of liver energy throughout.

The female period is part of the physiological process, not a pathological one; it is part of life's process, as is pregnancy. Physiology is often mistaken for pathology in the modern world, where so much sufferance is associated with natural processes, due to mind-identity's dominance. The female period cycle is heavily associated with blood—and so the liver system— but the cycle itself and its rhythm are under the control of the kidney yang energy interiorly. It is in coordination with the cycle of the Earth-moon exteriorly, which is why this is associated with age, and Jing (yin) of the kidneys which is derived from the ovaries, but the ovum is considered more in association with the yang–liver blood. It is not very effective to separate these aspects out; in fact, it causes fragmentation. The main principle is that the woman's body and internal sexual energy must be warm and produce yang, whereas the male's sexual energy must be yin and cool (testes are exterior to the body, to do just this). This combination creates the opposite of fusing that occurs at conception. The blood and kidney yang are very close to each other energetically, and the liver blood of the liver organ contains the kidney yang energy, so we can say that liver blood is synonymous with the period cycle.

The female cycle is usually fully established at about the age of 14. At this time, the woman has been through two full cycles of growth (7 x 2). The period essentially indicates that a woman has become fertile, so this is often around the age of marriage for some ancient cultures. The woman's body creates blood that can be used for reproduction. However, this blood is also within general circulation, so it is not only used for reproduction, but in fact all workings of her body. If there isn't enough food or energy in the woman's body, she will stop having a period in order to retain her nutrients and the body will be infertile. This is very important. In today's sexually addicted society, we have lost touch with the natural rhythms of nature. Most mammals actually reproduce at a time of opulence of the energy within their body. This is usually found seasonally, in the summer/late-summer period. At this time, there is very little energy needed, because the body has a lot of warmth and energy present, provided by the environment. This means that energy can become overall excessive, and the energy that is just over that which the body needs to function can be used for physical labour or sex, both of which will drain the body's energy. If conception occurs during this time, the woman's body energy starts the process of going internally during the autumn and winter months, which is perfect to tonify and strengthen the body of the foetus for birth in springtime. However, in the "civilised society", this is something, which is not felt, and sex occurs all the time, as we have lost touch with the primordial energy of the Earth and the seasons and do not protect the energy. Thus, often we die prematurely or do not live as long as the potential of the body could allow. Sex is utterly regulated by the nature of the harshness of the environment i.e. reproductive energy is only possible if there is enough energy to sustain the body exuberantly, if nature is left without resistance.

When there is opulent energy sex will occur, when there is not it will reduce, just in parallel with the seasons, no need to think about it, its already intrinsic.

If the woman does not become pregnant, she has a monthly cycle, and the cycle should be (ideally) directly in tune with the moon cycle. A full moon indicates the yang within the yin. This is the most yang moment of the cycle and indicates the hottest part of the female cycle, which is just before the actual period, when the moon begins to wane; then the period will come. The opposite side of the cycle, therefore, when the moon is covered (entirely or partially), is when ovulation occurs, as the yang then returns after the darkest point in the cycle. The cycle also relates to the woman's sexual cycle and sleep, as she has more sexual energy when the moon is bright and there is yang within the night, and she is able to sleep less. When the moon is waning, there will be less sexual energy and the body will return to stillness. Again, this cycle is not observed in "civilisation", as the natural light has been regulated and unnatural light is used constantly, which is the aspect that influences the woman's cycle. Light is yang, so too much yang being shone at her means perpetual light, which means the cycle cannot return to the yin and therefore recover back to the yang with fullness. This is very commonly the yang dominance of society; this can be seen literally, looking at city lights (yang) at night and noticing perhaps what this is like in more natural, rural areas where the only light is from the stars and moonlight.

Women naturally follow this seasonal cycle within themselves, the period being a time of fullness being emptied, and so of death, associated with autumn season. This is vitally important because it is the female connection to death within life that allows life to come again, therefore rendering life and death meaningless as concepts.

(fig. 2.3)

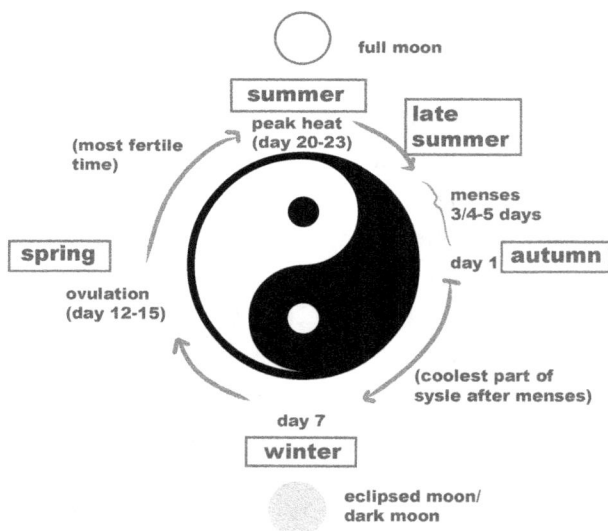

The ovulation aspect of the period is a very yang, pushing out of the ovum (within modern fragmented ideology). This is kidney yang. The blood supporting and nourishing the uterus is the liver's blood, as the uterus is like one big muscle. The uterus blood increases waiting for ovulation. After ovulation, it still increases, making the area very full of blood and nutrients, giving the best chance for pregnancy. This process is perhaps 2–3 times more powerful in the summer and late-summer periods than in the winter, where the woman's body must use some of this energy to heat her system. The blood collecting around the uterus, once there and not used, is not put back into circulation unless the body is very low on blood. The blood is released, because this blood has been static for a little while and its vitality has been reduced. The body gets rid of it with the excess of heat through the period. The period therefore is a cleansing, or better, a renewing process for blood in the body of women. The blood clears away the materials made for the possibility of pregnancy; these are all considered part of the liver blood energy.

Just a week before the period, the woman's appetite will increase and she will often crave sweet and salty foods. To a mild extent, this is natural, as the body wishes to make up blood that is now lost from circulation, as it is collected around the uterus, but critically BEFORE this blood is released, so that there is never a time of deficit and there is an overlapping of the beginning of one cycle with the end of another. At the same time, she will feel full and slightly bloated below, because the blood is about to be released. This of course can change the mood, and there can be a frustration feeling with this. All these symptoms should be very mild aspects of the cycle. When there is heavy bloating, excessive cravings, and so on, this is pathological and usually due to blood deficiency (we will look at this later). The liver, as we can see, is affected by this constant bleeding and the production of blood. The healthy woman will have a period in line with the moon cycle, a cycle of 28–30 days. This timing of the earthly moment with the woman and the period of light and day is very important, as the blood and the yin of Earth are one, whereas qi and the Heavens are the yang. Hence the female body is connected deeply with the Earth and with blood cycles driven by the kidney yang. The ovary is associated with kidney energy in women, but the uterus with liver and blood. It is natural that the woman will be mildly blood deficient after the period, and towards the end of the cycle, it is natural that the liver energy feels full. The liver pulse will also reflect this feeling; this will arise just before the period comes, as the blood is engorging the uterus, which as we have said is part of the liver organ energy. During the ages of 14-49 (7x7) approximately there will be a period cycle, this is considered as more yin time for women physically because they are constantly building up the yang of the liver blood and then discharging it or having children. Prior to 14 and after 49 therefore have a commonality in that the female body is not bleeding and therefore holds the blood and as a result can be considered to be more yang. It is important to always note the place a woman is within this spectrum of change when understanding a CP or DP.

The liver organ is very much linked to the lung meridian, and the liver meridian to the lung organ. This will be explained further in the metal section.

2.2 The Lung Meridian System
(fig. 2.4)

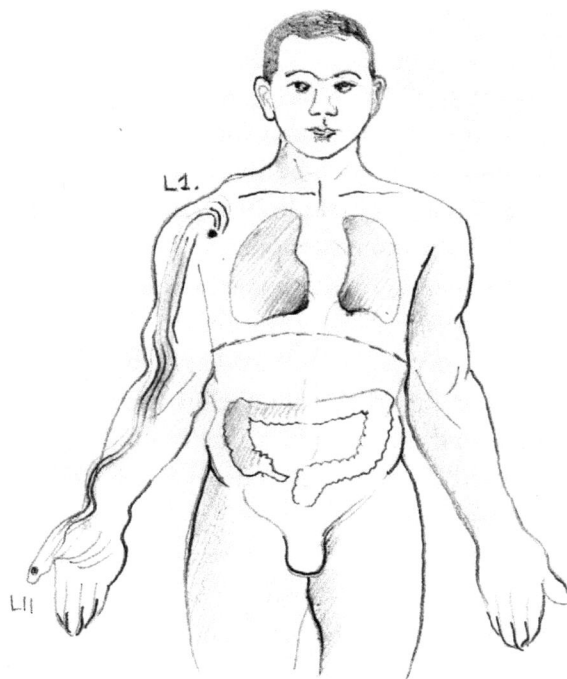

The lung meridian is linked to the qi of the liver organ blood. It is pungent and expressive, but on the surface rather than internally. The quality of the interaction of the lung meridian is explained in the metal section, but we categorise the lung meridian as part of the wood phase. The qi of the meridian is called wood but this meridian is labelled metal, this is only a labelling associated with the connection of the meridian to the Lu organ, however this connection and labelling is less profound than the energetics which it expresses. However in the further sections on meridians and points it will be important to understand the label as well as the qi.

If we consider that the pungent flavour of the wood phase tonifies the fire phase, we can see that the pungency of the liver organ internally and of the lung meridian externally radiates the energy to the exterior and upper aspects. This is the upper body organ of the heart (the lung organ is not part of wood or fire), the upper most yang, and the meridians of the upper body and yang surfaces—fire and the yang regions of the body, which once it reaches this peak starts to disperse, cool, and condense and come back down the yang meridians to the yin again and the lower body. This is a general flow, but it is important, as it shows the nature of the diffusing property of the lung meridian and the pungency to take the energy to its peak before it condenses down to the base. Notice that pungency therefore tonifies the right kidney, which is the root of the heart network or fire.

i) The Production of Post-Heaven Qi from Air*

(*Air (or breath) literally means qi or ki in translation, but here we are differentiating air-qi from food-qi and then the yinyang of this energy when inside the body forming yin qi (ying/ nutritive qi) or yang qi (wei/ protective or surface qi), note that Air-qi is intrinsically pungent in flavour and effect generally.)
The process of qi production from food follows this process:

1. The right kidney triggers the in-breath (pungent air is drawn in), and the left kidney anchors this process, drawing in air to the lung organ.
2. When the qi is full, it is drawn downwards towards the Ming Men. This aids the conservation of pre-Heaven energy.
3. The air qi is quickly absorbed into the body and rises up as a combination of salty (R-kidney energy) and pungent flavour, to join with the food qi to help to make blood (ying/ "nutritive" aspect or yin of qi), but most goes to the exterior and the surfaces of the body (Wei/ "protective" aspect or yang of qi), along with the salty aspect of the flavours from the blood, gathered from the food.
4. The qi permeating through the system moves the blood and helps the circulation through the body of energy and also regulates the heat of the body. It lets out energy and with it circulating fluids from the system if too hot (sweating), or, if too cold, the pores close and the out-breath is less, conceding expression of ki to the exterior.

ii) Understanding the ki passing through Pores of the Skin

The lung meridian is responsible for the pores of the skin. This associates very importantly with the connection to the lung organ. Understanding the pores of the skin also helps us understand the pathology of the exterior, so let's look at this ….

The following diagram expresses the opening and closing of the skin pores in relation to the various actions within the organs.

(fig.2.5)

out- breath   in- breath

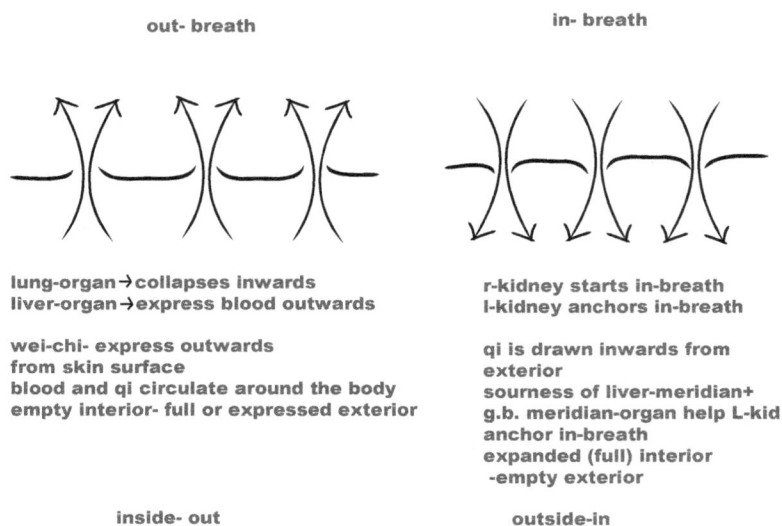

lung-organ→collapses inwards
liver-organ→express blood outwards

wei-chi- express outwards
from skin surface
blood and qi circulate around the body
empty interior- full or expressed exterior

r-kidney starts in-breath
l-kidney anchors in-breath

qi is drawn inwards from
exterior
sourness of liver-meridian+
g.b. meridian-organ help L-kid
anchor in-breath
expanded (full) interior
-empty exterior

inside- out   outside-in

The pores of the skin never actually close, just like the nostrils of the nose never close. There is always exchange and connection to the exterior. So often, I see people describing the skin or the body as a barrier or defence. This is true, but it is more of a radiation and exchange of interior with exterior, rather than a defence, which seems to cut off interior from exterior; this is not the case, there is no exterior really we don't begin or end anywhere! This is just a linguistic idea ONLY. The emptiness of the lung organ interiorly without breath (qi), i.e. the out-breath, is opposed to the fullness of radiation and expression on the body surface. The fullness and strength of the pungency of the air within the chest, i.e. in-breath, anchored by the pre-Heaven yin of the body in the form of the sour and bitter flavours, is full, but exteriorly the air is being drawn in, so this too draws in the pungent surface qi. We could say that the in-breath benefits the egocentricity, and the out-breath is the reality of unity!

The point is that this is a seesaw process of full and empty within the body. The pores of the skin blocks and cools/contracts only when the exterior cold affects them. The cold blocks or damages the yang of the pores to stop their function, whereas the in-breath is cooler than the out-breath for the skin. The skin belongs to the nature of the Lung organ, which as a accumulative function by nature, but here we are looking at the nature of that which moves through the pores of the skin, not the pores themselves. The in-breath draws energy from the exterior inwards, when this happens the lungs expand. This works against the accumulative energy of the inner body of the kidney and liver yin function, which is also that actual pores of the skin. When the lung exhales then the heat energy form pungency is pushed to open the pores, so the pores, or one could say the yin of the body never gets a break, it is always being worked against, this is because the body is yang overall. The time when the body recovers more is when breathing is at a low level naturally for example at night when one is asleep.

The lung meridian is said to originate from the middle Jiao. This region is dominated by the stomach; in fact, ST and spleen, SI, LI, and TB all connect to the stomach either through direct connection of their internal meridian pathways to the region or via the stomach meridian (He-sea point association). This simply shows the post-Heaven ki's production process interlinking to form air and food ki for distribution throughout the body. Pungency derived from the stomach (which is the originator of all the flavours from food through digestion, and the source of warm fluids for the body) can easily flow to the lung meridian in this way, just as pungency from the lung meridian can move and dry out excessive fluids in the stomach/spleen region through sweat and breathing out.

2.3 The Large Intestine Meridian Organ System
(fig. 2.6)

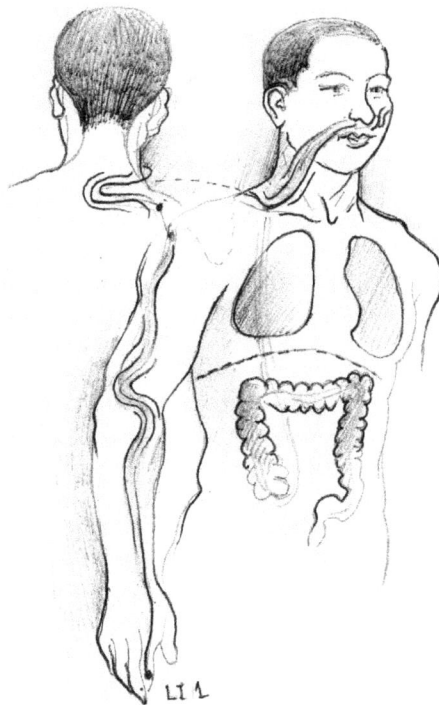

The large intestine has the same function for its organ as its meridian, following the meridian function of the lung meridian, just as the gallbladder has the same function for organ and meridian derived from the meridian function of the liver. The large intestine's flavour is pungent/spicy/acrid. Again the LI meridian is labelled metal although its functional qi is opposite to this designation, this will be further explained later. The large intestine's main function is to express unwanted products outwards. Second to this, it is to absorb and redistribute leftover fluids. The main expression of the large intestine is outwards, and this is exactly the effect of the pungent flavour—to open outwards and push from the centre out. The energy of the large intestine does this and judges the process of severing something from the body. Though this is sometimes considered to be metal energy, the L.I is actually wood within the metal of the body, it is the action and the expression and evacuation, it goes outwards not inwards as in autumn and true metal energy, hence when we are looking at L.I this is very connected to the liver organ and lung meridian energy or wood.

2.4 The R-Kidney Ming Men and Triple Burner-Bladder/Small Intestine Organ-Meridian

It is important to note that the initial burst of life that ignites the circulation system at the first in-breath of a baby, and the energy that weakens and dies at death, is a form of pungent energy, and so the R-kidney energy and Ming Men (including the bladder organ-meridian) associate with both wood and fire. The Ming Men and R-kidney are both a fire-starter and initiator of movement, as well as a burning, glowing, warming energy. The primary is the wood function, the latter is the fire function. Hence Ming Men and the triple burner, and also the bladder/small intestine organ-meridian, must be considered to have both a salty and pungent effect. This becomes very clear when looking at the very close connection between the distinction of dis-ease patterns of the lung meridian where the bladder/small intestine/triple burner energy is also involved; they are essentially one and the same thing. Generally we can say that the Tai Yang of the bladder and small intestine are a bit more superficial (more pungent-salty), and the TB is a little deeper (more salty-pungent), but all this comes through the skin, so it's all pungent, mixed with salty energy. As we will see in the next chapter, these organs and meridians are more clearly categorised by their warming function in the body and so the fruition of pungency or the fire or saltiness. However, it is important to note that there is an energetic crossover, which must be taken into account. This will become more and more obvious later, but if we think of birth spring and the initiation of fire of the "spice of life" at the Ming Men, we can see that there is an important connection here. True or pure salty flavour must be ascribed to the heart and its meridian, the pericardium (see next section), which is saltiness personified through the body. Otherwise, the salty flavour associated with the bladder and TB meridians therefore (including water points and source points, see page 350) have association to primarily salty flavour but also with pungency. As a general principle, where there is salty flavour in the body, it is usually accompanied with pungency. They feed into each other and wood-fire are one (perhaps the heart organ-pericardium meridian is the exception which is more salty overall).

2.5 Liver and H'un-Chinese / Kun-Japanese

魂

The 3 and 6 energies of the 9 energies, relate to the liver organ and, depending on the balance of spirit energy of the constitution, relay the information of vision/planning and more decisive action. The liver organ's energy (3/zhen associated) is often projected through the lung meridian, which is the expressive path of the liver's energy to the exterior. Notably the lung is not usually expressed; as the decision maker of the body, it is usually the GB in association with the liver, because it delivers heat and energy, which is passed to the GB to express to the exterior. However, the GB is cooling in itself; it carries the liver's heat and therefore is associated with the judicial calming of the expression of the liver. However, the expression of the liver associates with the function of the lung meridian and LI organ-meridian and the 6 energy of the nine energies.

The important concept is that all humans have both ethereal and corporeal spirit aspects. They are both yang and yin respectively. What is strong makes the other aspect weak, and vice versa. Those with a strong ethereal nature of such as the 3/ Zhen quality (6 has both yin qualities and yang expression, much like the nature of the lung organ and its meridian) will have less strong corporeal aspects. They will tend to be more mind orientated, less grounded, and more inspirational speakers, as we discussed in the constitutional aspect of this work. The Hun is described as the "cloud-spirit", indicating that it has the nature to connect with the sky and the Heaven energy, and rises from the body. This differentiates from the P'o, which is earthbound and dense.

Part 3

The Fire Phase

Part 3: The Fire Phase

The fire phase is associated with all the organs and meridians associated with the salty flavour. These are: SI, TB, BL organ-meridians, PC meridian and heart organ. This section is the most complex, as we will look deeply into the understanding of the true nature of the fire phase in Chinese Medicine to find its true meaning. This is one of the most important aspects of this book as it is brings into total clarity the concept of the 11 meridians, the nature of the heart-meridian and fundamental shifts out understanding towards the Classical picture of Han dynasty medicine. It is basically the keystone theory to understand this.

3.1 The Small Intestine Organ Meridian System
(fig. 3.1)

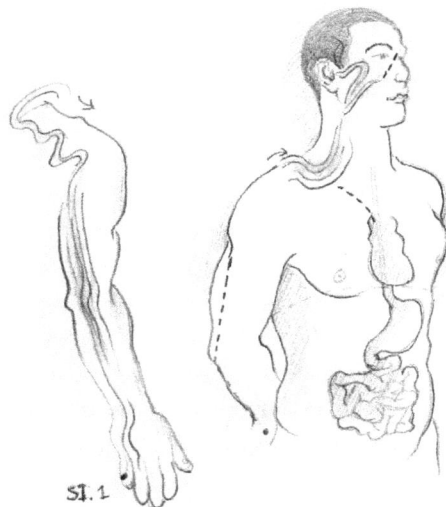

SI.1

The small intestine organ is part of the peristaltic ki, which is described as stomach Qi in most textbooks. The small intestine as an organ is simply a follow-on of the digestive system. It is hotter than the stomach, being deeper and closer to the Ming Men fire that arms it. It separates pure and impure energy and siphons off energy to be used for creating blood and more ying qi. It has an important function in post-partum milk production for women, as the SI is the most refined aspect of the system to create the heart blood, drawing the salty flavour from foods and using this for creating milk. The meridian follows its organ, as all yang meridian organ systems do. The ki of this organ system is salty, as it shares the yang energy of the R-kidney. It is part of the heart complex and as a result is powered by the R-kidney.

The link of the Shen or heavenly life spark to the digestive system indicates that good and clear digestion leads to a clarity of the heart and therefore a clarity of Oneness.

3.2 The Triple Burner Meridian and Right Kidney/Ming Men "Organ"

(fig. 3.2)

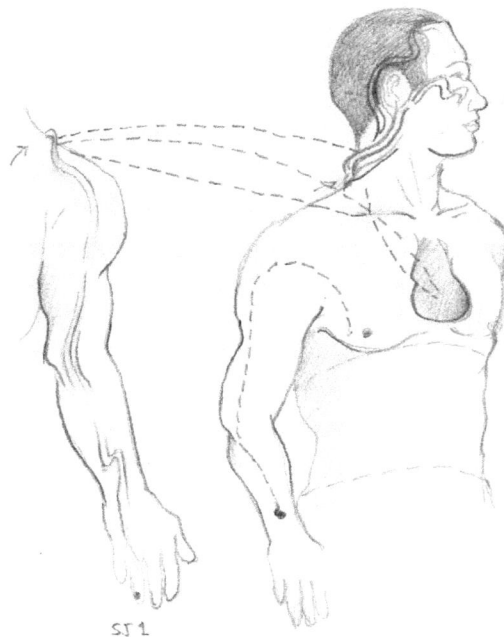

SJ 1

The triple burner is a meridian. It is said to have no organ, or an organ without form. The organ it is associated most closely with, that does have form, is the right kidney. The right kidney creates the Ming Men, which is the "gate of life" opened by the first breath and the gateway for the spirit's spark, therefore, when the first breath arrives. The triple burner energy's root is therefore in the right kidney, and this is its organ, but the Ming Men is the organ without form. The Ming Men is the impetus to give the heart its beat, and this energy is said to be the root of the heart energy and the Origin or source of life given by Heaven.

The TB is said to be an organ with "name but no form", and essentially, as we will see, the TB

296

is a full organ, full of yang energy, from the fullness of the Ming Men or R-kidney energy which too is full. This makes the heart the fu organ of the TB. Alternatively, the whole fire system of the body can be described as the heart complex or fire (see explanation below). This makes the TB-Ming Men an extraordinary system; it is both full and yang and within the yin. This energy we must call the yang within the yin of the body meridian-organ network, and as we will see, this joins with the GB to form the Shao-yang of the body and the GB being the yin within the yang. Hence balance is maintained in the body at all times. The following diagrammatically explains the connection of Ming-Men-TB to the Fire network:

(fig. 3.3)

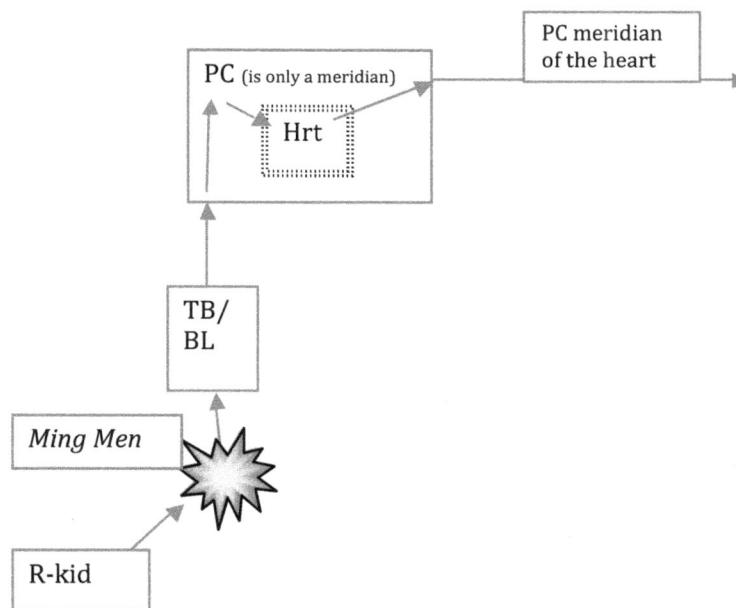

This sequence is only part of the expression of the TB, but this is why it is said to have name but no form. This means it has energy but no substance. This means too that the R-kidney can be thought of as an energetic kidney really. The physicality of the R-kidney could be said to be owned by the left kidney. The concept of "right kidney" and Ming Men merge here.

It is vitally important to understand that the TB is the only yang meridian that feeds the yin of the PC. The other yang meridians are fed by their yin counterparts. This is important, as TB and PC are the Shao Yang/ Shoyo/ 少陽 and Shao Yin respectively; this indicates pivot, between yin and yang. TB is yang within the yin, whereas PC is yin within the yang. (here yinyang is associated with inner-outer, where as T.B and GB are upper-lower and also qualitatively hotter-cooler respectively)

Note also that the Ming Men is pungent as well as salty in its explosive initiation (wood function) as was well as heat and warmth (fire function). This makes the flavour associated with fire energy both salty and pungent.

3.3 The Bladder Meridian and Organ System

The bladder and triple burner are very closely related:

(fig. 3.4)

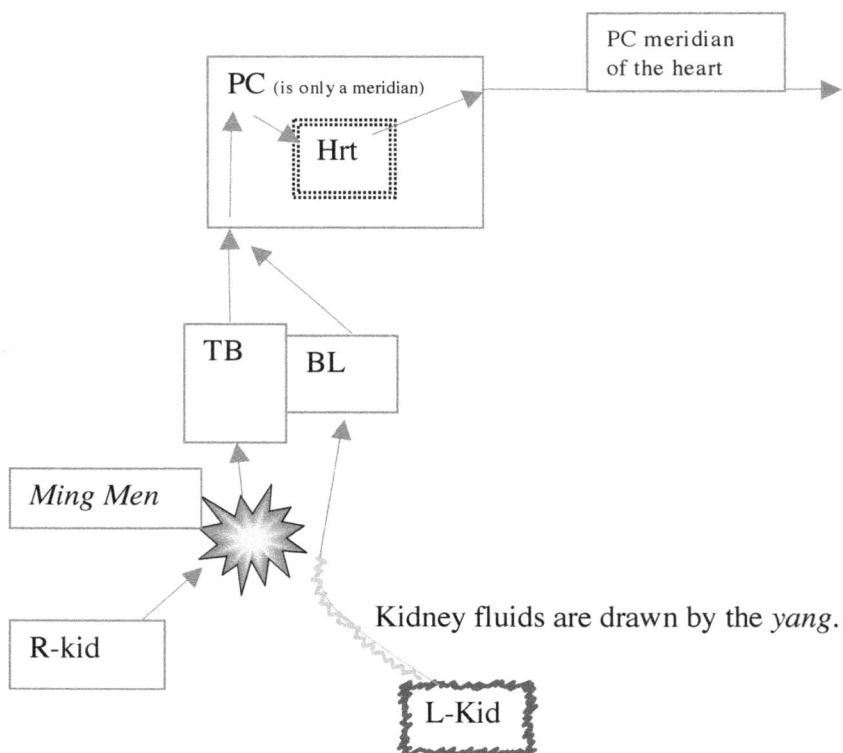

It is very important to note that the fluids of the left kidney are DRAWN by the fire network. This is to say that water controls fire, but only through the action of fire. Water in itself is still and yin and therefore will not move on its own. The bitter flavour is therefore drawn by the pungent saltiness, as expressed above.

(fig. 3.5)

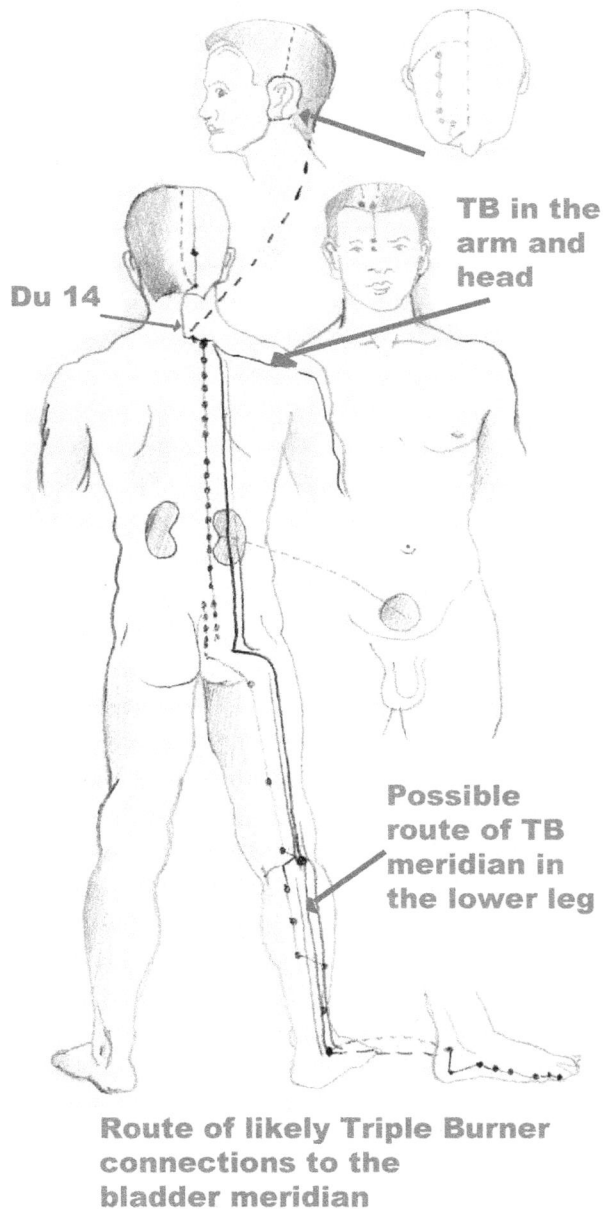

Du 14

TB in the arm and head

Possible route of TB meridian in the lower leg

Route of likely Triple Burner connections to the bladder meridian

The bladder and TB are the same meridian ostensibly. The bladder's route is derived from the L-kidney much like the R-kidney, and connected to the Ming Men and is especially the exterior of the L-kidney, which is especially the R- kidney on the surface, which is exactly the same as the triple burner. The triple burner is actually the outer line of the bladder meridian on the back. Also, the little known TB meridian in the leg is an extension from BL39 down the lower leg between BL and GB, which is simply an extension of the meridian down to the feet. This is seen as a branch of the main meridian in Ling Shu 10. Its importance is overlooked. It is perhaps more important to see the TB as both a leg meridian and a hand meridian, although its superficial vessel is also in the hand and it therefore is definitely affiliated more with being a hand meridian. The place where the bladder meridian joins up with the TB in the back is Du 14:

The bladder organ essentially functions on the heat of yang, as an impetus. The ability to pass urine is especially a yang quality. When urine cannot be passed, this is often an illness having to do with cold in the bladder meridian, or leakage of urine is again a cold issue. In health, the bladder functions by releasing and opening the gates of the body to the exterior, and this is a yang function. The bladder clears unwanted fluids or distributes the fluids to the correct place. It pushes fluids along, and this is the key function. This meridian along with TB and PC make up the fire of the body, and all this energy has a summer, salty flavour to it.

299

The triple burner energy is that which warms the digestive system, like a fire beneath a cauldron; it is the energy that lights the heat for digestion. The pericardium brings the heat back down to be recycled. The process is as follows; notice that the yang meridian pathways internally go upwards here, as on the outside the meridians go down. The yin meridian pathways go downwards on the internal route and upwards to the sky on the exterior.

(fig. 3.6)

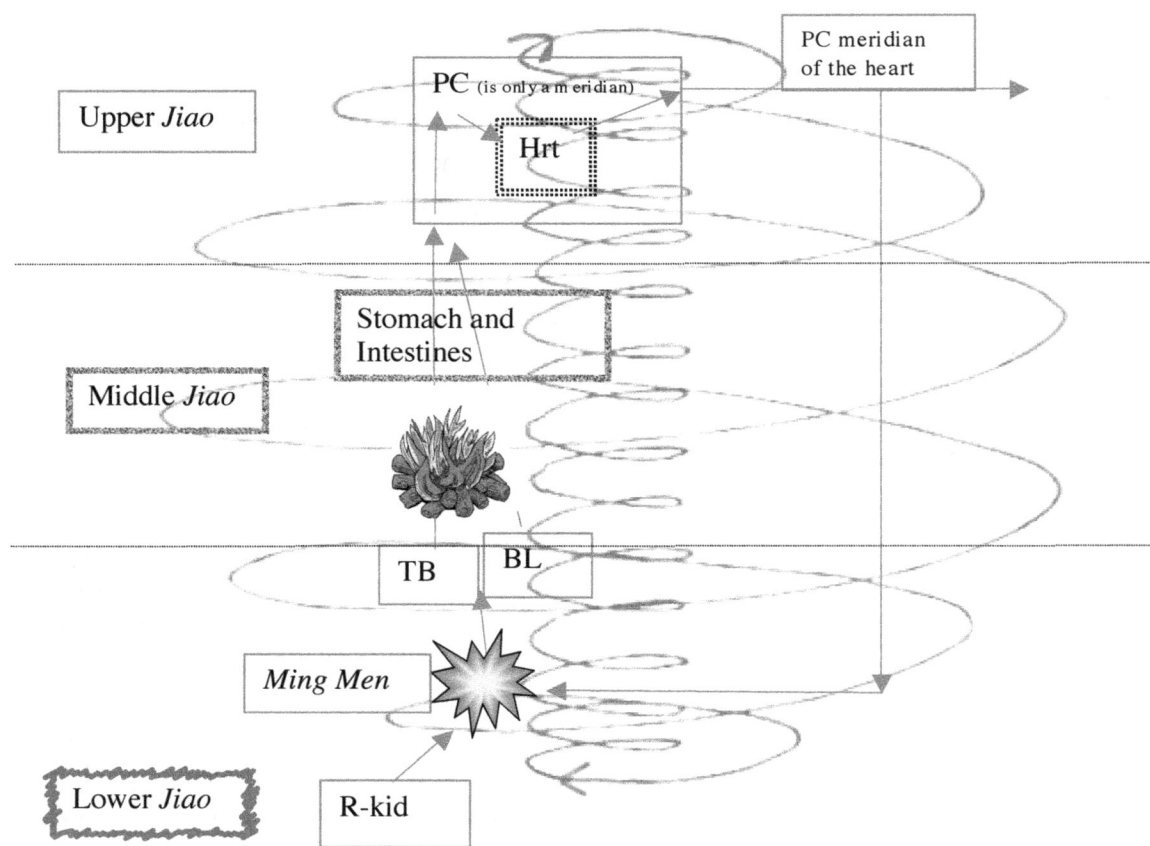

The expression above is simply a convention current of heat rising and then slightly cooling, sinking and condensing as hot air would follow in nature. Above, the fire is the stomach, which is often represented like a cauldron of stew (being the place where drink is taken into the body as well as food). The steam from this cooking process rises and circulates up to the top of the body, being diffused by the lung's out-breath and the heart's beat. The pungent (wood phase) and salty (fire phase) flavours together allow the circulation of the energy to the exterior. The TB is the route of the whole circulation process through the 3 Jiao of the body and thus governs the movement of blood and fluids. This is why it is considered in Su Wen 8 that ...

The triple burner takes the office of dredging water in the watercourses of the whole body; it takes charge of the activity of the vital energy of the body fluid and the regulation and dredging of the fluid. (based-on Wu N. L and Wu A.Q, (1997)).

This relates to the function of all of the above organs and meridians, being both salty and pungent in quality.

Please note that in this section we are talking about the actual ki within the meridians not their labelling. The Bladder meridian is considered as one of the water meridians due to its connection to the L-Kid but it is as we can see, part of the fire network. TB is so closely associated with the Bladder that it too is really a Water meridian, or the two are one really as we can see from the meridian. This labelling will become important when it comes to discussions of meridians and 5-phase points, but for now it's best to realize that the label, due to connection of meridian to its paired Zang, verses its actual qi quality is different and sometimes opposite in energetic nature.

3.4 The Heart Organ and Pericardium Meridian Complex

The heart is the most complex and safely guarded secret of Chinese Medicine. It seems that over many centuries, distortions and criticisms have raged about the heart and its interconnected vessels. I will now attempt to express the full and true understanding of the heart.

The main classical references for this discussion are:
- I Ching, interpretation by Wang Bi/王弼 (AD 224–249), the following chapters: "Explaining the Trigrams–Shuo Gua" and "Hexagram 61 Zhong Fu"
- Classic of Moxibustion, the following chapter: Moxibustion Classic of the Eleven Yin and Yang Vessels, MSI.b.10,11(CM 67–71)
- Ling Shu 1, specifically lines 32 and 33
- Ling Shu 2
- Ling Shu 10, specifically lines 8, 9, 10, 11, 12, and 13
- Ling Shu 41
- Ling Shu 63
- Ling Shu 71, specifically lines 6, 7, and 8
- Nan Jing 66
- Nan Jing 8
- Mai Jing: Book 2, chapter 2, and Book 6, chapter 3

The heart and its associated network of meridians is one of the most complex issues in Chinese Medicine, causing debate throughout the centuries amongst practitioners basing their practises on the methodology of the Su Wen and Ling Shu. This is because several points made in reference to the heart and its meridian make it stand out as the one of the 12 that is very different ... why is this?

If we first look at the I Ching, the base classic of all works in the Taoist arts, we look at the trigram for fire, Li. Fire is represented with a yang line, a yin line, then a yang on the top:

(fig. 3.7)

```
  _____

  ___  ___

  _____
```

The associations made for this trigram are as follows, from the original Classic of the I Ching:

> *Li is fire, is the sun, is lightening, is the Middle daughter, is mail and helmet, and is the halberd and the sword. In respect to men, it is those with big bellies. It is the trigram of dryness. It is the turtle, is the crab, is the snail, is the clam and is the tortoise. In respect to trees, it is the hollow ones with tops withered.(Lynn R.J, 1994)*

This is as opposed to water, Kan:

(fig. 3.8)

```
  ___  ___

  _____

  ___  ___
```

> *Kan, is water, is the drains and ditches, is that which lies low, is the now-straightening and now-bending, and is the bow and the wheel. In respect to men, it is the increasingly anxious, the sick at heart, and the ones with earaches. It is the trigram of blood, of the colour red. In respect to horses, it is those with beautiful backs, those that put their whole hearts into it, those that keep their heads low, those with thin hooves, and those that shamble along. In respect to carriages, it is those that often have calamities. It is penetration, is the moon, and is the stealthy thief. In respect to trees, it is those that are strong with dense centres.* (Lynn R.J, 1994)

These are very important. They tell us the nature of fire versus water. The key aspects are that the heart seems to be hollow, and water seems to be full and dense. A vitally important phrase from above is that fire in respect to men are those with big bellies. This indicates the possibility that it could be relating the essential emptiness of the belly of men (especially after sex), which is filled with the ki of the "immortal foetus" in Taoist practises and observance often of sexual abstinence. the opposite is true of women, holding the red/yang-blood foetus of a child within her womb. Usually the Li trigram represents 9 which is yin and the Kan 1 which is yang, but if we see the pregnant woman's belly, verses the space where the ovaries would have been in men, then in this picture the energetics are reversed.

This idea associates with the Kan, the water trigram representing blood and redness. It is interesting that the trigram of Li is in itself an open-hollow vessel, much like a blood vessel. The associations of the Kan to men who are anxious goes with the idea of men with yin deficiency, due to loss of Jing fluids. This is a common pattern of fullness in the heart, because of an initial fullness in the constitutional kidney energy and therefore a wish to use it, sexually or physically overworking, and it is associated with sickness.

The references and connections to the heart being hollow and the water being powerful and dense are endless here. A picture starts to emerge about the true nature of the heart.

If we then look at the I Ching, Li is doubled to form the Hexagram 61, called Zhong Fu or "Inner Truth, Innermost Sincerity":

(fig. 3.9)

```
━━━  ━━━
━━━  ━━━
━━   ━━
━━   ━━
━━━  ━━━
━━━  ━━━
```

As we can see, the representation of the heart is of an emptiness at the central point or a place of inner yin and outer yang. Yin associates with the state before yin and yang, which is Wu or emptiness, nothingness. This is very important for the understanding of the heart. If we look at the heart organ and its area of control, the blood vessels, we can see that even though it is a yin organ, it is hollow in all aspects. The heart is hollow, as are the blood vessels. In essence, the heart is a vessel of the spirit. It holds the Shen/spirit but holds it like a cup holds water; the holding is essentially a resting of the spirit in the physicality of the blood.

The heart therefore simply is there, or is a branch of something; it isn't the root. It doesn't do anything of its own volition. It is powered by all the other aspects of the body and simply radiates the spirit out. The spleen makes the blood, the right kidney gives the beat of the heart, the left kidney cools it, and the liver supplies the blood pressure or the muscular actions of the heartbeat. The lungs add the ki to the blood. Hence if the heart is to rule the body, it is basically as a flower of the spirit. A flower doesn't do anything; it just opens and spreads its existence. In order to do this, the physical structures of the heart are open for the spirit to reside and dictate the body's direction. The heart holds the yang of the Ming Men, and the Ming Men siphons off/draws the ki of the left kidney Jing. Hence the heart must merely follow the dictates of the root in order to express the character of the person.

A key point is that the heart represents the sovereign. However, as all societies are the same, is it actually the sovereign that rules? The sovereign is actually controlled by the minister fire, the triple burner—the prime minister of the body. We can see this in our own societies, and it was a common thread in the Chinese Imperial court, especially if we look at the lives of Confucius and also that of King Wen, the compiler of the I Ching! Note that in the trigram Kan, water, the water is associated with the drainage ditches, of which the triple burner is the master in the body. This is the root of the heart.

This being the case, the emptiness of the inner heart is vital to understand. This leads us to the explanations of the heart and its meridian in the ancient texts. Firstly, let us look for the heart meridian in the Mawuang Dui Literature, found in 1973, tomb 3, in China, often known as the Classic of Moxibustion. In this text, the heart meridian is not found on the body. In fact, what they call in this text the "Shao Yin" meridian is a meridian that runs close to the classical location of the pericardium meridian.

303

Hence some scholars say that it is actually the pericardium that is missing. However, if we think that the people who wrote these texts had some insight, it doesn't make sense to regard this as just a simple version of the more modern Su Wen and Ling Shu. Perhaps they were pointing at something. I have tried to look into this. In Ling Shu 2, line 2/3, the answer is revealed. Here there is an explanation that the arm Jue Yin channel (PC meridian) can actually be represented in the text as the arm Shao Yin channel (Heart meridian)! This is very important. It explains that the ki of the arm Jue Yin is the heart ki, but that also in the Mawang Dui this explains why the heart meridian itself is not presented. Why is it not presented? Looking at the actual meridian of the heart, all the points described in modern texts are actually pulse points of an artery, not points associated with energetics. Why is this? In the Ling Shu, perhaps the most important chapter in this investigation is that of chapter 71, lines 7, 8, and 9.

In these chapters, we have a look at the nature of the heart meridian and how it is used in dis-ease states. Here it clearly explains that the heart meridian has no five-phase acupuncture points. Hence if this meridian has no five-phase points, can we assume that it too has no flavour associated with it? Why is this? If we then go to Nan Jing 66, lines 2 and 6, we can see that the source of the heart meridian actually emanates from PC-7. And again in Ling Shu 71, lines 7 and 8, it is explained that the pericardium actually takes the ki of the heart and hence the PC owns the heart ki.

Nan Jing 8 corroborates the fact that the root of the heart is actually the Ming Men and that checking the wrist pulse is not an indicator sufficient for telling signs of death. The lower abdominal pulse (Ming Men-ki) indicates the root of the heart—i.e., the root of the TB ki. This is vital. If this is the case, it can explain why the Mawang Dui scriptures point out that in fact 11 functional meridians exist. This also explains that the TB must be the source of the heart's ki (heart's beat)—i.e., the TB is the only yang meridian to feed the yin. It feeds the PC, and being the meridian of the Ming Men /right kidney—i.e., the source of the heart—is at the TB point on the PC. In Nan Jing 66, line 6, we have the explanation that the hand meridian of the Shao Yin has its source point at Hrt-7, but this is not the source point of the heart ki itself, as this is associated with the PC-meridian, hence it just the meridian of the heart. However, line 13 of the same chapter indicates that it is the "stream" points that constitute the TB energy in all of the 12 meridians. Hence, how can there both be a functional source point, yet the heart meridian itself is non-active?

In all of the 12 meridians, apart from the Shao Yin meridian, there are actually sets of five-phase points. However, the heart has no five-phase points, questioning how the TB energy could actually emanate from HRT-7 at all, when it cannot be described as a "stream" point or an Earth point. Hence the connection even of the TB to HRT-7 is brought into question. We know there is a point at Hrt-7 (this is also stated in chapter 71 of the Ling Shu), but due to the fact it is not a five-phase point, isn't it stripped of its energetic property? If we consider that although the heart meridian is not a "meridian" in the same way as other meridians, it is still a meridian.

The TB connects to ALL the meridians, and so this, like all the other 12, is associated. However, its association is not the same as the other 11 meridians, and this is the key point. I do not believe that the Hrt meridian has the same energetic power as the other 11, but is a badge or ghost following/ tracing the general identity of the Jue Yin meridians, which is the true meridian of the heart. As Hrt-7 is the only point on the channel, it is used for treatment of the Shao-yin channel itself, but the energetic property is not present, so it has no impact on the rest of the system, or very little.

One must see the assignment of "source-point" to the HRT-7 point as an assignment of labelling and connection, rather than an assignment of actual energetic power. Chapter 66 of the Nan Jing is actually about the interconnections of the TB system rather than an exploration about the heart network itself, so one must infer the fact that this HRT-7 is somehow assigned to the TB. However, one must also include the information from the rest of the Nan Jing, the Su Wen, and Ling Shu to get the full picture, rather than taking one part out of context. This then indicates that there is some cut off between the heart and the heart meridian. Why?

The pericardium pulls all the heart ki to it. This is so that wherever the heart ki is at the exterior of the body, it is protected by the pericardium, which surrounds the heart and protects exterior energy (particularly cold) from actually getting to the heart organ itself. If this was not the case, then via the heart meridian, exterior/ pathogenic energy could get in directly to the heart organ and thereby end life. By exterior/ pathogenic energy, we mean cold here, as cold kills fire. The pericardium therefore holds all the heart ki and encloses the heart so that nothing gets to the heart other than via itself. Where does this leave the heart meridian?

Well, this is the difficulty. The heart meridian is actually the most intimately connected to the heart, but it has none of the heart's ki. In Ling Shu 10, line 10, we learn that the heart meridian itself starts from the very centre of the heart (the centre of emptiness). In fact, throughout the system, it follows the heart's ki in the PC meridian. Hence the main channel pathways of heart and pericardium basically go to similar places. The heart is especially following the ki of the heart energy, but it itself has nothing; it is a ghost meridian. It is empty.

This is the key point: in fact, the heart meridian is empty. In Ling Shu 71, line 9, it explains that the heart meridian can be tonified and shunted/dispersed, dependent on the changes of the key organ meridians of the upper body—the pericardium meridian and also the lung meridians, i.e., the controllers of the heart beat (PC holds the heart ki) and the breathing. Hence if pathogenic ki affects these areas, the heart meridian would follow their problems. It acts similarly to an extra-vessel or Luo vessel, to absorb pathogenic ki and hold it. It has no flavour to act on the system; it just sits there and acts as an overflow to either the lung or PC, but the PC is closest.

Hence the meridian can be dis-eased, but the heart organ itself will not be. The heart meridian acts like a Luo vessel that absorbs, like a reservoir, the ki that is in excess and sends it out when surroundings are deficient. It acts much like a pressure valve in the system, stiffening off excessive heat and stagnation to keep the heart safe and not overheated, or it takes on cold but can't get near the heart-ki itself.

What this leaves us with is a meridian which has no points, except for a badge representing that the meridian belongs to the TB system, i.e., Shen Men (Hrt-7), and therefore it is out of the five-phase points loop, it has no flavour/energy. The flavour of the spleen meridian is sweet, the liver meridian is sour, the kidney is bitter, and the lung is pungent. The TB is salty, and this represents the TB and Bl meridians, which are the kidney yang. The flavours all exist in the points. The water points are salty, the fire is bitter, the Earth is sweet, the wood is sour, and the metal is pungent. So why is it that although the heart meridian has no ki, fire, representing the yin organ of the heart, has a bitter flavour suggestion?

This has to do with the fact, again, that the heart itself is empty. Hence the heart cannot have a substance it makes. But the flavours are derived from the meridians … so the question again, why isn't the heart meridian bitter?

The answer lies in the root—the kidneys. The kidneys are the roots of the heart. The left kidney is cooling. Its meridian and organ are the same. They are both bitter, and they cool. The kidney meridian spirally wraps the heart ki; this is explained in Ling Shu 10. There is debate about what this means, but if we look at it clearly, the "heart" here is likely to mean the pericardium-meridian, or the heart's ki. The pericardium is in fact only a meridian; its solid organ is the heart, and most importantly, it holds the heart ki. There is no reason to connect the heart to the Shao Yin of the kidney directly, because this would go against the whole principle of the heart-enclosing network. There is no meridian that gets close to the heart other than the pericardium, so the cold of the kidney is the meridian most likely not to need to be in direct contact with the heart.

I feel that it is the pericardium that the kidney organ spirally wraps to cool the heart ki, and so one can see the anchoring effect the kidney has on the heart ki. The right kidney is the Ming Men and is salty, which means that the water points and source points have saltiness in them. The water points associate more with the bladder, the source points more with the TB. These represent the heating of the fire. Hence it is in the lower abdomen that we have the sources of both fire and water, and so two flavours reside here. Although we associate the heart with fire, in fact it is the TB that is the source of fire, and its opposite is the left kidney yin. Hence the TB and PC are both fire in organ and meridian, balancing out the system. The heart therefore is allowed to be at peace and empty and to be in connection with the Tao of Heaven.

Another interesting point is that the heart is filled with a great deal of energy. In fact, one could say that although the heart itself is empty and a vessel, it is the vessel that has the hottest (fullest) substance in the body—radiating blood. The opposite of this is the heart's actual meridian, which is totally empty again, but helps the overflow of the heart when it becomes too full, via the PC meridian. So we have an organ which is very full (although empty in nature) and a meridian (heart meridian) that is very empty. If we see all aspects, it is clear that the heart and its meridian are actually one and the same emptiness, but the heart is being filled via the TB to the PC and then to the heart, whereas the meridian of the heart is truly empty.

In Ling Shu 10, the triple burner shows us that it has direct connection to the pericardium meridian, which it spirally wraps also. However, this is the point at which the heart ki is actually transmitted to the heart to give the heartbeat. Hence the heart-enclosing network is like this:

(fig. 3.10)
TB:

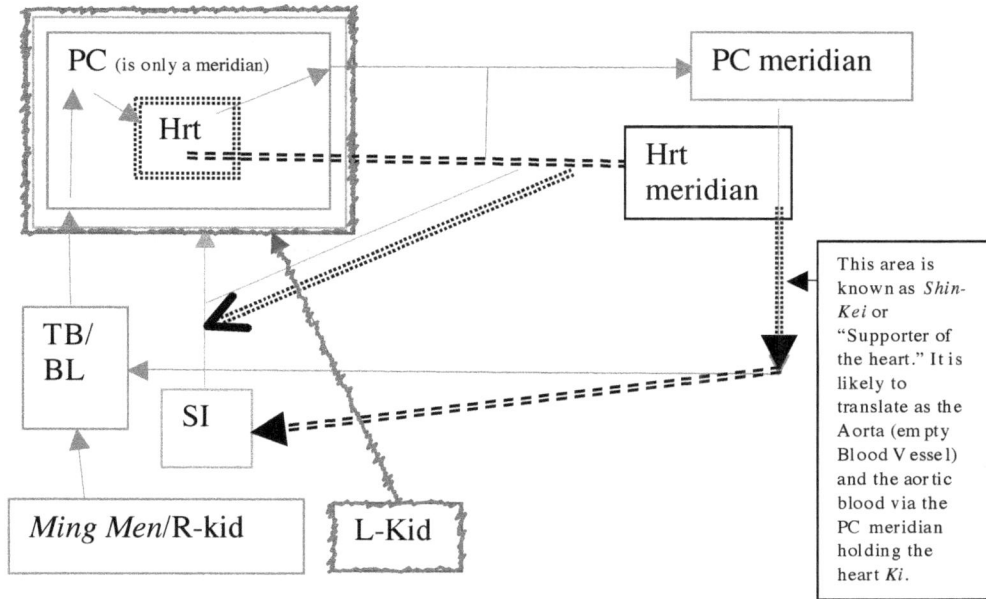

PC (is only a meridian)

Hrt

PC meridian

Hrt meridian

This area is known as *Shin-Kei* or "Supporter of the heart." It is likely to translate as the Aorta (empty Blood Vessel) and the aortic blood via the PC meridian holding the heart *Ki*.

TB/
BL

SI

Ming Men/R-kid

L-Kid

(fig. 3.11/ fig 3.12)

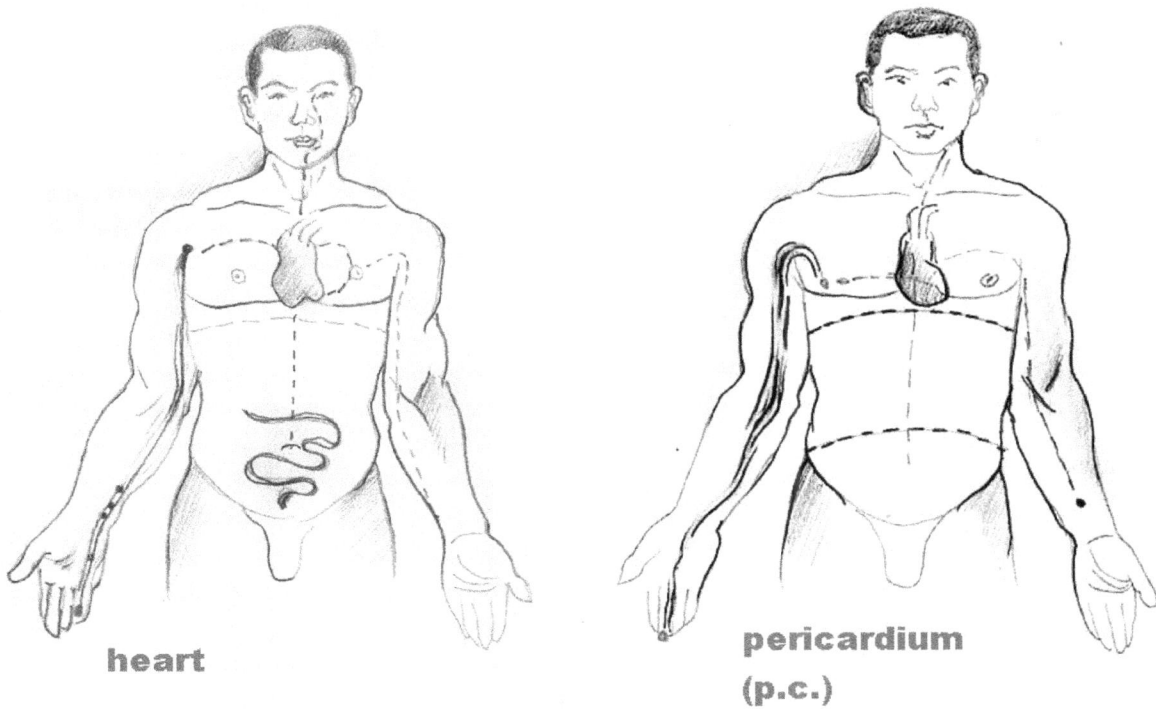

heart

pericardium
(p.c.)

As we can see, the heart organ is wrapped at least three times by the other meridians involved. There may be more, but the most important point is which order or which is the closest to the heart and which is furthest away. This is understood by the relative importance of the organs to the heart system. It is likely that the kidney and SI are around the same level of wrapping the heart, as the SI is the connected meridian, but the kidney heart wrapping is the only Zang-Zang wrapping in the whole body, so it is very special and has a similar level of importance as the paired yinyang meridians of heart and SI. Again this is the connotation of the Spiritual-Pivot meaning Ling Shu as explained previously.

The difficulty in understanding the importance of different types of connections of meridian to organ comes in the characters of the text. There are mainly four different types of connections (Chinese/Japanese): He/ Go 合 Zhu/ Chu 注, Luo/ Raku 絡, Shu/ Zoku 属 or 屬

He/ Go 合 type is like two sides of a coin or a lid and a pot; they go together as two different parts. Zhu/ Chu 注 type is the connection of one thing flowing into another. Luo/ Raku 絡 means to spirally wrap around something like string around a bobbin. Shu/ Zoku 属 or 屬 is perhaps the deepest connection. It means to be one with or to be joined intimately and seamlessly together, as if one, a "mating" connection is represented here. He/ Go to Shu/ Zoku represent a spectrum of interconnections to show the level of importance the body places on specific connections. Or another way of saying it is the quality of ki involved in the connection. The connection points provide very important information about the nature of the heart.

The person follows Tao by following the ki of their heart. Hence the heart must be open and free and clear and empty for this to be the case. It is true of the heart organ and meridian. Although it looks like the heart is active, it is being ACTIVATED, acting like a capacitor in a circuit, not the battery itself; that is the key difference. One doesn't look at the heart to see if signs of life have gone; you feel the moving ki between the kidneys in the lower heater to see if signs of life have disappeared. Then the Ming Men is no longer active and there is death. The Ming Men is probably the place at the centre of the egg and sperm joining (from the fragmented perspective), and it represents the source of life—the gate of life in fact.

So what of the SI and its relation to the heart?

Of all the yinyang pair connection, the heart and small intestine seems the weakest. This is due to the lack of ki found in the heart meridian, hence the lack of influence over the small intestine. The small intestine therefore finds more influence from the stomach and spleen as its root. However it has a connection with the heart itself; hence the connection of SI to the heart is actually stronger than the other way around, which is why constipation can affect sleep and so on. The spleen, however, is the ruler of the digestion, including the SI. Overall, the spleen governs the ST and SI as one unit.

The key points, therefore, are about the idea of true fire being actually the yang line within the yin of water:

(fig. 3.13/ fig 3.14)

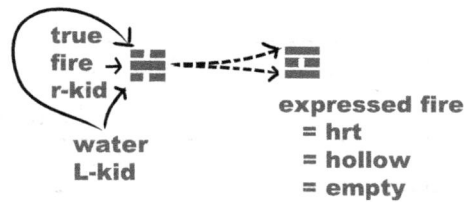

On the other hand, fire of the heart is actually a representation of emptiness. This gives us the power to understand the mechanisms of the body and use the flavours of the points correctly in the system. This also interlaces perfectly with the use of the five-phase points in Nan Jing and Ling Shu in relation to various conditions and also with the mechanisms of flavours in the Shang Han Lun.

A representation of the heart-pericardium main channel in relation to pathogenic ki is given here. It is a basic explanation of the Ling Shu 71, lines 8 and 9, which hold the key points to uncover the mystery:

(fig. 3.15)

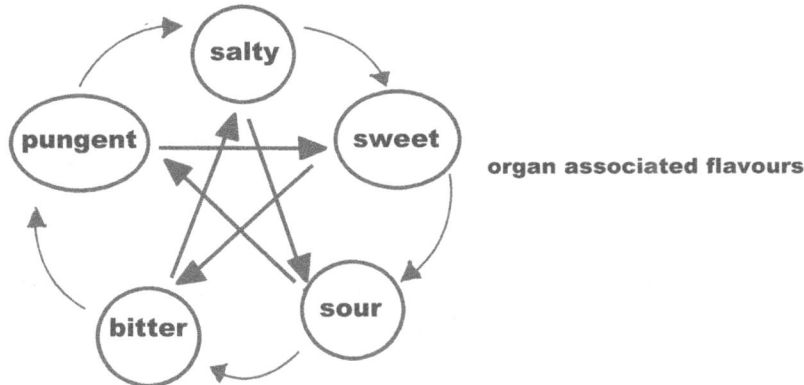

organ associated flavours

There are numerous references in the classical literature of the early Han upon which to base these concepts. When looking at the classic books, the importance of understanding the base perspective of the chapter and the area of the text one was looking at, in order to get clear perspective of what one was reading, was revealed to me. It is easy to read things into the literature, because one has a different perspective interest and wants therefore to embrace one part and discard the other. If we see the classic books as a completed whole which is the work of objectivity and therefore likely to be written by a group of clear-minded individuals, then we can investigate reality without going off course.

In Nan Jing 25, there is an explanation of the organs and meridians being of different numbers. This is the reason:

309

(table 3.1)

No.	Zang-Fu	Meridian
1	Sp	Foot Tai Yin of the spleen
2	Lu	Hand Tai Yin of the lung
3	L-Kid	Foot Shao Yin of the kidney
4	Liv	Foot Jue Yin of the liver
5	Hrt	Hand Jue Yin of the HEART (pericardium is meridian only)
6	St	Foot Yang Ming of the stomach
7	LI	Hand Yang Ming of the large intestine
8	GB	Leg Shao Yin of the gallbladder
9	BL	Foot Tai Yang of the bladder
10	SI	Hand Tai Yang of the small intestine
11		("Hand Shao Yin")
12	R-Kid, Ming Men	Hand Shao Yang of the triple burner

The reason for the problem is that the heart owns two meridians, and the PC's Zang is the heart. Also, the TB doesn't have its own "registered" separate Zang—i.e., it is the right kidney of the Ming Men that is at its root, so it is said to have no organ. There are really only 10 organs and 12 meridians, but if we count the Ming Men/R-kid, then it is 11 organs and minus the heart meridian makes for 11 organs and 11 meridians, a balanced match.

To continue, Ling Shu 63, line 1, explains that the tastes entering the body "settle" in the various tissues. This is not the same as desirous of the effect of ...

- Bitterness goes to the Bone
- Saltiness goes to the Blood

Here we have a perfect explanation of what is occurring. The bitter taste enters the body and is attracted to go to the place it is deficient, which is the place of its opposites residency. Hence bitterness going to the heart is an explanation of this. However, the actual residency of the bitter taste is in the bone, which is of the kidney yin, and the residency of the salty taste is in the blood, which is what fills the heart. Hence we can understand that when a taste is consumed from food, part of the taste's resident meridian in the body will be activated more, as the taste has powered up the flavour of that organ or meridian.

The taste of the organ and that of its connecting meridian are opposite in only two cases—for the lung and the liver. The other areas are the spiritual pivot, which are the heart and kidney and the spleen (centre). The lung organ has a sour taste; the meridian is pungent. The liver organ has a pungent taste, but its meridian is sour. The left kidney is bitter and it governs the lower body; the right kidney is salty and it governs the upper body. The saltiness mixes with the bitterness and vice versa. The spleen is constant and central and sweet.

Ling Shu 10, line 10, explains that the layout of the kidney meridian actually wraps the PC which surrounds the heart, so not the heart itself. This further pushes the kidney and bitterness away from the heart, so to say that the heart meridian is connected to the kidney Shao Yin is very unlikely.

In the Mai Jing, Book 6, chapter 3, there is an explanation of the heart and pericardium relationship and the separation of symptoms of the hand Shao Yin meridian. Again they refer to Ling Shu 71, line 8, to explain the lack of points on this meridian.

What all this means is that we have to consider the heart organ and the pericardium meridian to be one unit, and the heart meridian to be an "empty" meridian or an overflow meridian, which has no connection to the heart's actual energetics. This is the most important point of this debate. Note therefore that the pericardium organ does not exist; the pericardium is purely a meridian that wraps the heart, and the organ of the pericardium is the heart.

If we look at the triple burner and kidney, we will see this pattern switch, with the TB having a meridian but no organ, so attaching to the right kidney. There is always balance in the body. It is important to note here that there are 11 organs (separating the right and left kidneys, otherwise it's 10) and 11 true meridians, so 11 is a very important number in looking at the body. (This was discussed in the general construction section also.) In this book, we will no longer refer to the heart meridian, and we will consider that it is the "hand Shao Yin meridian" and has nothing to do with the heart ki. In essence, it is not part of the energetic balance, so it is not necessary to discuss it further (similar to not discussing further the extra-ordinary vessels). I will refer to the hand Jue Yin as the hand Jue Yin of the heart, incorporating PC and Hrt into one completed unit. I feel this is the original methodology, which combines perfectly with the stems and branches construction from Ling Shu 41 presented earlier. This in fact is a absolute key that in counting the meridians involved in Ling Shu 41, and in the points counting in Su Wen 58 and Ling Shu 2; 25 points are associated with the Zang organs and 36 are associated with the fu organs. Divide by 5 for the Zang 5 x 5 = 25, divide by 6 for the fu 6 x 6 = 36. Or we double 5 making 10 double 6 making 12 then we are back to the stems and branches of Ling Shu 41. This is therefore measured through the texts in numeral counting as well as other explanations for the overall balance of 11 as the key.

If we consider that the Ming-Men-TB-BL-SI are all the expression of the yang of the R-kidney, then we must say that the flavours of the organ meridians associated are all primarily salty with pungency, a mix of fire and wood energy. However, the pericardium meridian and heart form the yang within the yin, and the energy of this has less pungency and more salt; it is deeper as is the effect of saltiness. Hence we associate only the pericardium, meridian of the heart, and the heart itself, both of which are rooted in the Ming Men energy and are more salty, whereas with the surfaces of the body, pungency also arises with saltiness, mixing with the air drawn in from the breath and the initiating spark of the Ming Men all pungent in quality.

The heart meridian is empty and this emptiness associates with yin. The deepest yin in the body or the coldest is the water phase of the energy of the L-Kid (see section of the water phase). There is a relation therefore of the heart hand Shao Yin and the kidney foot Shao Yin, in that one carries bitterness (Kidney) the other is empty (heart). This connection will be looked at again when we come to look at points , labelling of meridians and associated connections.

3.5 Other Aspects of the Fire Network

The right kidney powers the whole of the lower abdomen. It warms the digestive function as the heat rises from below, so the intestines and the bladder and the fu in general are warmed by the lower burner. The lower burner also powers the process of ovulation in women and the cyclical process of the female period in coordination with the exterior change of the Earth and moon. The liver organ and the blood of the liver is therefore an extension of the kidney yang. The blood is drawn out of the body, and this is why it is not an entirely internal process but also has to do with the exterior-interior connection point. The right kidney powers the heating of digestion of the stomach and also rises into the upper body to connect to the lungs. As we can see and as expressed in chapter 66 of the Nan Jing, the TB permeates all aspects of the body functions, which is the Ming Men/R-kidney, and the TB function is literally the "life " of the body through the three burning spaces.

3.6 The Fire Network and the Lung

The fire network or the power of the kidney yang is the driving force for the heart organ, the driving energy of the PC to recycle back to the Ming Men (the heart organ being almost the fu-like (empty) organ of the Ming Men in the chest), so pumping the yang chi of the blood. In addition, the kidney yang is the power of the in-breath and so the driving force of the lungs' movement. The drawing inward of air from the exterior to the lower abdomen is the drawing in of the mother energy (pungency-spring/wood organ-energy) to the Ming Men (salty-summer/kidney yang).

(fig.3.16)

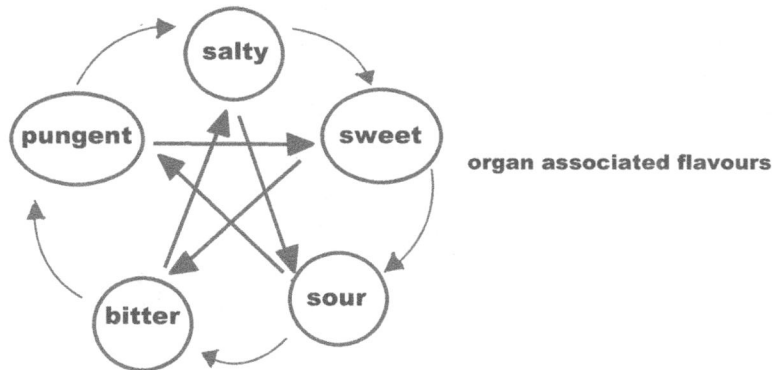

organ associated flavours

This is the true picture of the flavours in association with the five-phase chart. Where the sour and pungent, bitter and salty flavours are swapped over from the above, this CANNOT be used, as the flavours will then not be in line with the seasonal changes and with the change of human energy of five-phases sequences; the swapping is in association with meridian flavours and therefore not associated with the 5-Zang organs which are the inner origin of he 5-phase ideology (please see later sections).

Hence the power of the air can mean that vital pre-Heaven energy is not lost if air is breathed deeply and well; this is the nature of Qi gong practise. However, the power of the pungency (air-ki) drawn into the body is based on the mother of the spring energy (bitter-kidney-yin/winter), which acts as an anchor from the impetus of yang to draw the breath in. Hence the bitter-pungent-salty connection is very important for the organ energy to be provided for. Notice that the lung organ isn't part of the picture but is simply the mechanism of the kidney drawing air in. It is a "dead" organ in the sense that its activation is as a result of the kidneys, and its collapse (out-breath) is as a result of letting go of something or the cooling of something. It is a non-active energy, much like the L-kidney organ or liver meridian; the coolant is moved, but doesn't act itself. Hence the Wei chi is intricately connected with the power of the kidney.

The actual acquisition of the air to be used in expression of energy from the surface is dominated by the kidneys. The fluid of the lung is derived from the kidneys. The lung is at the kidney's command, and as a result, the Wei chi and the opening and closing of skin pores and cycles of breathing are constantly unified between the lung and kidney. Just as the heart beats at the deepest level, so the body breathes at a more superficial level above the heart, hence while the deep yang is dealt with by the Ming Men/TB/PC/ heart connection. The exterior of the BL/SI and Wei chi is in total union with the lung meridian (being the functional aspect of the lung and the expression of wood energy which creates "fire"/Wei chi on the surfaces—i.e., pungency plus saltiness), pumping the heat and warmth on the exterior. One route deals more with ying qi blood, the other with Wei-qi or qi. This shows why the kidney yin and yang energy are the fire and water sources of life.

The expanded lung full of pungency (air-ki) is therefore considered akin to the liver organ, and the expression of the lung organ and liver organ will occur together on the out-breath. The actual letting go of the lung in the out-breath is considered part of the liver meridian anchoring and heaviness interiorly plus/ forming lung organ collapse inwards, and the lung meridian is in coordination with the BL/TB/SI meridians exteriorly, to express the yang ki at the surface. Note that these meridians are associated with the Tai Yang stage of the Shang Han Lun dis-ease (see later section).

ONLY the potential to draw the breath is derived from the kidney yin and liver meridian. The actual impetus/activity of drawing in the breath is considered part of the kidney yang. Hence, unlike the other organs, the lung is two things, full and empty. The other yin organs are constantly full (the heart constantly has heat in it although it too has a full and empty cycle with the energy moving out towards midday and then back inwards at night). We will look at the changing nature of the lung organ again in the metal section for a deeper understanding of its expression and the reason for saying "collapse" associated with the lung.

3.7 The Heart and Shen – Chinese/ Shin – Japanese

神

The heart organ houses the "Shen" which, in this case, is an aspect of the ethereal soul. In macrocosmic terms, Shen is the radiance or energy of the body, while macrocosmically Jing is the constitutional-physical nature and solidity. Microcosmically, as in this situation, Jing is stored in the L-kidney and is the fluid coolant of the system, the platform with which the Shen can spring from, whereas Shen is the aspect of spirit, which relates initially at birth to the ignition of the spirit with the first breath and the Ming Men. The constitution balance of the spirit energy is forged at this time, but exuberant Shen spirit is the 9-fire energy of the nine spirit energies, and this represents the heart itself, where the spirit manifests. This energy is very ethereal and expressive and finds itself rising and being associated with the upper body and the heart, although its origin is in the Ming Men/right kidney. It is flower of the body's expression energetically. Again this expression exists in everyone but the 9 energy if emphasised by the constitutional time of birth will mean this expression is more profound and obvious or less so in the overall expression.

3.8 Zhi – Chinese/ Shi – Japanese, and the Right Kidney

志

Zhi is more associated with the 1 energy of the 9-energies and the left kidney than the right kidney, BUT the yang line within the Kan trigram indicates that the right kidney is also involved in Zhi. Zhi is considered as the base of will power from which the Shen can act. Rather than the base-nature which is still and cold like ice water, and associated with the L-kidney, the action itself is associated with the R-kidney. Hence the right and left kidneys, dominating the lower Jiao, combine to form the energy of Zhi within the system. Again the balance of spirit energies determines the relative power of Zhi. However, as Shen belongs and is representative of the fire within the body, so Zhi could be said to belong to the left kidney, but is also the root of the right.

Part 4

The Earth Phase

4.1 The Stomach Organ Meridian System

(fig. 4.1)

ST 1

St 45

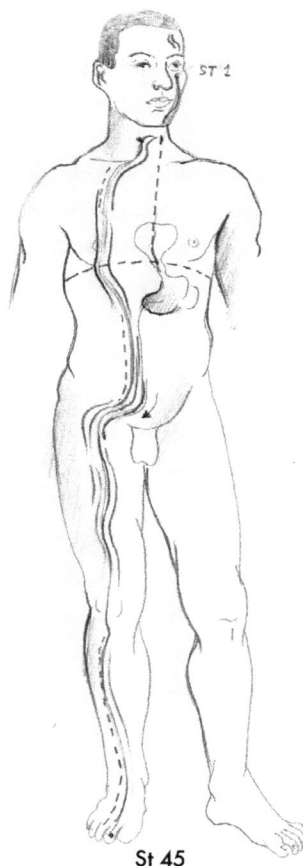

The stomach represents the heat of the middle burning space of the body. It is the place where food is broken down and fermented into energy that can be used later to create blood. The stomach is a hot organ, so it is dry and needs fluids to supplement it. Hence the absorption of fluids is the main desire of the stomach. The stomach's organ and meridian are of the same function and are of the sweet flavour. The sweet flavour is warming and at the same time fluid accumulating; if one eats a lot of sweet food, then one gets thirsty, because the sweetness attracts fluid to it and nourishes the body fluids. Notice that this fluid is warm fluid, not hot and not cold. Cold fluids belong to the kidneys, and hot fluids would be associated with the right kidney/ bladder TB system.

As the heat from the Ming Men rises up the body, it cools a little. The stomach is not as heated as the Ming Men but it is warm to digest, churning rather than burning the food to get out its vital energy. The stomach is the first place for food to reach, so it deals with coarse substances, hence it needs powerful heat. In fact, its heat warms the spleen a little, so while the stomach provides heat for the spleen, the spleen provides fluid for the stomach. It is basically like the Kid-yin and Kid-yang of the lower burner, just further up the system. The Ming Men draws yin fluids with its heat, and this is why the spleen and stomach are very united. The fluid aspect of this "steamed fluid" IS the spleen, and the heat aspect IS the stomach, hence they are one and treated as such. The spleen is not cold but requires the Ming Men and then the stomach to warm its energy to function effectively. It cannot, however, be hot; otherwise it loses too much fluid and dries out, just as the stomach cannot be too wet or it cannot burn the food easily if too cooled. As the spleen's fluid sinks the stomach energy and the stomach heat raises the fluid of the spleen, the mix creates a warm rising energy that is the digestive process and distribution of warm fluid throughout the body. The spleen and stomach are in total balance and harmony, mirroring the role of left and right kidney but within the post-Heaven energy.

The temperature expresses the flavour and the energy attached to the physical water itself. In numerous herbal treatments, the water used is one of the actual ingredients, and warm, hot, or cold describes the place that the energy of the decoction will go. Hence in the stomach and spleen they are in charge of the sweet flavour and warm fluids. In order to differentiate, yin fluids are cold and belong to the kidney, while the fluids of the stomach and spleen (often called "body fluids" in TCM) are warm and post-Heaven related and are called body fluids, not yin fluids; this is an important differentiation. The stomach cools itself by the exterior energy being taken in; it also draws fluids from the spleen, and the spleen from the L-kidney yin.

The stomach is open to the exterior through digestion and is therefore vulnerable to substances that can affect the Fu organs and then the Zang organs directly from the exterior. However, it is a Fu organ, so it has the strength of yang energy to protect itself. Yang Ming/ Yomei/ 陽明 is the brightness meridian, which means Ming: double sun or a place of greatest brightness. The heat of the Yang Ming is very great as it is the yang meridian on the front of the body, the peak of yang before it goes into yin.

The stomach is where the five flavours of food come into the body. Unlike the air, which is essentially of one flavour, pungent overall, one can derive all five flavours from food, and so the stomach introduces these to the body. However, it is only at a certain stage of digestion that the flavours are absorbed. Pungency can be absorbed directly by the lung meridian and later by the LI meridian organ, and some ends up within the liver blood. The sweetness is absorbed by the spleen and stomach, and the saltiness is taken in by the SI and TB/BL. The sourness is taken by the liver meridian and the GB meridian organ, and the bitterness is drawn in by the L-kidney meridian organ. The cooler the substance, the further down in the digestive tract it must go to be absorbed. Hence bitter and sour are drawn in right at the end of the digestive process, whereas pungency is almost immediately absorbed. The stomach connects with all the organs of post-Heaven production of energy, namely those associating with pungent, sweet, and salty flavours. This is because these combine with the pre-Heaven to form blood and qi to be circulated throughout the system. (Salty flavour is associated with the R-kidney which is started with birth, and so is a flavour between pre-Heaven and post-Heaven, but because it is warm, it can be easily supplemented in the body). The spleen is said to be the controller of the stomach because it is its Zang organ and contains vital essences. However, it is the stomach, which is said to control the Fu organs of the body, and this means all digestive processes of the post-Heaven energy.

The stomach and spleen are both associated with the sweet flavour but are highly different. The Stomach energy is sweet (earth) and hot almost pungent (fire and wood- yang), the spleen flavour is sweet (earth) and cool almost moving into sourness (metal -yin). It is important to realize that the nature of the spleen and stomach are sweet but do have this very different expression within their unity at the central region of the body. If we look at the nature of the organs and meridians and their opposite nature in the metal-wood and fire-water expressions, we can understand why spleen and stomach have this quality too, for each other.

4.2 The Spleen Organ Meridian System
(fig. 4.2)

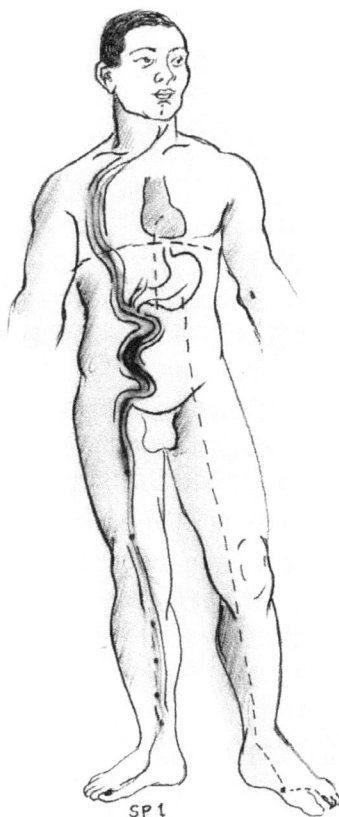

SP 1

318

The spleen too has a meridian and organ with the same function. The spleen function is ingestion and circulating the food energy, whereas the stomach is about digestion. The spleen is the root of the stomach's yin energy. It has the quality of being the supporter and supplier of the sweet flavour to the stomach, to nourish it and so its yin fluids and energy. The spleen is a source of the sweet flavour within the body. It has the capacity to create gu-qi (food-qi) from the broken down material of the stomach, and it has the capacity to spread this energy gently upwards, mainly to the heart and chest, where all the post-Heaven and pre-Heaven qi combines and makes blood with ying qi and Wei qi also. The spleen separates the flavours and sends them either above or below, where they can be absorbed. As most food that enters the stomach is either sweet, salty, or pungent, as these are the flavours of the post-Heaven energy, and of these, sweet is the main flavour of food, this tends to have a rising action. If the spleen can't process something, it is usually too cold. For example, there can be a sour or bitter flavour, some of which it might be able to process, but too much will drain its energy, as it's too cold for the stomach to keep up with. This shows why these flavours are so difficult to put back into the system via internal medicines like herbs, and also how external therapy is perhaps better at attempting this, or at least regulating yin. We will look at this later.

The spleen is the central point of the whole body. Like the Earth being the centre of human life, so the spleen is the centre of body-life. Its function generates all things, like Mother Earth. It creates shape and form and flesh, attempting to obtain from breaking down "fertilizer" in the form of food that then goes to make more of itself. It is warm and overall yang, due to the stomach's heat and expansive energy, but because it is not hot, it doesn't dry, so it contains water. Salty flavour dries, so this is very hot. The spleen itself is a yin organ and holds some of the yin of the kidney raised to a warmer temperature for use. Hence it works best with a relatively dry environment because it has a lot of fluid. The stomach and spleen organs balance each other's requirements, and the stomach uses the spleen's energy, while obtaining more food to generate fatty flesh and substance for growth of the whole body.

All fatty tissues of the body are part of the spleen network. There is some distortion of facts as far as the spleen's dominant areas in Chinese Medicine are concerned. The spleen dominates the "flesh", which has to do with the roundness and shapeliness of something. This is associated with the fatty tissues. The muscle, muscle belly, and tendon are much more to do with the liver. If we look at the colour of these tissues, the yellow is fatty tissue, the red is muscle or liver blood, and the tendon is liver yin, so it is opaque. The stomach deals with the yang surfaces of fatty tissue, and the spleen more with the deep tissues and around the organs and the front of the body.

The spleen is said to dominate the four limbs, just as the kidney dominates the back, the liver the flanks, the heart the face, and the lungs the chest. This is interesting as the 4 represents the 4 corners of the body, the representation of the Earth (body) being square. The abdomen as a whole is also related to the spleen as the centre of the body from this wider view. The connection to the four limbs is in relation to the sweet flavours' movement of warm yang energy through the limbs of the body, giving them fluid movement of blood to nourish and use them. The spleen particularly relates to the fluid distribution and the joints, which often is where fluid can accumulate if the spleen energy is not functioning effectively.

319

i) The Production of Post-Heaven Qi from Food

The qi production from food follows this process:

1. Ingestion of food-qi and fluids into stomach.
2. Absorption of food-qi and warm fluids by the spleen (sweet flavour formed).
3. The process of absorption is continued in the SI (including hot fluids) and LI (salty, also via the bladder [R-kidney] and pungent flavour formed). Also cool fluids are absorbed at this point via the L-kidney and liver meridian functions (bitter and sour flavours are absorbed and formed in this process, but not very much, perhaps equal or a little less than the output of the internal fluid required to follow through the above process).
4. The qi absorbed rises up the body gently and collects around the chest.
5. This adds to the air qi drawn from the lungs (pungent).
6. The rising process of the qi is aided with the drawing of internal TB energy—salty flavour, drawn from food and connected to kidney yang (drawing also with it some pre-Heaven yin energy—sour and bitter flavours) up the body, which is the origin of the stomach's power.
7. All the energy combines in the chest and is circulated by the lungs and heart.
8. The heart organ/pericardium meridian carries the heavier ying qi and this is, or forms, blood (sweet, pungent, salty, bitter, and sour flavours combined).
9. The lighter and coarser Wei qi comes to the surfaces of the yang meridians and the skin surface, where it is regulated by the lung meridian function and the TB/BL meridians. Pungent flavour and salty flavour are formed via air (pungent) and food qi (salty).
10. The kidney-lung and heart functions propel the blood and Wei energy around the body.

Note that the body uses up energy to digest and ingest, so in the process of digesting, the spleen uses sweetness, the lung meridian uses pungency, the liver meridian uses sourness, the kidney meridian organ uses bitterness, etc. Hence the aim is for one to put enough energy into the system to prevent loss of pre-Heaven energy and sustain the body. If too much of one or other aspect is taken, the reserves of the body draw in one or more of the internal organ five flavours to power the system. The yin reserves (bitter and sour) are the most precious of all, being pre-natally introduced and therefore very difficult to replenish post-natally. The basal yang of the R-kidney (salty) can be tonified, but without a base in the kidney yin, this heat simply rises right out of the body. The salty, sweet, and pungent flavours are more easily tonified through post-Heaven means, but the pre-Heaven of sour and bitter is hard to absorb, and this inevitably results in the process of death.

4.3 Yi - Chinese/ I - Japanese, and the Spleen

意

Yi is said to be the spiritual aspect of the spleen. The spleen is Earth within the body; it is therefore more corporeal than the liver and heart spirit aspects. This energy relates to the 2/5 and 8 of the 9 energies. It can be translated as the "intuition to follow the heart's/spirit's feeling". So it means to follow one's senses, far from the usually translated meaning of there somehow being a mental association, which is a very modern representation. It is far more about sensory following and being a part of the physical body dictated through the intuition of the Shen of the heart. It is therefore a more physical-sensory spirit, rather than a very ethereal aspect, like 3/4/9 would be. Essentially, to do this is to go with the natural movement and response to the unity of life—going where you need to go, doing what you need to do, etc. No intention is behind this natural movement; all intention is based in mind-identity, and mind-identity is fragmentation and dis-ease.

Associated with the Yi are three other aspects of the process of "intuition to follow the heart's/ spirit's feeling". They are the following (from Ling Shu 8):

1. 思 Si- Chinese/ Shi- Japanese
2. 慮 Lu- Chinese/ Ryo - Japanese
3. 智 Zhi - Chinese/ Qi - Japanese

These are expressed in the above order in chapter 8 of the Ling Shu. Previously, these have always been translated as part of the Westernised mental way of viewing, so are associated with aspects of thinking. If we consider the nature of the Earth energy and the fact that mind-identity in itself is a fragment and a source of dis-ease, we need to start looking at this from the concept of Taoist clarity.

Si/ Shi 思 depicts what looks like a rice-field character above a heart. However, an older view is that the rice field is not a rice field in this case and actually depicts a head with a hole at its centre. This relates to the soft material between the bones of a baby's forming skull (the fontanel) . The heartbeat can actually be felt here on a baby. This makes clear the expression of heart and head. Also the rice field associates with the spleen and granary, so a connection is made of the head, heart, and spleen. The depiction however is that the heart underlies the picture and is at the root of it. Hence it describes the interaction of the head or brain in connection with feeling from the heart—the clearer the brain, the clearer the heart, and vice versa. This is the first part, a clear mind-heart connection. Note that while the brain is associated with the kidney-yin and marrow, it is at the upper part of the body and so is the most yang marrow and is also associated with the fire network and is fed by the BL/TB and associated meridian, the yang of kidney. The heart and head/brain therefore are very much associated, which is why mind-identity, or the pathology of the head, is associated also with yang.

Lu/ Ryo 慮 can then be thought of as analysis, the process of overseeing myriad parts and forming a whole from them. So this isn't analysis using the brain. Again, the heart whole of the Si/ Shi character is within the Lu/Ryo pictogram, expressing that this is clear-minded or intuitive analysis. So again, the clearer the heart and mind, the clearer the intuition and ability to see the whole within the fragments.

This leads us to the last character Zhi/ Qi 智 meaning spoken wisdom or spoken truth. The importance is the understanding of "spoken". The mouth is a part of the picture here, and a directly spoken image is expressed of wisdom or truth. The process of clear-mindedness, of the connection of heart and mind and the intention through this to follow the heart's desire, allows for clear viewing and analysis of a situation and the ability to intuitively speak wisdom or truth.

This is what is meant by the spleen and Yi in association with wisdom of spoken truth. The connection of spleen being formed by fire, or the expression of the heart dictating the spleen function (mother-child relationship), shows how fire and Earth are energetically united throughout the system. Hence these connections give us a far better understanding. The clear-mindedness from a clear heart or un-fettered heart-Shen can be called wisdom or truth. The writers of the great classics understood this and wrote their works from this way of speaking the truth of wholeness.

The spleen organ associates with 2, and the stomach organ with 8 and 5 mostly.

Part 5:

The Metal Phase

5.1 The Lung Organ System

The connection between the lung meridian and liver organ, and liver meridian and lung organ is very important. The lung meridian is pungent in its quality and is generated by the out-breath, which is the dominance of the lung organ. When you breathe out, the pores of the skin open, and when you breathe in, they draw inwards. Hence the in-breath relates to the kidney's yang of the Ming Men, anchored by the kidney yin in the lower burner, but the lung organ is the out-breath.

The lung meridian has a pushing-out effect on the skin and the exterior. This matches on the inside of the body with the liver blood, which is pungent also and pushing outwards. Hence with every out-breath, the liver organ energy releases to the surface. This is often used in relaxation exercises and letting go of emotions, when sighing or grieving—letting go of the liver organ's stuck-ness/withholding of blood. This pumping effect of the energy pushes fluids around the body, essentially from the inside to the exterior. Part of the lung organ's function is to help the heart pump blood around the system, as well as body fluids and difuse them throughout the body. Note however that the lung meridian concerns qi, not fluids; though it moves fluids, it does not relate to fluids and their metabolism. This is more the function of the spleen (post-Heaven fluids, body-fluids) and the kidney (pre-Heaven fluid, yin).

In contrast, the liver meridian has a yin nature and "collapses" or collects inwards. This is matched by the nature of the lung organ when there is an out-breath. The lung actually collapses inwards. Notice that this is not a forced action, but a process of letting go of action; it is essentially a "death" within the system. This is very interesting for several reasons. The lung is associated with death and grieving and the autumn season of letting go of yang, especially, or life. Metal is a very heavy substance, so when the lung is full of air/qi, it will float on water. It will be light, like wood (consider this). However, if the air lets go, then the lung will draw inwards and form a mass, which is hard and solid like metal—sinking in water. This is why the lungs are expressed as metallic. This is expressed clearly in Ling Shu 8 where the description of the P'o is:

> The faculty of physical motion that formed from the coming in and going out of the motion of the refined essences is called P'o. (based-on Wu N. L and Wu A.Q, (1997)).

This means that the P'o is associated with the essences as it FOLLOWS their motion. The physical body follows the essences or energetic qualities of the being. The following of the essences means that the P'o follows the processes of life, but it itself is not involved in it.

Hence one can say that the effect of yin accumulation is associated with the liver meridian and lung organ, and the yang expansion is dictated by the lung meridian and liver organ. The anchoring of the in-breath would also be associated with the kidney yin and therefore liver yin, but the actual impetus to take the in-breath is the kidney yang. It is the Ming Men that dominates the in-breath. Consider that the first breath is dominated by the in-breath, which is the Ming Men's impetus; the last breath is out, and this is the lung and death. It is important to note that whenever something is moving or is impelled to do so, it is the yang that is involved. The lung organ collapses not through force of action but through acceptance and due to the fact that there is no yang. This is the way of all things. When Heaven's change ceases, then yin will be all that is left, forming Void or the Original Mother again.

Therefore, we can say that the out-breath is only as strong as the in-breath. The power of the lungs' own pungent flavour is dependent on the impetus and anchoring of the kidney yang in the lower burner. The air is the pungent flavour, just as food is essentially the sweet flavour, as a general identity. When this is drawn into the body by the kidney, the out-breath takes over. It diffuses the energy throughout the skin and the whole body. This makes a pump. The kidneys pull inwards, and the lung organ lets go and so presses outwards. Hence "effort"/yang is used from the in-breath, but there is no effort/yang in breathing out. The lung simply follows the kidneys' energy, being more or less dependent on the kidney function.

To some degree, a weakness of the lung function (lung meridian) is always due to a weakness of yang kidney function at the origin of the in-breath. Spleen energy is essentially a layer below that of the kidney yang (salty) and lung meridian yang (pungent) flavours, so it constitutes the leg aspect of the Tai Yin. The lung meridian relates to the upper body, skin- pores (not skin), the LI meridian, and the functional aspects of the BL/SI and TB meridians more. Salty flavour is closer associated with pungency than sweet. They represent the body yang within yang flavours. This again shows how important the fire and water in the body are, as the fundamental system that creates function for the rest of the body.

So where does this leave the actual function of the lung organ? What we find is that the lung itself is a non-functional entity in the body. It doesn't have any essence of itself to derive function. It has no quality of actual accumulation and no quality for expansion. In fact, it has no yin or yang. What does this mean? If we consider that the lung's out-breath is simply a letting go and an expression of the weight of the physical body pushing out the air when we let go, and also that this is the very last thing we do in life, we begin to see that the lung organ is representative of the physicality of the body—not its energetics. The energetics that power the body are yin and yang energies in forms of the pungent, sweet, sour, bitter, and salty flavours. The lung organ acts similarly to the liver meridian and GB organ-meridian in that it accumulates inwards. However, this drawing in is due to letting go of energy; it is not due to a fullness of the sour flavours. So although it is synonymous with the autumn and sourness, it is not actually sour. Also the lung is associated with qi and energy function and to the pre-Heaven energy; the post-Heaven energy is sourness and bitterness in the body. This is derived from the kidney Jing fluids, which are only added into the system pre-birth. This means that lungs cannot be a yin holding organ, being post-Heaven energy related (lung meridian) and not storing the yin essences. This is why the lungs can't become fluid deficient, or only via the kidneys' yin weakening and the whole body drying out or the post-Heaven fluids drying out. The lung organ cannot store a pure yin fluid. So then why is the lung organ strengthened in the autumn and winter, and the nature of it seems so yin? The answer is that it does not contain yin essences of sour or bitter, or in fact yang essences of pungent, salty, and sweet; it is death within the body. Just as life has a beginning, which perhaps we can call the kidney yang or spring, so there is an ending in the last out-breath of life, autumn. The lung organ is exterior to the process of the body, or one could say the pivot point that the in-breath and out-breath move through. The collapse of the lung in the out-breath is exactly the same expression of the body's mass falling downwards to the Earth at death; each out-breath is a small version of this.

The seesaw effect of in-breath and out breath is literally an expression of the microcosmic cycle of life. The lung organ function therefore is absent from the body, and in medicine, the function of the lung organ is not talked about at all, because it is irrelevant in the process of understanding, and understanding is always focused on the meridian function, which is truly a function.

Hence the lung organ relates to yin but is not yin, and nor is it yang (in the context of the body energy). Therefore it is synonymous with Wu or the exterior of yin and yang, and this makes it associated with yin. This is important because yin is often associated with deficiency, whereas yang is to excess. This is because yin is associated to Wu or the exterior of the universe—nothingness. This nothingness underpins the moving universe, it is AS IF the lung is nothingness in relation to the rest of the body systems. It is always important to see the relation to yin and Wu. This gives us a context to discuss what is meant by a fullness of yin energy, which is a cooling, sour-bitter energy. Associated with diagnosis, yin excess will always mean blood stagnation (see later sections). Another important issue to note very clearly is that all FUNCTION is associated with yang aspects of the system. These are aspects that actually do something and usually lead the processes. The yin aspects, such as sour or bitterness, are not functional; they have a quality that cools and gathers—this is yin. Yin and deficiency are perhaps the deepest, most important concepts we discover in Chinese understanding, because they are the antithesis of the yang of life as we understand it; mind-identity only understands things in yang format.

The associated tissue of the lung organ is the skin. The skin relates to membranous tissues of the body. While the facial tissues (the white tissues of the tendons) are more associated with the liver and GB meridians and GB organ and relate to the sour taste, the skin also seems to be a sheath-like tissue. I believe that we should consider the skin itself as a form or tendonous sheath wrapping the body. The membranes all over the body have this same connotation, and so skin is in relation to the energy of the lung organ. There is a distinct connection therefore with the bone or internal skeleton and the skin or exoskeleton, and also with the skin and marrow or the brain and nerve tissues. This exterior is where the interior ends and where the exterior begins, so it belongs to both worlds. Hence, again, its connection is to the lung and P'o (see below).

Please note that we previously spoke about the lung meridian in the wood section , we did not speak about the actual skin. Wood tissues associates with muscles and blood, the red tissue of the body , heat and pungency. The Lung meridian therefore is simply an expression the heat in the blood to the surface it is to do with the energy that comes through the skin not the skin itself, the skin itself belongs to the lung organ. Hence as with the all other aspects of the system there is a balance , the air in the lung and therefore the meridian energy is pungent, the skin is sour. The Liver blood and its organ and the muscle bellies of the body are pungent but the tendonous tissue that wraps the muscle fibres and the meridian that wraps the liver is sour. With the liver and its meridian it's internal with the Lung meridian and the skin it's external. As we said before the pores of the skin are like the nostrils- they are holes through which energy manifests and expresses, but the skin itself has a tendency to hold inwards rather than express outwards, it is a shell. Hence the pores of the skin have a tendency to move towards drawing-in but the energy pulling through or pushing-out has no such sensibility.

i) The Importance of the Out-breath

The out-breath is representative of death within life. During an out-breath, the body becomes stiller and quieter than in any other experiences of existence. Hence it is in the out-breath that one has the opportunity to realise the nature of reality versus the nature of the movement of the body-spirit interaction. The out-breath is a gateway for connection of the Stillness within you to the Stillness behind the exterior world—or Wu. The out-breath is the most sacred of bodily expressions. All deep meditational practises from ancient times focused deeply on the out-breath, simply allowing the in-breath to come of its own accord. The out-breath was the main part of the cycle to which one was aware, for within it is the ability to become unified with Nature. When one realises that the essence of the out-breath is non-resistance, letting go rather than doing anything, then one learns the true nature of reality and is no longer deluded by deriving a sense of self from attachment to the moving aspects of the body (e.g., the heartbeat or the in-breath). If one instead derives a sense of self from the out-breath, this provides a realisation that the self is not just within the body-spirit form. Hence one goes beyond mind-identity's limitation/perspective. Shizuto Masunaga made the understanding of the out-breath in self-therapy and shiatsu a mainstay of his work, an important expression (See "*Meridian Exercises*" by Shizuto Masunaga, see bibliography).

As expressed earlier, the metal energy is very close to Wu, being the Central pivot or empty-hub and too, background upon which movement occurs and the mother of water. If water is "close to the Tao" as it says in the Tao Te Ching (see next chapter), then the mother of water is also the implication of metal and Wu's connection. This connection to the out-breath expresses how important this understanding is in Oneness. Metal is very much associated with clear awareness, and P'o (see below) is intricately involved in the understanding of the impermanence of the form/physical, but is death within life, and relates to the calmest of all aspects of being. Water is Stillness within the world of yinyang, but metal is almost between the worlds of Wu and yinyang. Observe this yourself. The supreme yin is in fact Metal in the body.

5.2 The Liver Meridian System
(fig. 5.1)

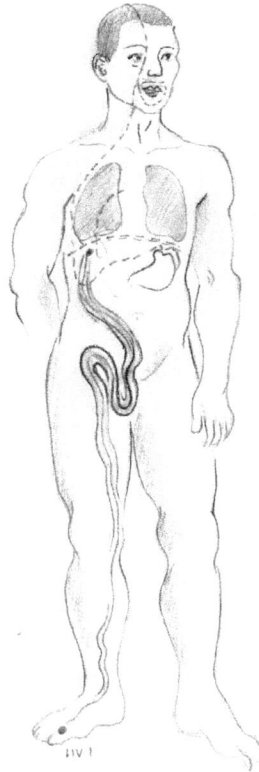

The liver meridian is sour in flavour, as explained earlier. The liver meridian is labelled wood due to its connection with the Liver organ, but it is energetically metal. The liver meridian wraps and cools the liver organ and has the yin function in coordination with the lung organ and GB organ-meridian. These are considered the aspects of the metal phase, which act to tonify and draw in water energy. Its associated tissue is the deep tendonous connection from bone to muscle body and also the facial material. Interestingly the Gao-huang as it is called (Matsumoto, 1988, p97-129), was the primordial tissues of the body was a kind of tendonous mass originally during the forming of the body at foetal stage. The interesting aspect of the first two meridians of the body activating the initial aspect of gestation the liver and GB meridians also indicates the importance of the facial tissues at this early stage. The basis of the main-meridian system is still seen as the tendonous meridians which underlie the pathway of the main meridians throughout the body, and are likened to the nature of the Liver and GB yin energy throughout the body they form the deep and superficial tendonous strands that are actually a foundation substance for all body processes. This however doesn't make them exclusive. Meridians energy doesn't run

328

through these exclusively but throughout the various structures of the body each with different quality in the system. The tendonous meridians are really all part of the GB organ meridian and liver meridian throughout the body.

The left kidney is bitter and tonified by the sour flavour, so the sour flavour of the liver meridian helps to condense fluids into the tendon and bone. Hence it is said to tonify the left kidney, but if we consider that bitterness is cold fluids and sourness is cool fluids, they are very deeply intertwined, so really, kidney yin is the store of fluids that liver fluids are part of.

5.3 The Gallbladder Meridian Organ System

(fig. 5.2)

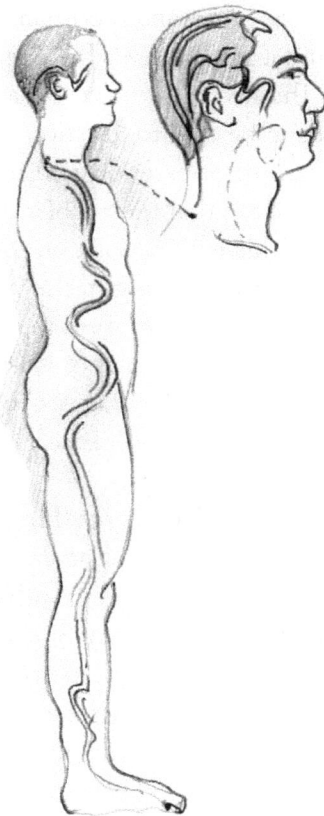

The gallbladder is associated with the yang energy of wood. The Gallbladder organ and meridian are closely associated with the liver meridian, rather than its organ. The GB meridian-organ is labelled wood when it energy is that of metal. Please be sure to understand the reason for this in the section on meridians and points.

The liver meridian is sour, and so the gallbladder organ is full of sour bile, which is cooling and oiling to the system. The gallbladder meridian, too, follows on from this and is the coolest of the exterior

329

meridians, the yin within the yang, so to speak. Therefore GB organ is considered as one of the six "extraordinary" Fu, along with the uterus (belonging to the wood energy mainly and to water), the brain, marrow, bones, and extraordinary vessels (belonging to the water energy). The tendon-sinew meridians and structures all over the body are too part of the GB-organ meridian and liver meridian expression.

The gallbladder meridian is at the sides of the body; it pivots between the front and the back of the body. This area is associated with balance and with the ability to turn to see side to side. It is the decision aspect of the spirit. The ethereal soul is associated with the liver, and therefore the action to move towards a vision or an image of something is very closely associated with the gallbladder. While the liver organ creates the vision, the gallbladder acts on it. Hence it is the gallbladder energy that one would expect to be hot and sparky. However, the organ and meridian are cool, and this is because in order to make good decisions, one needs to be fluid and objective. The energy of the liver is very hot, and the gallbladder balances this with its coolness. The liver heat or vision is given to the gallbladder, and so it takes on a massive amount of heat and needs to be cool to function under this temperature.

The coolness of the GB is an entry point of exterior cold into the liver and therefore into the internal organs; we will look at this later. Notice too that the gallbladder is extraordinary because it is full. It is not an empty organ like all the other Fu; this is in opposition to the heart and PC meridian where the heart is a Zang organ but is empty. The heart is the yang in the yin, whereas the GB is the yin in the yang. The sourness-bitterness of the bile is a reservoir of yin from the origin of the kidney yin and adds to the yin of the liver. Also note that sour tonifies bitterness; hence it is actually the sour flavour that helps to draw in the kidney energy and condense the "ice" of the kidney coolant. Therefore, although the origin of Jing is the kidneys, the sour flavour at the end of the five-phase cycle helps to draw energy back there, recycling the yin energy.

The vital essences of sour and bitter flavours are the pre-Heaven essences. Even the salty flavour is associated more with post-Heaven or the light of life—Ming Men at birth starts the salty flavour in the system. Saltiness can be tonified, as can sweet and pungent in life, because they do not require the cooling and gathering process that is the function of pregnancy. Sourness and bitterness are only put into the system at one point, pre-Heaven or pre-natally. This is then used up during life. To reclaim it from the sour and bitterness of food is very difficult (this was the attempt of using decoctions of "immortality" using very yin ingredients (such as heavy metals) to give eternal life or youth) as it must pass through the post-hence-qi systems, which are all based in warmth. Salty, sweet, and pungency are easier to take in. The lung's in-breath, remember, is based in the Ming Men fire's actual action; this doesn't benefit the yin. Often the best way we can benefit the yin is actually by using less of it, which means a slower lifestyle. Acupuncture can also be beneficial (we will look at this later).

There is a similarity of nature in the heart and the lung. The heart is empty, yet it is a yin organ, and it is life-force that powers it. The lung is also empty, but of life itself, though it is denser than any other organ of yin when in a collapsed state. It is powered by the physicality of being, or death.

The gallbladder makes up the exterior coat of the liver, and so the liver commands its function. The gallbladder system therefore includes all surface and tendons (including facial tenuous tissues) of the whole body. Note that this material is all white in colour. Muscle is blood-red and associates with the liver-organ (blood) and heart.

One often associates the GB with wood energy, but this is very similar to the 4 energy of the nine energies, which is what the GB and liver meridian are associated with. The key is that the GB cools the liver, as well as the liver meridian, and so this energy is a wrapping. This is clearly expressed in Su Wen 8 when the description of the organs of the empirical court are divulged:

> *The liver is a vigorous organ, its emotion is anger, it is like a general who is valiant and resourceful. The gallbladder is like an impartial judge who makes one able to judge what is right and what is wrong.* (based-on Wu N. L and Wu A.Q, (1997)).

(Another reference to the explanation of judgment related to the GB is the Pulse Classic Chapter 2 of Book 6.)

The liver is the organ of heat, which passes its heat to the GB, but the decision process is derived from the judicious and cool impartiality of the sour flavour. This actually tempers the liver's energy and so allows one to have clarity in judgment rather than merely explosive expression. The GB being a cooling aspect is very important to understand its extraordinary nature. It is the yin within the yang of the body, the Ming Men/TB being the yang within the yin.

5.4 P'o – Chinese/ Haku – Japanese and the Lung

魄

P'o is considered the most corporeal soul aspect of the spirit energies, described as "white spirit". Opposite to both the wood and fire energies 3/6/9, the corporeal soul relates to the 7 and 4 energies of the nine spirit energies. The out-breath is dominant in this phase, and so when the lung has no air in it and becomes hard and solid, this is the organ energy of the P'o. It is a very still energy with great potential to change to its opposite, such as when an out-breath reaches its maximum, then an in-breath comes. This is just as life and death. The corporeal soul therefore governs the ending of the life process or the monitoring of the life process, until death necessarily must come in order for life to continue in a broader spectrum. It is important to understand that here the "white" aspect of the spirit nature of P'o is probably associated with the white of a skeleton of dead bones. P'o is dense and heavy and very yin. It is not an ethereal energy or what we would commonly think of as a "spirit" but is actually the Earth, structure and form. It is non-ethereal, unlike Hun, the yang spirit form that we commonly identify in English as "spirit".

Relative to all the other spirit energies, it is the one that has the most solidarity about it, and this too relates to death, each out-breath a cessation of yang, showing simply that we are dying and living every moment—it is One, and so takes away the fear of death itself, as we are part of it. This is an important part of the spectrum of existence. The white colour has to do with the reflective property of metal. Metal is hard and solid, but it reflects bright white light, like a mirror. When something has become very dense and is about to change to black, it first looks bright; this is the same, and the brightness is just before the darkness of winter. Hence, although it is bright, it is cold light and solid-hard light, meaning death. Death and grief within the body is not aliveness, but quite the contrary, a deficiency of life.

In relation to the flavours, we could relate the P'o to sourness. However, as discussed above, although the liver meridian, GB organ-meridian, and lung organ make up the autumn energy within the body, the lung organ itself is essentially bound to the body, the physicality, not to the yinyang essences within the body. The flavours denote the essences, so the lung organ has no flavour in reality but is associated with sourness through its sense of being. In fact, one must associate the P'o with the physical body structure that causes the collapse and the density of energy within the chest (i.e., drawn in lung) in its gravitational down, and therefore is associated with the end of the body. If there was no lung, there would be no death, and without death, there could be no life, so it is a vital structure within the body. Notice the structural nature expressed here and the cold, hard nature of the lung that is without air in the chest. This is an expression of metal, buried in the Earth, and that's where it stays. This is a vital expression of the energy of P'o. The P'o therefore is that which goes back to the ground after death, and with each out-breath gets us closer there—just as each in-breath lifts the spirit energy in the body upwards to Heaven. To understand death is to understand the lung organ. The out-breath is a leaching of the energy internally to the exterior; it is becoming one with the exterior and blurring the exterior with the interior. It is borderless. The skin, the energetic expression of the metal phase associated with the lung organ (the lung meridian is associated with the pungency within the skin), is also representative of this which both belongs to the interior and exterior at the same time, as do all the meridians.

Notice too that the P'o is said to have seven parts. As the 7 energy is associated with yin-metal in relation to the five phases of the body, this would be lungs. Hence there is a relation to the seven aspects of the P'o and the Lung energy also. The 4 energy is also associated with metal, but this is more of the surfaces of the liver and the GB meridian and the GB organ particularly. This makes up the sour flavour also, so this is an actual fullness of energetic flavour, rather than the P'o, which is outside of the human yinyang energy. It is yin in relation to the yin and yang of the body; therefore we call it death within life.

Part 6

The Water Phase

6.1 The Left Kidney Organ Meridian System
(fig. 6.1)

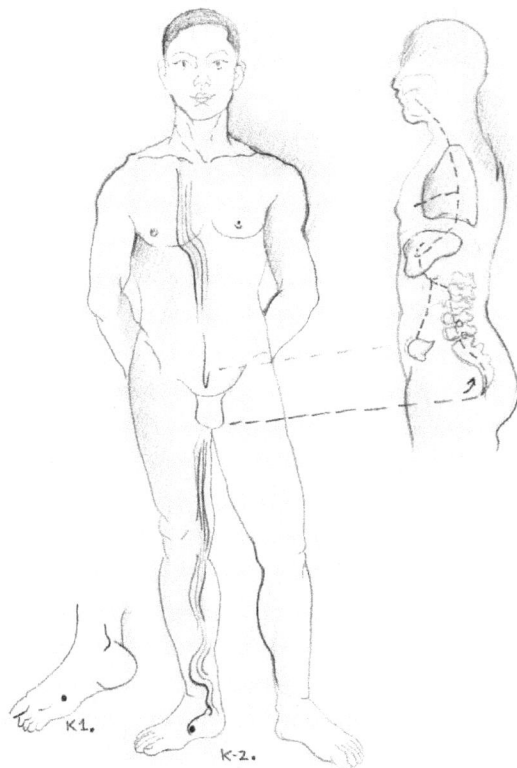

This phase actually has no yang partner. The bladder is part of the fire energy of the body, hence this is the only true aspect of water within the body. The nature of water is cold and deep; this is the nature of this organ meridian system. Aided by the liver meridian, the GB organ meridian, the lung organ, and the sour aspects of the body, the left kidney holds the bitter flavour within the body, and this is the root of the cooling system. This energy is called the pre-Heaven or pre-natal energy given to the baby directly from the mother's own essence at the ninth month of pregnancy.

This energy is very difficult, if not impossible, to supplement after the umbilical cord has been broken, because the body then reverts to post-Heaven energy. That means that supplementation through food and air will be difficult because the lungs and spleen are higher up in the energetic spectrum. They are within the yang in relation to the kidney Jing, and so placing very cold things into the upper body simply destroys the yang of their functional systems, making it hard to supply via the digestive system. It may be impossible to supplement this energy, as this would be literally reversing the aging process. The only way the Chinese have understood it is by slowing the aging process, which is not so much supplementing what is there, but making the best of what there is, the efficiency of this being the art of longevity in Chinese philosophy—not that this is an aim, but a product of letting go of mind-identity naturally. This will be touched upon later.

The key point is that this is the coolant and the pre-Heaven energy. This energy is very precious. It is drawn off via the fire of the Ming Men and used to moisten and enhance all body processes, dominating and regulating all functions of the body having to do with timing and the aging processes. This energy is stored in the bones and bone marrow of the body, the fluid nature of the bones, giving them density and flexibility. The marrow fills up the spine and the brain and spreads through the nervous system, hence the nervous system is dominated by the kidney yin and works best at cool temperatures. The testicles in men and the ovaries in women, and their connected systems, are dominated by the kidney yin. All accumulations of rarefied or powerful fluids within the system are kidney yin related, so in Modern Western physiology, this would also incorporate endocrine and lymphatics. However, this is a fragmented way of looking at the body systems, so be careful not to look narrowly here, but broadly; water course systems are the key.

Through the process of life, the yin of the kidney not only acts as fluid that cools but also as an anchor to the energy literally rising up and out of the body. The Ming Men/R-kidney energy, that is a density of energetic yang, is anchored down in the lower body, due to the kidney yin that it is in relation to. The weighty kidney yin and the sour flavour in the body help each other in the cooling and gathering of energy, but this is a non-function effect rather than a yang-functional change. Yin works through gathering energy, and this occurs through stillness, not through change. The foundation for the Ming Men expression in the in-breath impetus or the holding down of the Ming Men, so that the heart can be stabilised in its beat, is all due to the kidney yin. Without it, the yang would just drift right out of the body, and the body itself would fall downwards in death.

Differentiating again the difference here between yin and body fluid, the kidney deals with the true yin of the body. This is not so much the yin of blood or body fluids, but the energy of cooling given pre-birth and so stored and only used gradually if life is to be long, or used fast and furiously if life is to be short. This yin energy smoothes cycles and allows energy to be let out gradually; overusing the yin will speed everything up and make life very fast. Also note that while the yin of the bitterness is the coldest part of the body, the Jing energy of a human is relatively warm in comparison. The point is that it is always relative to the other elements within the human five phases. Overall, humans are more yang than our surroundings, but perhaps less yang than light itself. So again, when one uses yin and yang, it is important to understand the comparison one is making.

335

There are 11 meridians and 11 organs, and of these 22 aspects, only six aspects have to do with the bitter and sour flavours, which means that 16 aspects are yang. This is around 30% of the body which has yin relation and two thirds that have yang relation. Again one can see the balance of power here is so much warmer for humans.

Included in the left kidney's associated energetics are the extraordinary vessels. These eight vessels are all part of the cooling influence of the kidney energy throughout the body, being reservoirs of this energy from all aspects to draw on, but also acting as buffers to collect energy that would be pathogenic at a deeper level, or simply heat that is surplus. They are however cool and still by nature; without the flow of the other meridians drawing and affecting them, they would be, as the L-kidney, still and calm as winter. They are also non-directional, and their use in meditation practises is more to allow the person's mind-identity to respond to stillness rather than to attempt to "do" something with them, which is commonly the error.

Kidney yin is the most associated expression to Wu within the human energetic spectrum. The second closest is the metal energy of sourness. Hence these energies of pre-Heaven sourness and bitterness have great importance; these energies are the source of life processes. From these essences, life can be nourished and cooled, or if ignited, this energy can expand and re-form and re-generate the body. Hence most meditational understanding relates to drawing the attention of the body away from the head and down to the lower body and lower back, filling this area which is the dominant area of the kidneys. The kidney yang goes up from here, but the kidney yin stays as an ever-present anchor, till yin is drawn completely by the yang, which then turns to death. This anchor is the anchor of mind-identity. It is also therefore the deepest place within the being and the Stillness place. From here, one can get the notion of what this energy connects to, which is beyond the quality of cooling. Because of its coolness, it connects to Stillness-Wu and so through the process of awareness meditation, we notice that the lower abdomen, or centre of our being, is where Stillness emanates from within us and so connects to that which is Stillness in the exterior. Hence we can say that the Jing essence is the deepest root of being that connects back to Wu. We can then say that the Jing essence is as close as one can get to Wu within the body. (note that the lungs and skin are P'o, which is not of the body, but almost belong outside of the yinyang of the human energy expression). This is why it is held with such reverence, as are the lower abdomen and legs and feet which root the being, calm mind-identity, and allow for life to be as long as it can be (i.e., the art of longevity, which is not so much prolonging life but accepting it in the moment). Note that Jing is not Wu because it is still part of the cycle of yinyang. However, it is close and so trains mind-identity to observe Stillness and dissolve/ draws inwards mind-identity from its exterior and fragmented viewing. This is expressed in the famous line from the *Tao Te Ching* in the first few lines of chapter 8:

> *Profound is the Nature of water!*
> *Because water benefits all life and does not contend*
> *And settles in places where even human life cannot*
> *So it is an example of Naturalness.*

6.2 Zhi – Chinese/ Shi – Japanese and the Kidney

志

Zhi is often translated as the will power. However it can be considered to be the base from which the Shen of the heart can take direct action. The consideration is of the will being a solid base. The kidney yin is the solid base from which the Shen is anchored in the lower body, and from this place, the Shen springs up into the heart to take expression in the heart. The solid base gives a firm footing for this, and this is why the potential power of the kidney yin is associated with this energy. This relates to the 1 energy of the nine energies. If we consider that the yang line within the Kan trigram is the yang of the right kidney, then the two exterior lines of the Kan energy can be considered Zhi. Again we can use this diagram to explain: (fig. 6.2)

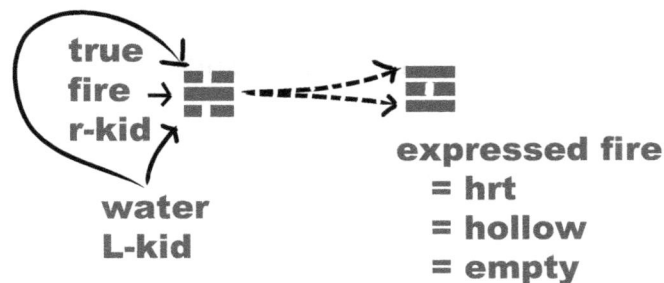

Using the 9 energy itself as the heart enclosing network association means that we associate the 1 energy with winter and the L-kidney organ. However, please notice the fact that within water there is fire, and within fire there is emptiness. If we associate the Shen with the R-kidney and heart network, then the Zhi would be more associated with the L-kidney, winter, and Kan trigram and cooling, although it is important to still realise that the energy of 1 is a combination of fire and water and has the root of fire in it as well as water; hence it is complex.

Part 7

The 5-element Interactions of the Zang, and Understanding the 10 Celestial Stems of the Nan Jing

The 5-element Interactions of the Zang, and Understanding the 10 Celestial Stems of the Nan Jing

As we have seen, the five phases originate from humans. The origin of the five phases within the body is the 5-Zang, which is the root of the energy. The Fu derived from the Zang and the meridians, derived from the Zang-Fu, make up the exterior and are one with the exterior. Only mind-identity differentiates that we are separated; we are borderless in fact. The organs of the Zang-Fu are not in contact as directly with the exterior, hence the interactions relate to the internal environment of the five phases' balance, according to that person's constitution. The exterior, of course, has influences, but there is also an internal change of the five phases going on at the same time. When interior and exterior match and balance, the person is completely one with their exterior. This is a sage-like existence, as explored in Su Wen, chapter 1.

The interactions of the five phases therefore are often associated with the Zang in the following way:

(fig. 7.1)

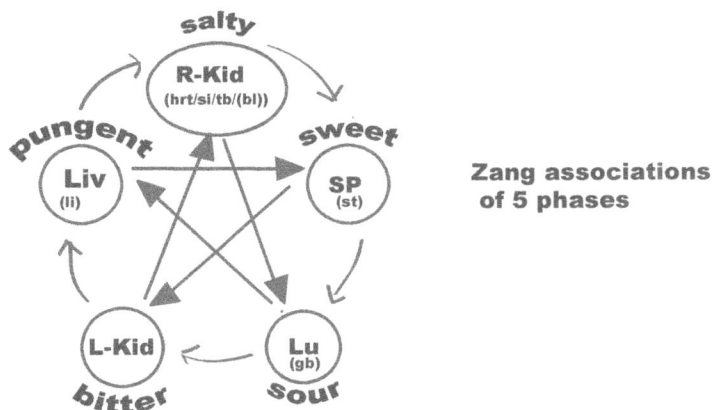

Zang associations of 5 phases

Notice that the Zang here include, or it is understood that the Zang hold or control, their related Fu organ. Notice, too, that the bladder is part of the heart network, and also the PC is a meridian, but it isn't an organ, so it isn't included. The overall picture is one where we can understand the Zang ruling the interactions. We can see that for each Zang there are two interactions with other elements.

In chapter 33 of the Nan Jing, the interactions are associated with the 10 celestial stems of the calendar (pre-Heaven), splitting the five phases of humans into yin and yang partners. Previously, in modern accounts, this has been associated with splitting off the Zang and Fu of the five phases and becoming involved in this interaction. However, this is NOT stated in the classics at all. There is, at no time in the classics, a correspondence made for wood yang to mean GB, or for metal yang to mean LI. This is, in fact, an interpretation of the classics rather than a deciphering of the meaning as clearly expressed without the need for interpretation. The yang aspects of wood control the Earth; the yin aspects are controlled by the metal. If we look at the actual text in Nan Jing 33:

> *The Thirty-Third difficult Issue:*
> *The Liver is green; it reflects the wood energy. The Lung is white; it reflects the metal energy. When the Liver is brought to water it will sink; when wood is brought to water it will float. When the lung is brought to water it will float; when metal is brought to water it will sink! What are the respective issues here?*
>
> *It is like this: The liver is not pure wood; I which is [resonates with the musical note] chiao, constitutes the soft [opposition partner] of keng. In macro-terms, I and Keng represent yin and yang; in micro-terms, they constitute husband and wife. [The liver] has subtle yang [influence] and absorbs subtle yin [influence]. [Wood's] sentiment is joy of metal. Furthermore, it proceeds mostly through yin's Tao [yielding]. Hence when liver is brought to water. It will sink.*
>
> > *The Lung is not pure metal. The hsin, which is [resonates with the musical note] shang, is the soft [opposition partner] of ping. In macro-terms these represent yin and yang; in micro-terms, they constitute husband and wife. [The lung] has subtle yin [influence]; through marriage it is drawn towards fire. [Metal's] sentiment is the joy of fire. Furthermore, it proceeds mostly through yang's Tao. Hence [in this case] when the liver is brought to water it will float.*
>
> *When the lung is mature it will take a turn and sink; when the liver is mature it will take a turn and float. Why is that?*
>
> *It is because we know that hsin must return to keng, and I must return to chia.*
> *(based-on Unchuld P., 1986)*

This explains that the 10 stems are of the pre-Heaven and are being looked at through the five phases. In the exterior/ macrocosmic perspective, they are I and Keng of the pre-heavenly 10-stem energy. HOWEVER, within the body (i.e., the microcosmic perspective), these are within the five phases and represent a husband-wife relationship. Without this perspective of the five phases, they would be simply opposite yin and yang aspects, two sides of a circle, but with the five-phase interaction they associate with the identity of husband-wife energetics.

Here is a better expression of what occurs. In the 10-stem picture of the change as seen on the calendar, we can see it like this:

(fig. 7.2)

the 5 phases and 10 stems
wan jing 33

(fig. 7.3)

5-phase point interactions

As we can see, the above pictures give us the fuller account of the 10-stem variations. The differentiation being made as to whether we see 10 stems or 5 phases but when looking at one, the other dissolves. The macrocosmic is the 10-stems, the microcosmic is the 5-phases. If we are viewing from the 5-phase perspective then we see just the colours in the diagrams above and below. If we see the 10-stems we see the individual interactions. The nature of the interactions is different. 5 phase is more post-heaven , 10-stem more pre-heaven. In a sense the 10-stems are in a zone that they don't belong, being associated with macro-cosmic but being applied to the 5-phase micro-cosmic. The important point here is to see that when something is split up into pieces then the pieces that make up the 5-phase may be scattered somewhat in order to maintain balance within a more complex structure, the more complex the structure from 2 (yinyang) or 5 (5-phases) the more complex the interaction and interpretation of the original parts, like an ever expanding mandala of more and more complex patterning from the original core of Pure simplicity. Or a ripple in a pool.

In the 10-stems the yang aspect of the fire controls the yin aspect of metal, the yang aspect of wood controls the yin aspect of earth, etc. Hence this creates the following stem-branch husband-wife interactions:

1 to 6

7 to 2

3 to 8

9 to 4

5 to 10

Notice that the masculine aspect always advances, whereas the feminine aspect stays static. This is important because in the upper picture we can see that the interactions of change are all of fire, earth, and wood; these are all the more yang counterparts. The energy of metal and water don't seem to interact. If we look at the second diagram, however, we can see that indeed the reactive aspects of the yang of metal and the yang of water are actually part of the yang WITHIN the yin. Hence the 10 stems expressed like this shows their interactions. Taken to a fundamental level, the basis of all the five-phase interactions is the Zang energy, but in its more complex patterning of 10 stems or even 12 branches, there are variations in order to show the total interconnectedness of yin and yang within the system. This relates to the symbol:

(fig. 7.4)

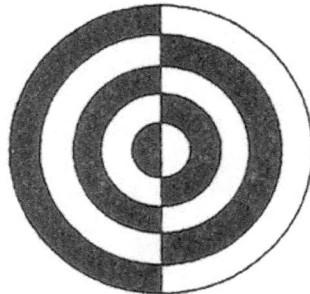

It is only here that we can consider the interactions of the five phases to emanate. The yin aspect of each phase is that it can be overcome by another phase; the yang aspect of each phase is that it can overcome its resonant wife partner. This simply means that the bitter taste overcomes the salty, the pungent taste overcomes the sweet, the sweet overcomes the bitter, and the sour controls the pungent. Also notice that the yang tonifies the yang, and the yin tonifies the yin, in the generating cycle, because the yang aspects are above and the yin below, and so they move at different strata in the organ interactions. These tastes express the seasonal internal nature of the Zang organs.

So Nan Jing10-stem interrelation verifies this further, saying that metal is not "pure metal", meaning that within each phase are all the other phases. Again, this calls upon our understanding of the classics. When an interaction is being explained, it is for us to determine whether the focus of the Nan Jing writers was to be more specific or broader in view. As far as the 10 stems are concerned, the interaction is broad, and the five phases are a simple and human overlay to highly complex/ more fragmented expression of the 10-stems in the Nan Jing.

The importance is to see the complexity of seeing all the different aspects of a system in relation to one another or the whole of an energetic phase. When we are splitting aspects apart we cause fragmentation and therefore complexity of attempting to balance these aspects into the whole again, and as such for example the liver meridian and its organ are different expressions of complexity. But if we see all of wood and all of metal then we are viewing from a much more non-fragmented perspective and hence the expression is simpler.

Just as at the peak, yin turns to yang and yang turns to yin, so the dots within the eyes of the yin yang is what is represented in Chapter 33 of the Nan Jing and especially in relation to wood and metal which are an example of this mix within the body energy. This chapter is vital for our understanding of the mix of metal and wood, fire and water throughout the texts and so is highlighted here.

343

Part 8

The Acupuncture Points and their Energetics

The Acupuncture Points and their Energetics

There are said to be 365 acupuncture points, one for every day of the lunisolar calendar, governed by the cycles of 12/Earth. The five Zang organs are the seat of yin (interior) but hold yang (spirit of the nine energies/Heaven). The meridians are the seat of yang (exterior) but connect exteriorly to yin (Earth) and are subject to the changes of the calendrical change of Earth. Of course there is no separation of organ and meridian, BUT the five-phase (Jingshen) change of the interior has its own motion, as well as the motion of the meridian-Earth exterior affecting this interior.

The five phases of the organs are governed by the Heaven/nine energies but are affected/augmented by the Earth if we consider the ancient proverb to the unity of yang influence in the universe; "All under Heaven". The point of this is to understand that humans are totally in unity with the change of the Earth, utterly one with it. However, there is an interior to the human that radiates outwards (five phase of Jingshen), and just as the exterior affects it, so it affects the exterior. When the two aspects join, this is called harmony; when they are out of synch with each other, this is called fragmentation/dis-ease. Hence complete health is when one's internal spirit is unified with the exterior changes of things. This is the same as saying, Be "your"Self, go with the flow. What we are saying is that while Jing is of the Earth (12) and Shen is of the Heavens (9), Jingshen is of the organs or the internal nature of humans (5). The meridians are between the interior and exterior, hence they have the 12 branches and 10 stems, which are governed by the change of 12, which is of the Earth:

(fig. 8.1)

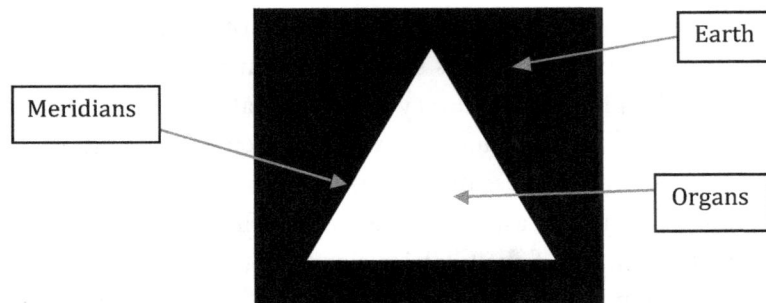

If the edges of the triangle represent the meridians connected to the Earth-exterior, then the five phases of the organs are the white triangle's interior, connected to Heaven/spirit and yang-life.

Just as all other phenomena of yang movement are ripples on the Stillness of Wu, from the centre of a human being, or the origin of life, energy ripples outwards creating all other aspects, then eventually going back to Stillness in death:

(fig. 8.2)

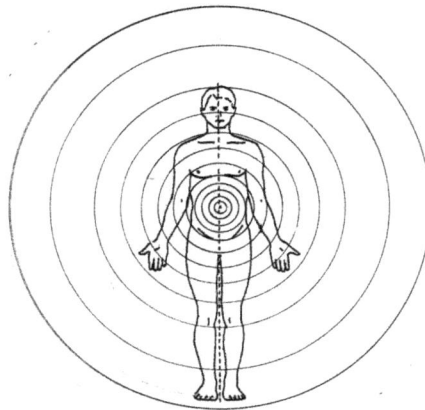

I will not be explaining the routes of all the meridians or all the point locations of the body. This information is available in many sources, most expertly through Ikeda Masakazu's work. When considering meridian and point location , the modern measurement system of locating points was never practiced by the Han dynasty practitioners. There was a learning of where points location was through feel and palpatary sense, the points not being in the same exact place on each patient and on the same patient variations would constantly occur with the various changes of the environment and the patients condition, like the woods of stringed instrument constantly being affected by environmental changes expanding and accumulating. Whenever one looks at a text of point locations it is very important that no matter how detailed the writer thinks the point is specifically located this is absolutely untrue. Precision and absolute accuracy of a point that is objective is impossible as even the subjectivity of the individual practitioner palpating will have a different effect on the patient. This is something that makes the Kozato method (from the Toyohari style of acupuncture) of pin-point accuracy at point location something of a fallacy in Classical understanding, as is similar with TCM style location via measurement . Location of points accurately is as subjective to the practitioners sense as pulse diagnosis, the key is for the practitioner to sense where the location of point is, is through feel and all aspects of treatment follow through from this instinctive sense of how to treat. This is what the maps of the Classics aim to do, not to produce a prescriptive method. Therefore any prescriptive ideology is modern in explanation and should be taken with handfuls of salt. Again I must integrate that this is intuitive sense based medical healing at heart. (Shudo Denmai has written a text on location of acupuncture points ("Finding effective acupuncture point") which is useful as it explains Mr. Denmai's own difficulties and eventual learned skill of sensing the point locations himself after much deeply interested practice. This gives a good idea , not to follow his code of points but to again see if one can sense this oneself).

What I will attempt to do in this section is deal with the most important points used in the treatment of the meridians and explain the general configuration of points in the following ways:

1. The meridian where it (the point) is found
2. Depth of the ki in that meridian
3. Flavour of the point
4. The effect of the flavour within this meridian

The aspect lacking in the understanding of points is that they are prescriptively looked at and applied to the body, like herbs. This is due to the modern Chinese approach to acupuncture (see appendix on the history of acupuncture and herbs in China and Japan). If one understands the nature of the energy of the points and meridians in terms of their flavours, as expressed in the Su Wen, Ling Shu, and Nan Jing, then we have the principle rather than the recipe. This will aid us in making this an art form, where one is forming various dosages for contrasting energies when treating, rather than prescriptions of points. Let us have a look at the points. It is always what you do with the point that effects the energy rather than purely the point itself. One point can offer 100 possibilities so it's more about connecting to the intuitive sense when following the general map of the Classics in order to come up with the appropriate response to a given situation.

8.1 Points in Relation to the Meridian where the Point is Found

The meridian is a flow of energy; therefore the points are simply aspects along this flow. "A point" is any place at all over the whole surface of the body. The reason specific points are chosen is that they are key to where energy changes in its state, like the five-phase changes, and these are the ones we will look at the most, as they are applied to the meridians very effectively here. From the distal areas to the inner body, there is a transition from the very narrow, fine, sensitive key of the fingers and toes and facial features to the broader, wider, and deeper energy. This is expressed in the classics as a flow of water from well points to spring points to stream points to river points to sea points … and then centres of energy seas, or oceans of collected energy on the fount of the body. Each meridian has a flavour, as was discussed earlier.

We must understand, when looking at the meridian the point is on, that it is the meridian, which is the large aspect, and the point, which is the smaller aspect; the point is in relation to the meridian, not the other way around.

8.2 Depth of the Ki in that Meridian

The body has several layers to it where the energy of the meridian, associated with particular flavours, runs.

(fig. 8.3)

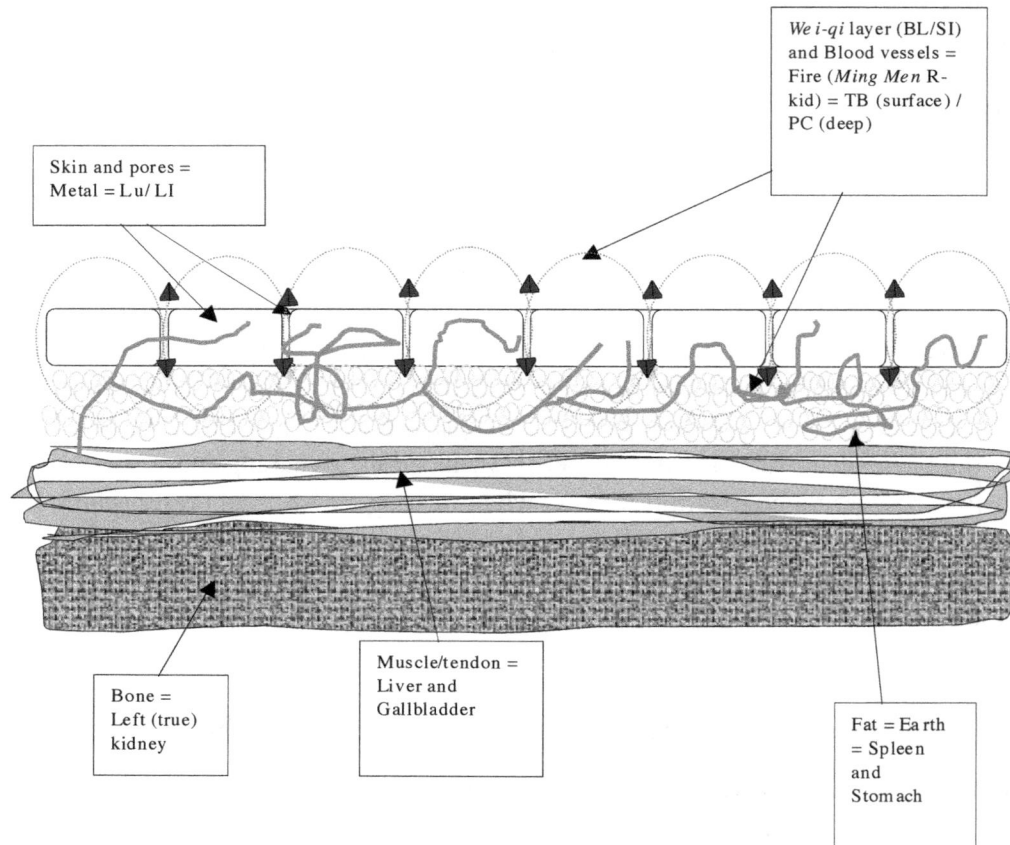

Skin and pores = Metal = Lu/ LI

Wei-qi layer (BL/SI) and Blood vessels = Fire (*Ming Men* R-kid) = TB (surface) / PC (deep)

Bone = Left (true) kidney

Muscle/tendon = Liver and Gallbladder

Fat = Earth = Spleen and Stomach

As we can see above, there are five meridian-ki depths, six if we consider the twin depths of the heart network. These depths of Qi (secondary to, but one with, the physical tissues) associate directly with the various depths. One will feel the radial and other pulses in the body. The tissue-ki is formed from the organ energy of the five phases; this is then where the meridian energy associated with each phase moves through the body:

i) Wei Qi/ Blood Vessel Level (Water (fire qi – salty/pungent)-meridians)

This is the source energy of the TB and the dominance of the heart network throughout the system. This layer is in fact two layers, above the skin and below the skin. Above the skin relates to the BL-SI Tai Yang meridians, the surface blood vessels, TB. Below the skin and the deep blood vessels relates to the pericardium in charge of the heart ki. This associates with the salty flavour. The Bladder meridian is labelled "water" as it is the yang within the yin of water. Hence this expression is associated with water meridians.

ii) Skin qi Level and Skin-surface itself (metal (wood qi -pungent)-meridians)

This is the level of the metal-**meridian** energy. More exterior and on the yang surfaces is the LI; on the yin is the LU. This associates with the pungent flavour (wood organ energy). Note that this is meridian energy; the organ energy is not represented here, as this is where the points are; on the exterior of the body, not in the organ region of the body. The tissue skin itself is associated with the metal energy, hence the Lu and LI meridians are wood qi in the metal region, labelled metal meridians but contain wood ki. This is important.

iii) Blood Vessel Level (Water (fire – qi - salty) meridians)

Between the skin and exterior are the superficial blood vessels that relate to the TB energy, and at the deeper levels to the PC systems. This is the deeper level of salty (and pungent = TB-Ming Men association) flavour that is regulated by the skin pores and mixed with the bladder energy on the body surface. The bladder energy too is regulated by the pungency of the skin pores. The salty flavour therefore is distributed to the surface with pungency. This level of meridians associates with that of the TB as the origin of the PC meridian energy so is fire within water and still therefore associates with the label of water meridian, although the ki it carries is opposite.

iv) Flesh/Fat Level (Earth - sweet meridians)

This is the level of the spleen and stomach, the stomach dealing with the more superficial layers, the spleen with the deeper aspects of this layer. It associates with the sweet flavour.

v) The Muscle-tendon Level (wood (metal qi- sour) meridians)

This is the level of the gallbladder and liver meridians , the liver meridian dominating the interior tendons, and the exterior is the gallbladder organ-meridian. The redness of the blood within the muscle is dominated by the deeper blood vessels of the fire energy, remembering that liver blood is not really owned by the liver meridian and is, in fact, opposite to its action, relating more to the pungent and salty flavours. This is the labelled wood meridian level qi BUT has the metal-sour flavour.

vi) The Bone Level (Empty-Fire (Heart) meridian (water-qi - bitter))

The bones are dominated by the left kidney. Most bones are deep within the tissue, or at least the deepest aspect of the tissues in a given region. The bones include the bone marrow, brain, and intra-fossa nerve tissues in more fragmented Modern Western anatomical terms. The bones are turgid with fluid, and this makes them strong and dense, not brittle. This is the associated region of the bitter flavour. This level of meridians is the deepest and coolest it relates to yin and so to emptiness hence it associates with the empty quality of the Fire and the heart as well as to the L-kidney, however the heart meridian doesn't contain bitter flavour, it is empty. The bitterness is of the L-Kid only.

What this explains to us is that the energy of the materials/ tissues of the body holds the ying energy of the various meridians. Accessing this level anywhere in the body will access the energy of this level with the energy at which it is located, creating a mix of association of that energy.

vii) The Flavour Level
The depth of the qi can also be described simply by the flavour at that level:

Wei Level: Pungent/salty
--------Skin level: Pungent--------
Blood Vessels: Salty (pungent)
Fat: Sweet
Muscles: Sour
Bone: Bitter

The pungent flavour tonifies the salty flavour (mother-child relationship of organ energy), which means spring transforms to summer. This is expressed in the Wei level and blood vessel level which are encouraged to open out to the surface by the opening function of the pungent flavour. When the pungent flavour is affected, the salty flavour can escape from under the skin and mix with/encourage the Wei energy of the BL-SI related meridians. This is how the LU/LI/BL/SI/TB all make up the surface yang energy, which is a mix of pungent and salty flavours, to protect the body from the exterior cold.

8.3 Flavour of the Point

When looking at the yin and yang meridians, the most important points used in treatment are the five transportation points and the source points. Other point categories are supplemental to this. The other two categories of secondary importance that we will look at along the distal aspect of the channel are Luo points and Xi cleft points. A note on these is below. However, the energetic points of importance are the five phase points. These points represent the energy of the other aspects of the system within one meridian. Just like the macrocosm in the microcosm, the fractal pattern continues in and in.

The chart below shows the points of all the 11 meridians and their five-phase interaction. Note that HRT-7 is associated with PC-7, but the hand Shao Yin heart meridian has no five-phase points, as expressed in Ling Shu 71.

Key:
Salty = Sa
Pungent = P
Sweet = Sw
Sour = So
Bitter = B

(table 8.1)

5 Yin Meridians	Wood-Well - So	Fire- Spring - B	Earth- Stream – Sw SOURCE – Sa-P	Metal-River - P	Water-Sea – Sa-P
Liver – So	Liv 1	Liv 2	Liv 3	Liv 5	Liv 8
Pericardium – Sa	PC 9	PC 8	PC 7	PC 5	PC 3
Spleen – Sw	SP 1	|SP 2	SP 3	SP 5	SP 9
Lung – P	LU 11	LU 10	LU 9	LU 8	LU 5
L-Kidney – B	KID 1	KID 2	KID 3	KID 7	KID 10

(table 13.2)

6 Yang Meridians	Well – Metal-P	Spring – Water- Sa-P	Stream – Wood- So	SOURCE- Sa-P	River – Fire- Sa	Sea Earth- Sw
Gallbladder – So	GB 44	GB 43	GB 41	GB 40	GB 38	GB 34
Triple Burner – Sa	TB - 1	TB 2	TB 3	TB 4	TB 6	TB 10
Small Intestine – Sa	SI - 1	SI 2	SI 3	SI 4	SI 5	SI 8
Stomach – Sw	ST -45	ST -44	ST -43	St 42	St 41	St 36
Large Intestine – P	LI 1	LI 2	LI 3	LI 4	LI 5	LI 11
Bladder – Sa	BL 67	BL 66	BL 65	BL 64	BL 60	BL 40

Note that the flows of the numbers are the direction of the flow. For example, in the yang meridians, the arm meridians will start at the fingers and end on the chest, as they flow down to the chest, and the yang meridians of the foot will go down to the toes. This is shown in the Ancient Eastern anatomical position in the last chapter with hands raised above the head, connecting the 10 meridians of the upper body (considering bi-lateral meridians) to heaven. Notice too that the toes and fingers always end with metal or wood points, which shows that the flux of polarity at the end points is very great, if we consider spring and autumn energy to be the most moving aspects.

As we can see above, the way of change of the yang meridians, from the well to the sea points, has opposite polarity, point for point, from the paired yin meridians. The exception is the fire point of the yang meridians (this is full fire (which is represented in the yin meridians as water inclusively), empty-fire does not exist in the yang aspect of the body), which is a key reason why the yang meridians and yin meridians are not used in the same way in treatment (see below). Also, there are six points on the yang and five on the yin, again expressing the obvious difference between yin and yang functional aspects.

The meridian is the seat of the flavour, so the point's effect is as Shudo Denmai writes, like an embassy within a foreign country. If the point is the same flavour as the meridian (Not necessarily its connected Zang organ), it is the home office! For example the wood point on the wood meridian (please note however that both these express the sour flavour, it is sour within sourness. This relates to Metal Phase or the Metal Zang, but the labelling of the point is wood due to the meridians connection to the wood meridian. The "point" is that these have influence over the governing of the country, through the influence of the other countries on their soil.

For example:

> SP 2 is the bitter point within the sweet meridian.
> Liver 3 is the sweet and salty (pungent) point within the sour meridian.

The main issue we are discussing here is that on the bitter meridian of the kidney yin, all the points will have the bitter quality, BUT at the fatty tissue level (Earth point), on this meridian, the bitter will mix with the sweet. On the skin level (metal point), the bitter will mix with the pungent; on the bone level, the bitter will mix with the bitter (fire point); on the muscle level (wood point), the bitter will mix with the sour; and at the very surface of the Wei qi (water point), the bitter will mix with the salty and pungency. This has obvious benefits of understanding when considering treatment possibilities. Ikeda Masakazu expresses why understanding meridian and point flavour is so useful in treatment in an article about point selection for the "North American Journal of Chinese Medicine", July 1997, p.16:

> *In Meridian Therapy all one has to do is to tonify and disperse the meridians, and any means can be used to accomplish this end. Although this position may be a little extreme, I think people can use practically any point. And when it comes to the symptomatic treatment, a person can do just about anything. Sometimes, in order to disperse an excess condition, we must resort to strong measures. If, however, we are overly concerned with the (effect of treatment on) pulses, we become timid and our treatments are compromised.*

This points out several things; seeing the broader-general picture and seeing the meridian before the point. The rationale for this comes from a realization of the mixes of energy that the point expresses based on the meridian energy. If one senses this, then literally any point can be used on the meridian as the method of treating will adapt the qi at each area of the body appropriate to what the body requires to relax.

The source points are the energy of the TB within each of the meridians. The kidney yang energy is that of the source, and this energy is heating. The representative of this within the system is twofold: the water points belong to the bladder meridian, and the source point belongs to the TB meridian. These both have a salty effect. However, the TB is associated with slightly more internal heating, and the bladder has more to do with urination and the external heating of the yang surfaces. The bitter point is that of fire.

This is strange, is it not? Why would fire be associated with the coldest of flavours? For an explanation of this, we have to identify again what the five phases are. The five phases associate with the five seasons and hence to the five organs of the yin. The organ energy reflects the seasonal changes:

- Liver organ IS Pungent, Spring
- Fire network (HRT) IS Salty, Summer
- Spleen organ IS Sweet, Late summer
- Lung organ IS Sour, Autumn
- L-Kidney organ IS Bitter, Winter

These we will call the Five Origins of flavours or Five Resident flavours. This is expressed as the arrangement of the Sheng cycle (please see Su Wen 23).

When one is in these seasons, one DESIRES the following in each season due to the nature of heat or cold at the time, to create balance:

- Spring desires Sourness (Liver meridian)
- Summer desires Bitterness (L-kid organ-meridian)
- Late Summer desires Sweetness (pivot point) (Spleen organ-meridian)
- Autumn desires Pungency (Lu meridian)
- Winter desires Saltiness (TB/BL organ meridian network)

These we can call the Five Desires of the five flavours (this is expressed in the Ke cycle interactions). The desires differ slightly as we move out to the yang meridians and interact yang with yin meridians, as we will see below, but these are all desire-based flavours of whichever organ is involved in the desiring!

Five phases are of the organs, not meridians. This is why the five flavours, when associated with the meridians here, and in some of the descriptions in the Su Wen and Ling Shu, are to give a picture of the five phases/flavours within the meridian's system, which tend to counter the effect of the season they are within. Hence salty and bitter and sour and pungent swap places as far as the organ flavour is concerned. Only the sweet flavour is still and centred. The meridians and point labelling of wood, fire, earth, metal, water are based on their actual qi but based on the 5-desires of the Zang they are connected to. Notice that the organ function of the lung and liver is opposed to the meridian energy anyway, so the sourness is associated with the wood point, and the pungency of the lung meridian with the metal point. If one makes clear consideration that the five desires of the meridians—and so points—are different from the five origins of flavour within the organs, and one sees these two as one, this is one of the keys of understanding the classical works. This is understood clearly if we look at the nature of meridian and organ in wood and metal. Also, if we consider fire and water, these two are both founded in the lower jaio at the left and right kidney organs; these balance each other, just as wood and metal balance each other. Earth is central or neutral, as the sweet flavour is balanced in itself, being warm (yang ST) and moist (yin

353

SP). Hence there is consistent balance of flavours through the meridians and organs, and this is reflected in the points on each meridian.

Hence when considering the five flavours, notice the difference in organ versus meridian energy we looked at before. The route to this understanding comes from a very important chapter in the Nan Jing that describes the use of each point of the five in treatment; Nan Jing 68. Notice that the yin meridian points are the only ones discussed here. The importance of this chapter along with Nan Jing 69, will be discussed further later.

> The sixty-eighth difficult issue: Each of the 5 Zang and 6 Fu has a [meridian point called] "well", "spring", "stream", "river" and "sea". What are there effects?
>
> It is like this. The scripture states: Where they begin/appear are the "wells", where they flow is the "springs", where they rush down are the "streams", where they proceed are the "rivers" and where they disappear/ disperse are the "seas". The "wells" resolve fullness below the heart. The "springs" resolve heat over the body. The "streams" resolve heaviness in the body and [accompanying] joint pain. The "rivers" resolve difficulty breath and coughing with chills and fever. The "seas" resolve qi-reversal (qi not going in the correct direction) as well as diarrhoea. These are the illnesses resolved by the "wells", "springs", "streams" "rivers" and "seas" of the 5 Zang and 6 Fu. (based-on Unschuld P. 1986)

Because the five phases of the points are expressions on the surface of the desires of the Zang organs, this can make the picture confusing—to know what to use—as fire in the Zang means heat, whereas in the point it means cold. This shows how one cannot simply use the concept of mother-child tonification using the points of the exterior, as they represent something different on the inside. The points can be converted:

(table 8.3)

Point	Flavour	Zang/ phase Tonified
Wood point	Sour	Tonifies Metal phase/ Zang
Fire point	Bitter	Tonifies Water phase/ Zang
Earth point	Sweet	Tonifies Earth phase/ Zang
Metal point	Pungent	Tonifies Wood phase/ Zang
Water point	Salty (pungent*)	Tonifies Fire phase/ Zang

(*The water points associate with saltiness and pungency as they connect to the function of the bladder and therefore TB meridians. This energy has both the pungency of the Ming Men and the warmth of it, salty. Hence it has both properties, but we categorise water as salty as an overall category.)

It is far too simplistic to associate the five-phase sequence to the five-phase point on the surface. We must, as practitioners, understand where we are looking, what we are looking at, and what we are treating as a whole. The five phases of the internal organs are the basis for treatment. We have to develop understanding of what is needed in treatment by completing the picture of chapters 68 and 69 of the Nan Jing together (see pages 354 and 560).

So why do the yang meridians have a source separate from the five phases, and why is the fire point salty here? These are important; they relate back to previous chapters where we explained that the meridians associate more with the yang and the organs more with the yin—hence the yang meridians are not the roots, and as such they are not regarded as important for "root treatment" (discussed later). Also, the energy circulating in the yang meridian is different from the yin. In the yin meridians, there is more of a connection to the Zang organs, whereas the yang have no such connection, only via their Fu. Hence the yang meridians follow very much the energy of their related meridian; they are much simpler.

Essentially, one doesn't tonify the Fu meridians because the Fu relate to the yang/meridians and so are exterior and tend to get full of heat blockage from yin or yang deficiency of the Zang root. The Zang meridians relate to the organ and interior, so even if one wants to affect the Fu organs, the best way is through the Zang. This is very important because the five functional flavours of the points are explored in Nan Jing 68 (see Treatment Principle chapter), and only the yin meridian points are explored. The yang meridians' point flavours are not indicated. As we can see in the above chart also, the points on the yang meridians' flavours change, as the yang meridians are not related to the bitterness of the kidney yin and are superficial, not deep. The five-phase points of the yang meridian's relevance to root treatment is insubstantial (as Nan Jing 68 implies). Hence these points are best denoted by their meridian flow—sea to well. The Zang are storages of the flavour, hence they collect, so when one does a root treatment, one wants the energy to collect. Therefore the yin meridians are always used to tonify the root. The yang meridians are associated with the surface and have connection to the Fu. Either way, they let go of the energy as fast as they take it on. The actual points on the yang meridians are not as functionally important energetically; the root is the base.

However, chapter 64 of the Nan Jing again reflects upon an important key (which has been reflected upon before in Nan Jing 33), which is the expressions associated with the points of the metal and wood, placed in association with the 10 stems. If we follow this to its logical conclusion, we can see that the energetics of the points of yin and yang meridians do in fact create a dynamic balance through the system, so it would be possible to use the five-phase points of the yang as well as the yin in root treatment. However, it is often best to use the yin as the root, as explained above. The diagram below shows the entire dynamic of the points' association for yin and yang meridians, as associated with the classics. Note that the flavours of the yang point and yin point are different, and this completely ties together with what we learned about the 10-stem associations in the previous chapters. Remember whenever the 5-phase colours are being used it is associated with the Qi of Zang energy base connecting to the meridian network, or what I've called the 5-Origins, NOT the 5-Desires which is the labelling of the meridians and organs NOT their actual Qi, I only relate to the label a point or meridian is given so you can see the difference and adjust to this in other texts, as for my own work I am only concerned with an understanding of the actual qi not the labelling process which I feel is a major confusion or attempted method to confuse, by the writers of the texts in order perhaps to enclose the use of the 5-phase points into a mechanism that only people who had clinical experience could follow which is why Nan Jing 68 is so vital. For the rest of the text, as from the origin, I will refer to the points or meridians NOT as their labelled variety but from the 5-Origin perspective, (I do refer to them as labels in Appendix 2 and I will assume knowledge of these sections material):

(fig. 8.4)

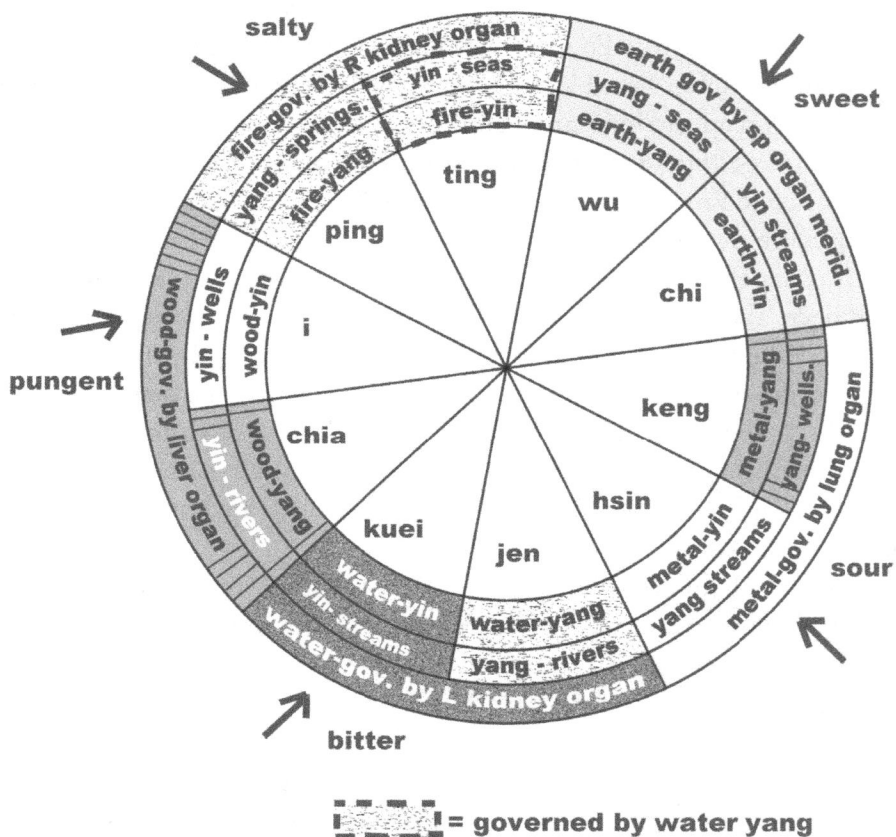

(fig. 8.5)

5-phase point
interactions

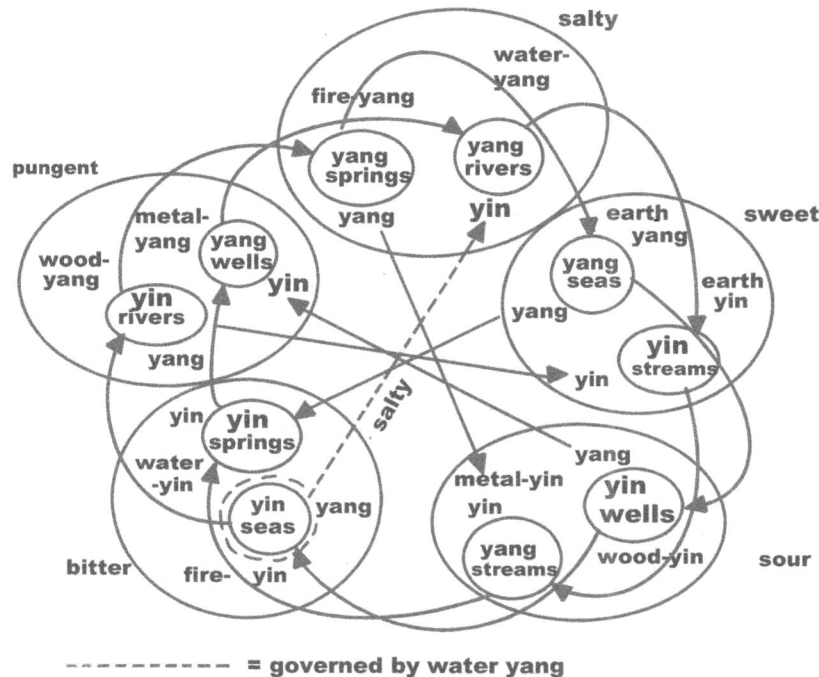

- - - - - - - - - - = governed by water yang

From the above, we can derive why the yang-river points would be salty (pungent); the desire of the fire-yin energy is to be filled with yang, because fire-yin is essentially empty without the interaction of the yang of the water (i.e., kidney yang energy, which is what is represented by the yang-river points). The fire yin seas are in association with the Heart meridian discussed previously which is why there is always an empty factor in this process, the dotted line above therefore represents the PC meridian governed by water yang. This is still a desire of opposites—full and empty. This very much relates to the understanding of the heart network, which we discussed earlier.

The above would seem to suggest that the yang meridians are just as important in the root treatment. However, we must recall that the above is a point-only expression; it does not indicate that the yang meridians are far away from the source of the organ energy and so are distant echoes of the root energy. This is why in chapters 68 and 69 of the Nan Jing associated with the root treatment, only the yin meridians are indicated. The above shows the nature of the points in relation to each other, but as far as the context of treatment goes, we also must consider their direct relation to the sources of energy. This is why we focus on the yin meridians and their flavours. Hence treatment based on the above would not yield as good results as yin meridian point association, due to the fact that the CONTEXT is overlooked; we must always see this in understanding the classics.

(fig.8.6)

**5 - phase point interactions
of nam jing 68/69**

salty

yin-seas
(+source points)

pungent

sweet

yin-
rivers

yin
meridians
ONLY

yin-
streams

yin-
springs

yin
wells

bitter

sour

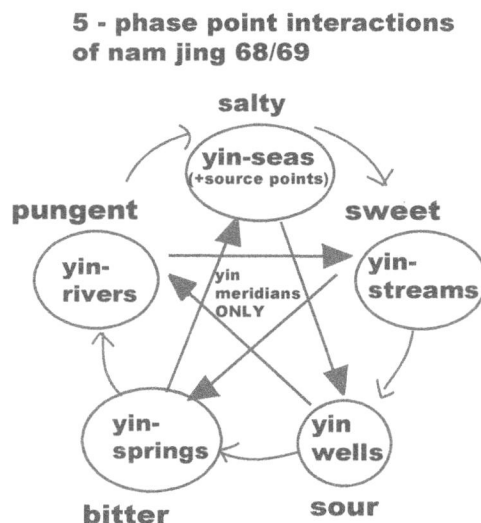

(again the above are the 5-Origins of qi of the point, rather than the 5-desire-labelling)

The source point is separate because there are six yang meridians and five yin meridians. Six relates to the six meridians involved. The source point in the yin is a doubling up of the salty flavour and represents the yang energy within the yin, and the fact that the Ming Men source energy is associated with all of the meridians.

This is clarified in Ling Shu 1:

> In each channel of the 5 solid organs, there are 5 acupuncture points of Jing, Xing, Shu, Jing and
> He, they are altogether twenty-five acupoints; in each channel of the six hollow organs, there are six
> acupuncture points of Jing, xing, shu, yuan, Jing and he, they are together thirty –six acupoints.
> (based-on Wu N. L and Wu A. Q., 1997)

This tells us several things. The first is that five yin meridians discounts the heart meridian, because it has no five-phase points, so there are not 12 meridians but 11 stated here, as 5 plus 6 is 11. Also it shows us that the Yuan source point is associated with the Earth point in the yin and is separated off into its own category in the yang. Why would it be that the sweet spleen Earth point would associate with the triple burner then? This is important because it relates to the nature of the TB being yang within the yin (depth of the body i.e. Ming men, its source). The TB's source is the right kidney or Ming Men, which essentially is a Fu organ. Although the Ming Men is solid energy, it is the organ that, according to the 25th difficult issue of the Nan Jing … "Has name (or essential nature) but no [physical] form".

Essentially the TB is yang, though originating from a deep place within the body. The right kidney is really the Fu of the left kidney or "true" kidney. However, it is "full" in its own right, full of yang not yin, so it is deep, but yang. This quality means that in the yin, the TB is a Fu meridian within the

Zang meridians, just as its brother BL meridian is associated with the salty (+pungent) flavour. Hence the TB here associates with the Earth and the spleen, warming the digestive system (Earth point) and making the association that the spleen's warmth is formed mainly by the stomach's energy giving to the spleen. Note that the yang tonifies the yang, and the yin tonifies the yin aspects of the system. Hence TB would normally warm the stomach and PC would normally help warm the spleen, but if we see clearly, the PC simply draws heat down from the upper body to the TB again, so the TB warms the stomach, and the stomach again warms the spleen. There is a continuous pattern. In the yang meridians then, the Fu meridians are all displayed individually, as they are not penetrating the yin to warm it. There are six points on the yang meridians, stomach (Earth) and TB (source) are differentiated, but again in close proximity. The yang energy of the body cycles in the yang, and the yin cycles in the yin; this crosses over at the TB-GB Shao Yang meridian-organs. The TB (note that the bladder is more superficial than TB, but has the same flavour, essentially salty-pungent, see below) takes yang to the yin (inside) of the body, whereas the GB takes yin to the yang (outside) of the body. This is interesting, as the TB/GB is considered the pivot point between the exterior and interior of the body as far as energetic layers go. This will be discussed in the chapter on pathology.

i) The TB Source Points

Very similar to the phases, the source points are representative of the right kidney energy within the flow. This energy on the meridian resonates with the source of the Ming Men energy, explained in Nan Jing 66. The TB therefore forms a phase in its own right, just as it has a meridian separate from the bladder but basically does the same thing. The bladder is the outside of the TB though, so the TB represents the fire of the body. Hence in yin and yang aspects, the TB energy is represented. Whereas the water points are salty in flavour and have more distinct connection to the salty expression of the bladder/TB meridians especially, the source points relate to the TB's origin (Ming Men) more directly and so have a mix of flavour, that of wood and fire, pungent and salt. This mirrors the energy of the Ming Men, initiating energy (pungent) and keeping warm deeply (salty).

In the yin, it resonates at the same place that the Earth points are, so these are special points carrying both the sweet and salty flavours. They also connect to both properties of warming and initiation of energy; hence they have a pungent quality to them also. This gives strong yang and mild yin qualities, so they are often considered as good tonification points for yang deficiency, and in this respect, because of the mild nature of yang tonification, they will cause no damage to a system if used in this way. For more exploration of use of these points, please refer to Ikeda Masakazu's work and to Appendix 2 of this book. The fire (saltiness) of these points tonifies the effect of the Earth nature of the point, so they are strong Earth points in reality. Essentially this is a mix of pre-heavenly yang and post-heavenly sweet/pungent flavours.

In the yang meridians, the TB points relate more to the combination of salty and pungent flavours.

ii) Luo Vessels

(fig. 8.7)

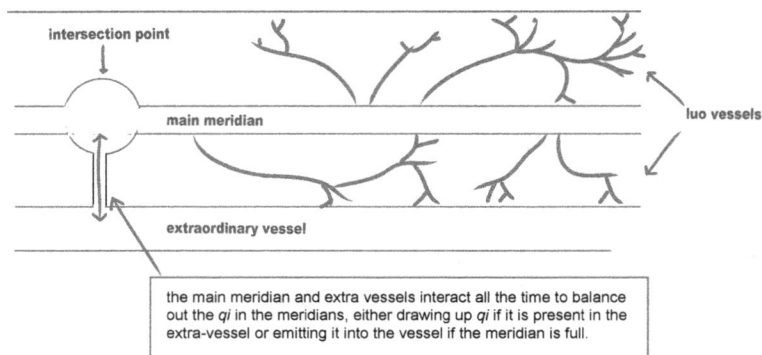

The Luo vessels are like tributaries of the main meridian. These tributaries connect the paired meridians (e.g., Sp-St, Liv-GB, Kid-BL, etc.) and spread throughout the body tissues, supplying the associated tissues with the meridian it is related to most.

These vessels represent the wefts in cloth, whereas the warps are the main meridian flows up and down. The wefts go side-to-side, and the warps up and down. This cloth is an interesting picture, as the associated descriptions of the characters used for methods of treatment in acupuncture often associate it with weaving cloth or sowing patches in the energetic system of the body, using needles. We will look into this more later.

(fig. 8.8)

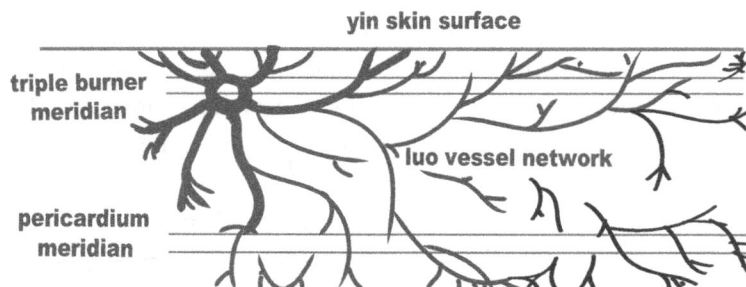

iii) The Xi Cleft Points

The cleft points are where qi can accumulate in the meridian. Often found at regions of the body where there is a dip in the bony feature of the tissue, these clefts act as areas where qi can "eddy" and accumulate.

iv) Back Shu Points and Fount Mu Points

Shu points are considered to be associated with the yang, hence are particularly used in the tonification of the yang of the body. The bladder and TB meridians run here, but the following points connect directly to their various organs via the connection with the bladder yang energy.

(fig. 8.9)

These points are:

Bl-13: Lung

Bl-14: Pericardium

Bl-15: Heart

Bl-18: Liver

Bl-19: Gallbladder

Bl-20: Pancreas

Bl-21: Stomach

Bl-23: Kidney

Bl-24: Triple burner

Bl-25: Large intestine

Bl-27: Small intestine

Bl-28: Bladder

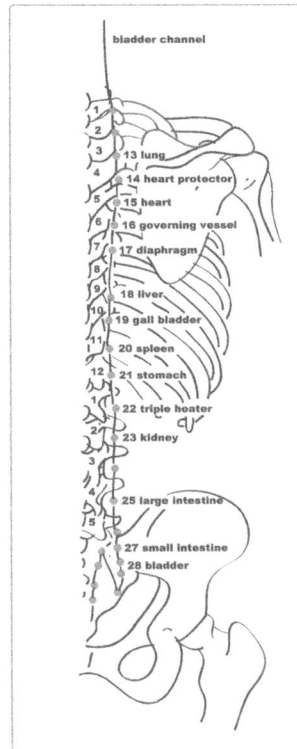

These points draw energy from the back to the front of the body, or draw energy into the body. This is their transport function; they are yang, and they move back to front. They have the potential to absorb yang energy drawing it inwards to the organs.

v) Mu-Collection Points

These points are similar to the Shu point, but instead of having a suction type effect, the points have a collecting effect. This collection is useful for yin accumulation, so yin energy can accumulate here more easily. The collection points have little dynamic function as they are in the sea of the yin aspect of the body, the front being full of the yin meridians, and the back being yang and full of yang meridians.

(fig. 8.10)

They are:

Lu-1: Lung

Ren-17: Pericardium

Liv-14: Liver

Ren-14: Heart

G.B.-24: Gallbladder

Ren- 12: Stomach

Liv-13: Pancreas

G.B.-25: Kidney

St-25: Large intestine

Ren-5: Triple burner

Ren-4: Small intestine

Ren-3: Bladder

I will not go through any more point categories in this book. These are the main point categories used in the Han Dynasty text. There are many more that came afterwards, obscuring the true simplicity of understanding the energetics, rather than really adding anything to the picture—for example, the "extraordinary vessel opening points", the "window of the sky" points, etc. Plenty of other books have been written with many more categorisations than this. My focus is on your understanding of the principle of energetic change through the system as a whole, not as fragmentary parts or points. With this clarity, you will be able to derive your own understanding of what point is "good for" what, or hopefully you won't use this terminology! This allows us to move away from what TCM calls point prescriptions (see Appendix 1), and also away from the fear of contra-indicated points and the possibility of "causing" illness using acupuncture. If the energetics are understood, each point becomes a versatile energetic energy that has properties rather than functions, which can be enhanced or affected by the practitioner. True energy medicine is always about the practitioner's intuitive connection to the above understanding, rather than what it says in a book learned by rote. Contra-indications become something laughable because one can affect a single point/meridian/person in so many ways—again, it's the context that counts.

www.ingramcontent.com/pod-product-compliance
Lightning Source LLC
Chambersburg PA
CBHW082129210326
41599CB00031B/5916